CW00430621

Inherently Safer Chemical Processes

This book is one in a series of process safety guideline and concept books published by the Center for Chemical Process Safety (CCPS). Please go to www.wiley.com/go/ccps for a full list of titles in this series.

Inherently Safer Chemical Processes

A Life Cycle Approach

Second Edition

Center for Chemical Process Safety
New York, New York

A JOHN WILEY & SONS, INC., PUBLICATION

A Joint Publication of the Center for Chemical Process Safety of the American Institute of Chemical Engineers and John Wiley & Sons, Inc.

Published by John Wiley & Sons, Inc., Hoboken, New Jersey
Published simultaneously in Canada

Library of Congress Cataloging-in-Publication Data is available.

ISBN 978-0471-77892-9

Printed in the United States of America

10 9 8 7 6 5 4 3 2 1

Inherently Safer Chemical Processes: *A Life Cycle Approach*

Is dedicated to

Trevor Kletz

Successful people are admired for their motivation and persistence to making a difference. It is to Trevor Kletz that we dedicate this Second Edition of "**Inherently Safer Chemical Processes: *A Life Cycle Approach.*"** Considered by many to have created the concept of inherent safety, Trevor's persistence, imagination and passion have truly taught us to look at safety in a different way.

Chemical plants worldwide are safer because of his influence, his writings and his teachings.

Contents

Foreword

About 1700 years ago, a 4th century theologian wrote that "... all great controversies depend on both sides sharing one false premise." In disputes about the amount of resources to be spent on safety (or the prevention of waste and pollution), the common false premise is that safer (cleaner) operations can only be obtained by spending more money. As the examples in this book show, this is often untrue. If we can *avoid* hazards instead of keeping them under control, the resulting designs will usually be cheaper as well as safer, for two reasons: less added-on protective equipment is needed and, if we can intensify, the plants will be smaller and therefore cheaper to operate. Intensification is therefore the preferred route to inherently safer design.

The disasters at Flixborough, UK in 1974 and at Bhopal, India ten years later demonstrated the need for inherently safer (IS) designs and provided good examples of their advantages. The material which leaked at Bhopal, killing thousands of people, was not a product or raw material but an intermediate, which was convenient, but not essential, to store. Since then, many companies have drastically reduced their stocks of hazardous intermediates, often manufacturing them as required and keeping no stock at all. For example, there is no longer any need to store phosgene, even if the usage is small, since small output units are now available. What you don't have, can't leak.

The inventory of hot, flammable liquids in the Flixborough plant was large as the process was a slow, low conversion process. Only 6% of the input was converted in passing through the plant. The other 94% got a free ride and subsequently had to be recovered and recycled. Reducing the inventory is not as simple as at Bhopal, but, nevertheless, new designs are possible. However, they have not been pursued new plants built since 1974 have used essentially the same process as that used at Flixborough.

This illustrates one problem with IS design. While its use has grown substantially since the first edition of this book was published in 1996, it has not caught on as quickly as other new process safety ideas, such as quantitative risk assessment, hazard and operability studies and new attitudes to human error. Why is this?

One reason is the view that innovation is risky. The thinking is that there may be unforeseen problems, so it is safer to stick to the designs we know—until someone else introduces a new design and we lose our market. A more important reason, however, is that innovation of any sort in design takes longer than repeating the old design and this time is usually not available. The new plant is wanted as soon as possible or we might miss the marketing window. The way around this is to look further ahead and think about the plant after next. When we are designing a new plant, we are conscious of all the changes we would like to make if only we had more time. We should note these changes and start work on them well in advance of the plant after next.

A change like this can come about only if the directors and senior managers of the company understand the advantages of IS design and are committed to it. Unfortunately many, perhaps most, of these individuals have little idea of its scope. If they have heard about IS design, they think of it as another safety gimmick.

So, this book is not one just for experts on IS design or for safety professionals. It is one for all those involved in plant and process design at all levels, including the most senior. Safety professionals should not just buy a copy to read. They should buy a few copies and try to get the senior people in their company to read it as well.

Outside the oil and chemical industries, the understanding of IS design is zero. It is beyond the ability of most newspaper writers and politicians to grasp that we can often get increased safety (and less waste and pollution) by spending less money, not more. It seems counter-intuitive, against experience and common sense, that we can get more for less money. It is true that we may have to spend more educating designers in IS design, but it is easier to train designers to remove opportunities for errors than to persuade large numbers of operators not to make errors.

Trevor Kletz

Preface

The American Institute of Chemical Engineers (AIChE) has been involved with process safety and loss prevention issues in the chemical, petrochemical, hydrocarbon process and related industries and facilities for more than 50 years. AIChE publications and symposia are information resources for chemical engineers and other professionals on the causes of process incidents and the means of preventing their occurrences and mitigating their consequences.

In 1985, the Center for Chemical Process Safety (CCPS), a Technology Alliance of the AIChE, was established specifically to develop and disseminate technical information for use in the prevention of major chemical process incidents. With the support and direction of the CCPS Advisory and Managing Boards, a multifaceted program was initiated to address the need for Process Safety Management systems capable of reducing potential exposures to the public, the environment, personnel, and facilities. This program includes:

- developing and publishing *Guidelines* and *Concept Books* relating to specific areas of Process Safety Management
- publishing a monthly newsletter, *Process Safety Beacon*
- organizing, convening and conducting seminars, symposia, training programs, and meetings on process safety-related matters
- cooperation with other organizations, both internationally and domestically, to promote process safety.

CCPS activities are supported by more than 100 corporations that provide funding and professional expertise. Several government agencies and academic institutions also participate in CCPS endeavors.

In 1989, CCPS published *Guidelines for Technical Management of Chemical Process Safety*, which presented a model for Process Safety Management characterized by twelve distinct, essential and interrelated elements. These *Guidelines* were refined over the ensuing years and updated most recently in *Guidelines for Risk Based Process Safety*

(2007), which expands the definition of Process Safety to twenty distinct elements.

This "Concept Series" book supports many of those twenty key elements of Process Safety, as identified in the *Guidelines for Risk Based Process Safety*, including Process Safety Competency, Workforce Involvement, Hazard Identification and Risk Analysis, Auditing, and Management Review and Continuous Improvement. The purpose of this Concept Series book is to demonstrate the application of inherently safer strategies throughout all the stages of the chemical process life cycle.

The intent of the Inherently Safer Design Subcommittee was to update and modify the existing text where appropriate based on progress in the field since 1996. As such, the Second Edition would better illustrate and emphasize the merits of integrating process research, development, and design into a comprehensive process that balances safety, capital, and environmental concerns throughout the life cycle of process. This volume provides examples and useful tools to help any company understand and employ inherent safety concepts. It also discusses incentives and barriers to IS adoption, providing technical illumination for policy debates on the role of regulatory bodies in encouraging IS. The Subcommittee believes this new edition will be especially helpful in:

- Clarifying the definition of IS through the introduction of First and Second IS orders
- Expanding the technical discussion of IS to incorporate the latest research
- Adding practical tools, such as evaluation methods and checklists
- Examining the incentives and barriers to the use of IS in regulation and for homeland security

Inherently Safer concepts have been well received by industry and there has been significant advancement in the concept over the last 10 years. Since the 2001 terrorist attacks, IS is receiving increased attention from industry, academia, advocacy groups and government as an element in process industry security assessments and countermeasures. The 1996 book is frequently cited as an authoritative source on inherent safety, and, as such, it is important that a new edition be published to reflect the most current knowledge on the subject. The Second Edition of the book is expected to be highly influential in the continuing discussion of the inherent safety.

Acknowledgments

CCPS wishes to thank the Inherently Safer Design Subcommittee for their efforts to create *Inherently Safer Chemical Processes: A Life Cycle Approach, Second Edition (2008)* :

Steve Meszaros	*Wyeth, Committee Chair*
Mike Broadribb	*BP*
David Clark	*DuPont*
Bob Conger	*Celanese*
Don Connolley	*BP*
Mark Davis	*Eli Lilly*
Art Dowell	*Rohm & Haas*
Bill Hague	*Honeywell*
Lou Higgins	*Rhodia*
Doug Hobbs	*Advantica*
George King	*Dow*
Bill Marshall	*Eli Lilly*
Craig Matthiessen	*US EPA*
John Murphy	*CCPS Emeritus Member*
Americo Neto	*Braskem*
Richard Pickup	*Schering Plough*
Randy Sawyer	*Contra Costa County Hazardous Materials Programs*
Jan Windhorst	*Nova*

Dennis Hendershot, CCPS Staff Consultant

CCPS also thanks David Moore, Dorothy Kellogg, Michael Hazzan, David Heller, Gary York, Lee Salamone, and Martin Rose of AcuTech Consulting Group for authoring the text and providing valuable input to this book.

Finally, CCPS would like to thank the following peer reviewers for taking the time to review the final manuscript:

Paul Amyotte	*Dalhousie University*
Nick Ashford	*Massachusetts Institute*

Jim Belke	*US EPA*
John Bresland	*Chemical Safety & Hazard Identification Board*
Cho Nai Cheung	*Contra Costa County Hazardous Materials Programs*
Jamie Conrad	*Conrad Law & Policy*
Jim Cooper	*National Petrochemical & Refiners Assn.*
Ted Cromwell	*American Chemistry Council*
Graham Dalzell	*TBS-Cubed*
David Edwards	*University of Loughborough*
Richard Gowland	*European Process Safety Center*
Jai Gupta	*Indian Institute of Technology Kanpur*
Anna-Mari Heikkila	*VTT*
Louis Higgins	*Rhodia*
Markku Hurme	*Helsinki University of Technology*
John Kindervater	*Eli Lilly*
Hungerbuhler Konrad	*Swiss Federal Institute of Technology Zurich*
Sam Mannan	*Mary Kay O'Connor Process Safety Center*
David Mansfield	*AEAT*
Michael Morris	*Bayer*
Tim Overton	*Dow*
Srinivasan Rajagopalan	*National University of Singapore*
Jari Schabel	*VTT*
John Sleigh	*Consultant*
Angela Summers	*SIS-Tech Solutions*
Jan Windhorst	*Novachem*
Malmen Yngve	*VTT*

1

Introduction

1.1 OBJECTIVES, INTENDED AUDIENCE AND SCOPE OF THIS BOOK

1.1.1 Objectives

The goal of this book is to influence the future state of chemical process evolution by illustrating and emphasizing the merits of integrating process research, development, and design into a comprehensive process that balances safety, capital, and environmental concerns throughout the life cycle of the process. The authors hope that this book will influence the next generation of engineers and chemists, as well as current practitioners and managers in the field of chemical processing.

In 1996, the Center for Chemical Process Safety (CCPS) published the first edition of its inherent safety concept book. Lessons learned since then, combined with the fact that inherently safer design (ISD) is becoming more widely accepted, prompted CCPS to update the book. In the intervening years, several jurisdictions have also mandated consideration of inherently safer design for certain facilities, and similar requirements have been proposed at the Federal level in the United States. Clearly, there is a need for more guidance, especially in practical, step-wise approaches to conduct inherently safer studies. This edition builds on the same philosophy as the first edition, but clarifies the concept with recent research, practitioner observations, added examples and industry methods, and discussions of security and regulatory issues.

The primary objective of this book is to provide a tool that can be easily used by any industrial company that handles hazardous chemicals to better understand inherent safety concepts. In addition, the book provides some tools and guidance on approaches to implement inherent safety.

1.1.2 Intended Audience

This book is written for chemical site managers, process safety managers, engineers, chemists, regulators, engineering educators, and others responsible for chemical safety and interested in the application of inherent safety to process safety management.

1.1.3 Scope

The book covers the history, research, and basic concepts of inherent safety. In particular, it includes guidance on how to conduct inherent safety studies, and to incorporate inherent safety into an organization's process safety management processes. The method described in this book may be widely applicable to inherent safety as it relates to safety, environment, and security issues.

1.2 INTEGRATION OF THIS GUIDANCE WITH OTHER CCPS GUIDANCE

Inherent safety is an integral component of process risk management. It is a foundation topic to managing chemical risks and has been referenced in other CCPS guideline books including:

- *Guidelines for Engineering Design for Process Safety*, published in 1993, includes a chapter (Chapter 2) on inherently safer design.
- The original edition of *Inherently Safer Design, A Life Cycle Approach,* 1996.
- *Guidelines for Design Solutions for Process Equipment Failures,* 1998, includes checklists with inherently safer design options for many types of processing equipment.
- *Guidelines for Implementing Process Safety Management Systems,* 1994.
- *Guidelines for Hazard Evaluation Procedures, Third Edition, with Worked Examples,* 2007.
- *Guidelines for Safe Storage and Handling of Reactive Materials,* 1995.
- *Layer of Protection Analysis, Simplified Process Risk Assessment,* 2001.
- *Guidelines for Identifying and Analyzing the Security Vulnerabilities of Fixed Chemical Sites*, 2002.

1.3 ORGANIZATION OF THE BOOK

The book is written with the key principles for inherent safety in the body of the book. Tools for implementing the approach, as well as worked examples and checklists, are included in the appendices. The contents are as follows:

1. Introduction
2. The Concept of Inherent Safety
3. The Role of Inherently Safer Concepts in Process Risk Management
4. Inherently Safer Strategies
5. Life Cycle Stages
6. Human Factors
7. Inherent Safety and Security
8. Implementing Inherently Safer Design
9. Inherently Safer Design Conflicts
10. Inherently Safer Design Regulatory Initiatives
11. Worked Examples and Case Studies
12. Future Initiatives

Appendix A: Inherent Safety Checklist

Appendix B: Inherent Safety Analysis Approaches

Appendix C: Applying Inherent Safety to Risk Based Process Safety

Chapter 2 introduces the topic of inherent safety. The key terms and the philosophy behind inherent safety are also described.

Inherently safer concepts will enhance overall risk management programs, whether directed toward reducing the frequency or consequences of potential accidents. The different ways in which inherent safety can be applied can be categorized into "strategies." These strategies—**minimize, substitute, moderate, and simplify**—are discussed in detail in this book in Chapters 3 and 4.

The process industry has recognized that a process goes through various stages of evolution. In this book, these stages are called life cycle stages as shown in Figure 1.1. The life cycle of a process begins with discovery at the Research and Development stage. Then, a process progresses through the stages of Conceptual Design, Detailed Engineering, Construction and Startup, Routine Operations, and Plant and Process Modification, including Decommissioning when the process ends.

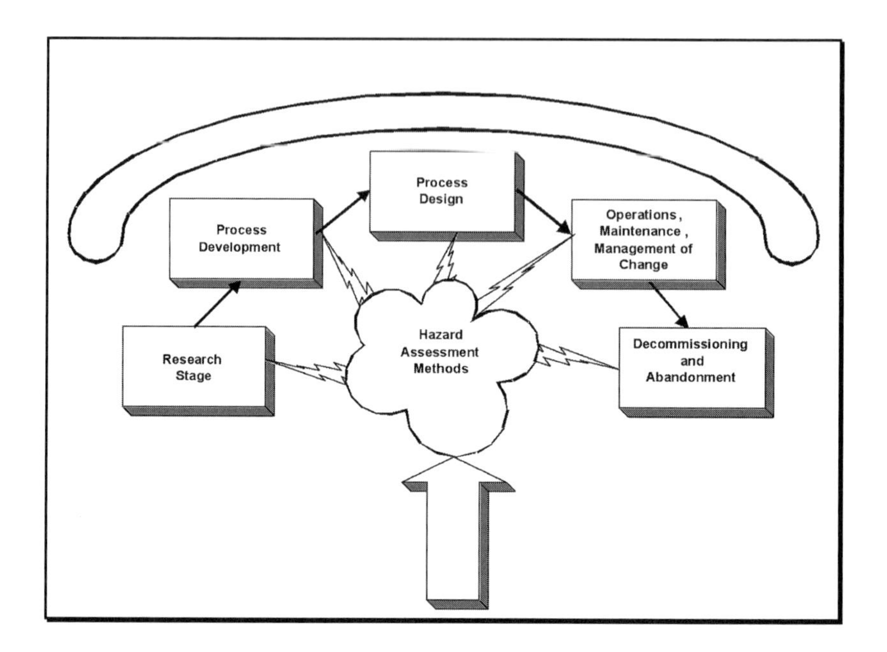

Figure 1.1 Process Life Cycle Stages

Exploring inherently safer alternatives may require more resources during the early stages of development than at other points. However, the resulting understanding will, in many cases, minimize or eliminate the need for appended safety mitigation devices and the costs of maintaining them, as well as reduce the possibility of incidents. Inherently safer considerations may reduce the life cycle cost of the process. In general, the earlier inherently safer thinking is applied to a process, the greater the economic benefits and, in some cases, the feasibility. However, it is never too late to use inherently safer concepts for existing facilities, as it is likely that some positive risk reduction can be achieved even after the facility is operating.

"Inherently safer" is a way of thinking and, to successfully implement it, this thinking has to be continually employed wherever possible. Improved understanding of the process may result in a better process and higher quality products. Processes should be reviewed for hazards and risks periodically. Chapter 5 discusses review methods to do this.

Human factors are an extremely important part of inherently safer concepts. Processes should be designed to reduce the opportunities for human error. Chapter 6 of this book presents a discussion of human factors as related to inherently safer design.

Chapter 7 discusses the role of inherent safety in chemical process security. Chapter 8 presents available methods for implementing inherently safer strategies. These can either be independent, special studies, done periodically or before a major project or change is undertaken, or opportunistically applied integral day-to-day process risk management strategies.

Chapter 9, entitled "Inherently Safer Design Conflicts," describes the conflicts that often develop between the various attributes of safety, operability, cost, and other risk parameters and the ways to understand and make decisions in light of those constraints. With the advent of state and local actual and proposed regulations that require inherent safety consideration or implementation, and proposed Federal regulations for inherently safer design, Chapter 10 was written to help guide regulators and industry through the various considerations and challenges of IS.

Chapter 11 contains examples of IS study methods and case studies to show the step-wise process that can be followed for an IS evaluation. It also gives practical examples of successful implementation.

Lastly, Chapter 12 describes potential future IS initiatives, including needs, research, expected practice issues, and regulatory issues. There is still work to be done to improve the tools available for the application of inherently safer concepts.

In this book, the terms and acronyms Inherent Safety (IS), Inherently Safer Technology (IST), and Inherently Safer Design (ISD) will be used interchangeably. All three terms are used in the literature and in the field.

1.4 HISTORY OF INHERENT SAFETY

Inherent Safety is a modern term for an age-old concept: to eliminate hazards rather than accept and manage them. This concept goes back to prehistoric times. For example, building villages near a river on high ground, rather than managing flood risk with dikes and walls, is an inherently safer design concept.

There are many examples of milestones in the application of inherently safer design. For example, back in 1866, following a series of explosions involving the handling of nitroglycerine, which was being shipped to California for use in mines and construction, state authorities quickly passed laws forbidding its transportation through San Francisco and Sacramento. This action made it virtually impossible to use the material in the construction of the Central Pacific Railroad. The railroad desperately needed the explosive to maintain its construction schedule in the mountains. Fortunately, a British chemist, James Howden, approached Central Pacific and offered to manufacture nitroglycerine at the construction site. This is an early example of an inherently safer design principle – *minimize* the transport of a hazardous material by in situ manufacture at the point of use. While nitroglycerine still represented a significant hazard to the workers who manufactured, transported, and used it at the construction site, the hazard to the general public from nitroglycerine transport was eliminated. At one time, Howden was manufacturing 100 pounds of nitroglycerine per day at railroad construction sites in the Sierra Nevada Mountains. The Central Pacific Railroad's experience with the use of nitroglycerine was quite good, with no further fatalities directly attributed to use of the explosive during the Sierra Nevada construction (Rolt, 1960; Bain, 1999).

Clearly, by today's standards, little about 19th Century railroad construction would qualify as safe, but the in situ manufacture of nitroglycerine by the Central Pacific Railroad did represent an advance in inherent safety for its time. A further, and probably more important, advance occurred in 1867, when Alfred Nobel invented dynamite by absorbing nitroglycerine on a carrier, greatly enhancing its stability. This is an application of another principle of inherently safer design–*moderate*, by using a hazardous material in a less hazardous form (Henderson and Post, 2000).

A milestone in process safety was the 1974 Flixborough explosion in the United Kingdom that caused twenty-eight deaths. On December 14, 1977, inspired by this tragic event, Dr. Trevor Kletz, who was at that time safety advisor for the ICI Petrochemicals Division, presented the annual Jubilee Lecture to the Society of Chemical Industry in Widnes, England. His topic was "What You Don't Have Can't Leak," and this lecture was the first clear and concise discussion of the concept of inherently safer chemical processes and plants.

Following the Flixborough explosion interest in chemical process industry (CPI) safety increased, from within the industry, as well as from government regulatory organizations and the general public. Much of the focus of this interest was on controlling the hazards associated with chemical processes and plants through improved procedures, additional safety instrumented systems and improved emergency response. Kletz proposed a different approach—to change the process to either eliminate the hazard completely or sufficiently reduce its magnitude or likelihood of occurrence to eliminate the need for elaborate safety systems and procedures. Furthermore, this hazard elimination or reduction would be accomplished by means that were inherent in the process, and, thus, permanent and inseparable from it.

Kletz repeated the Jubilee Lecture two times in early 1978, and it was subsequently published (Kletz, 1978). In 1985, Kletz brought the concept of inherent safety to North America. His paper, "Inherently Safer Plants" (1985), won the Bill Doyle Award for the best paper presented at the 19th Annual Loss Prevention Symposium, sponsored by the Safety and Health Division of the American Institute of Chemical Engineers.

Since 1978, interest in inherently safer chemical processes and plants has grown, and that growth was particularly rapid in the 1990s (Kletz, 1996).

Inherently Safer Design (ISD) received additional attention from the chemical industry as a result of a number of major industrial accidents. Two examples are:

- Mexico City (1984) – a series of explosions in an LPG terminal killed approximately 500 people and destroyed the facility.
- Bhopal, India (1984) – the worst disaster in the history of the chemical industry occurred when water entered a storage tank containing methyl isocyanate (MIC). The resulting reaction generated heat and pressure, lifting the tank relief valve and releasing highly toxic MIC vapors into the city. Exact casualty figures are disputed, but tens to hundreds of thousands of people were exposed. In 1994, the official Indian government estimate of fatalities was 4,000 (Hendershot, 2005).

In 1995 and 1996, there were more than 30 papers and presentations related to inherently safer chemical processes given at six different meetings, conferences, and congresses sponsored by the American Institute of Chemical Engineers and the Center for Chemical Process Safety. During these years, inherently safer design was also receiving attention from government and regulatory organizations in the United States and Europe (Ashford, 1993; Lin, et al., 1994; Mansfield, 1994; Moore, 2004), joint industry-government working groups, such as the INSIDE Project in Europe (Rogers, et al, 1995; Mansfield, 1996), and environmental and public interest organizations (Tickner, 1994).

In a similar manner, the work of Etowa et al. (2002) sought to clearly identify the inherent safety features of the Dow Fire & Explosion Index and the Dow Chemical Exposure Index—two commonly used tools in process safety management for relative risk ranking. Goraya, et al. (2004) developed and validated a methodology for incident investigation that makes explicit use of the principles of inherent safety in determining the root causes of an incident. By means of an illustrative example, Gupta et al. (2003) provided evidence of the linkages between inherent safety and the cost of process safety. Their work helps to establish a clear business case for the use of inherent safety principles in management efforts directed at enhancing process safety.

Further motivation for the current research is found in the comments of workers who have reviewed the field of inherent safety and inherently safer design, including Bollinger, et al. (1996), Gupta and Edwards (2002), Kletz (2003), and Khan and Amyotte (2003). For example, Khan and Amyotte (2003) have remarked that the various elements of process safety management can be seen to have at least a partial basis in inherent safety. This fact has been recognized by companies that have incorporated inherent safety as a "named feature" in their safety management documentation and have developed internal standards for the use of inherent safety principles. Yet the term *inherent safety* is typically not named as such in the general description of process safety management systems. According to Bollinger, et al. (1996), explicit use of inherent safety terminology within such management systems is a possible means of furthering the adoption of inherent safety principles in industry.

In recent years, particularly since the terrorist attack on the United States on September 11, 2001, there has been accelerated interest in inherent safety from government, industry, and the public. Inherent safety's promise has produced heightened expectations and it is seen almost as a panacea to reducing risks in the chemical process industries as the public becomes aware of the concept. Inherent safety is mentioned in European safety regulations (Seveso II) and is incorporated into safety and security regulations in parts of the United States. Inherent safety has been proposed as a leading requirement for chemical security regulations in the U.S. Congress.[1]

Some of the best compilations of information on ISD can be found in the works of Kletz (1978, 1998), Lees (1996) and the final report of the INSIDE (INherent SHE in DEsign) project in Europe (1997). Khan and Amyotte (2003) provide an excellent summary of the applications of ISD (2003).

---

[1] See transcript of Senate Environment & Public Works Committee, June 21, 2006 on S. 2145, *Chemical Facility Anti-Terrorism Act of 2006*, and related Senate and House debate on chemical security from the 107th, 108th and 109th Congresses.

2

The Concept of Inherent Safety

2.1 PROCESS RISK MANAGEMENT AND INHERENT SAFETY

The modern approach to chemical process safety is to apply risk management systems theory. This includes recognition of the hazards posed by the process, and a continual effort to analyze the risks, and to reduce or control them to the lowest levels practical, while balancing other business objectives. A **hazard** is classically defined as a "situation with the potential for harm," while **risk** is defined as "the likelihood that a defined consequence (harm) will occur." In other words, *risk* is a function of both *consequence* (hazard) and *likelihood* (frequency).

Safety can be defined as tolerable risk in comparison to the benefit of the activity. This considers who receives the benefit when compared to whom bears the risk. Inherent safety is one of the tools available to improve safety; it is a preferable method, if feasible, but other approaches are valid and successful depending on the specific situation. Inherent safety is used where it meets overall safety objectives, where it is sensible given various alternatives to manage risks, and where there is an application for the inherent safety principles.

Process risk management is the term given to collective efforts to manage process risks through a wide variety of strategies, techniques, procedures, policies, and systems that can reduce the hazard of a process, the probability of an accident, or both. In general, the strategy for reducing risk, whether directed toward reducing the frequency or the consequences of potential accidents, can be classified into four categories:

- *Inherent*—Eliminating the hazard by using materials and process conditions that are non-hazardous; i.e., substituting water for a flammable solvent;

- *Passive*—Minimizing the hazard through process and equipment design features that reduce either the frequency or consequence of the hazard without the active functioning of any device; i.e., providing a diked wall around a storage tank of flammable liquids;
- *Active*—Using controls, alarms, safety instrumented systems, and mitigation systems to detect and respond to process deviation from normal operation; i.e., a pump which is shut off by a high level switch in the downstream tank when the tank is 90% full. These systems are commonly referred to as engineering controls, although human intervention is also an active layer;
- *Procedural*—Using policies, operating procedures, training, administrative checks, emergency response, and other management approaches to *prevent* incidents, or to *minimize* the effects of an incident; i.e., hot work procedures and permits. These approaches are commonly referred to as administrative controls.

All four categories can contribute to the overall safety of a process. Ideally, the steps of analyzing, reducing, and managing risk will be considered in a hierarchical manner as shown in Figure 2.1.

Inherent safety uses the properties of a material or process to eliminate or reduce the hazard. The fundamental difference between inherent safety and the other three categories is that inherent safety seeks to *remove* the hazard at the source, as opposed to accepting the hazard and attempting to mitigate the effects. If implementing inherently safer approaches alone to meet project risk goals is feasible, other layers of protection—and their associated costs in time, capital, and expenses—may not be required.

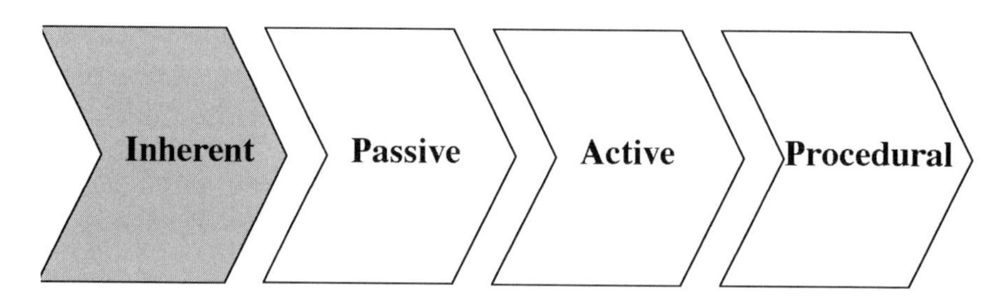

Figure 2.1: Hierarchy of Process Risk Management Strategies.

2.2 INHERENT SAFETY DEFINED

in·her·ent: Adjective. Existing as an essential constituent or characteristic; intrinsic. From the Latin *inharens, inhaerent-*, present participle of *inhaerere,* to inhere. [1]

What do we mean when we speak of "inherent safety" or an "inherently safer" chemical process? "Inherent" has been defined as "existing in something as a permanent and inseparable element, quality, or attribute" (American College Dictionary, 1967). **Inherent safety** is a concept, an approach to safety that focuses on eliminating or reducing the hazards associated with a set of conditions. A chemical manufacturing process is **inherently safer** if it reduces or eliminates the hazards associated with materials and operations used in the process *and* this reduction or elimination is permanent and inseparable. The process of identifying and implementing inherent safety in a specific context is called **inherently safer design**. A process with reduced hazards is described as inherently safer compared to a process with only passive, active and procedural controls.

An inherently safer process should not, however, be considered "inherently safe" or "absolutely safe." While implementing inherent safety concepts will move a process in the direction of reduced risk, it will not remove all risks. No chemical process is without risk, but all chemical processes can be made safer by applying inherently safer concepts.

2.3 INHERENTLY SAFER APPROACHES

Where feasible, reducing or eliminating hazards through changes in materials, chemistry or process variables is preferable to adding layers of safety to a process. While layers of passive, active or procedural controls *will* reduce the risk, they will do nothing to reduce the nature of the hazard itself.

A hazard is a physical or chemical characteristic that has the potential to cause harm to people, the environment, or property (adapted from CCPS, 1992). Hazards cannot be changed—they are intrinsic to the material or to its conditions of storage or use. They can, however, be removed or mitigated.

[1] Definition taken from *The American Heritage® Dictionary of the English Language, Fourth Edition* Copyright © 2000 by Houghton Mifflin Company.

Chemical process hazards come from two sources:

- hazards that are *characteristic of the materials and chemistry used,* and
- hazards that are *characteristic of the process variables,* or the way the chemistry works within the process.

Examples of Chemical Process Hazards

- Chlorine is toxic if inhaled
- Concentrated sulfuric acid is corrosive to the skin
- Ethylene is flammable
- Steam confined in a drum at 600 psig contains a significant amount of potential energy and may explode
- Acrylic acid monomer can polymerize, releasing large amounts of heat

---

Inherently safer design can either change the conditions under which a hazardous chemical process is conducted (change the characteristic of the process variables) or completely eliminate the hazardous agent (change the material and chemistry used in the process). Changes in the chemistry of a process that reduce the hazard of the chemicals used or produced can be considered **First order Inherent Safety**. Changes in process variables can be considered **Second order Inherent Safety**.

Such approaches can be implemented to address any of the three stages that most accidents follow:

- *Initiation*: the event and contributing factors that start the incident;
- *Propagation*: the events that maintain or expand the incident;
- *Termination*: the outcome of the events, including the sub-events that stop the incident or diminish it in consequences.

The most effective strategies will prevent initiation of the accident. Inherently safer design can also reduce the potential for propagating an accident or terminate the accident sequence before there are major impacts on people, property, or the environment.

Inherently safer design should be an essential aspect of any process safety program. If the hazards can be eliminated or reduced, the extensive layers of protection to control those hazards may not be required or may be less robust.

However, inherent safety is not the only process risk management strategy available and may not always be the most effective. A system of strategies that includes both inherently safer design and additional layers of protection may be needed to reduce risks to an acceptable level.

2.4 LAYERS OF PROTECTION

Layers of Protection[2] is a concept whereby multiple devices, systems, or actions are provided to reduce the likelihood of an undesirable event (Figure 2.2). This strategy works on the premise that multiple layers of protective features and countermeasures reduce the potential for the event to occur. Each layer should be independent of the other layers, such that each device, system, or action is capable of preventing the event from proceeding to the undesired consequence regardless of the initiating event or the action of any other protection layer. Independent layers of protection are inherently safer than layers of protection that share common elements. This is a general assumption and the actual reliability and robustness may vary.

The passive, active, and procedural process risk management strategies mentioned earlier are considered "layers of protection," as they involve the addition of safety devices or work processes to reduce risk. Passive safety devices do not perform any fundamental operation, but are designed to reduce the damage when a process upset occurs. Procedural safety measures, or administrative controls, utilize safe work practices and procedures to reduce risk.

Independent protection layers (IPLs) may include control systems, operator supervision, alarms, safety instrumented systems, and mitigation systems. Creating layers of protection between a hazard and potentially impacted people, property and the environment can be highly effective, and its application has significantly improved the safety record of the chemical industry. However, such an approach may have significant disadvantages:

- *The hazard remains, and some combination of failures of the layers of protection may result in an incident, thereby allowing the hazard to be fully realized.* Every layer has a certain likelihood of failure, due either to mechanical means or management systems failures, such as not maintaining or keeping administrative controls active. The outcome of the event may be limited to whatever passive or inherent layers have been applied. If the overall risk was considered low because of those layers, there could be substantial residual consequences.

---

[2] See *Layer of Protection Analysis: Simplified Process Risk Assessment* (CCPS 2001) for a more in-depth discussion of layers of protection.

- *Potential impacts could be realized by some unanticipated route or mechanism.* Nature may be more creative in inventing ways by which a hazardous event can occur than experts are in identifying them. Accidents can occur by mechanisms that were unanticipated or poorly understood.
- *The layers of protection can be expensive to build and maintain throughout the life of the process.* Initial capital expense, and costs of operation, safety training, and maintenance, along with diversion of scarce and valuable technical resources into maintenance and operation of the layers of protection can add significantly to plant operating budgets.

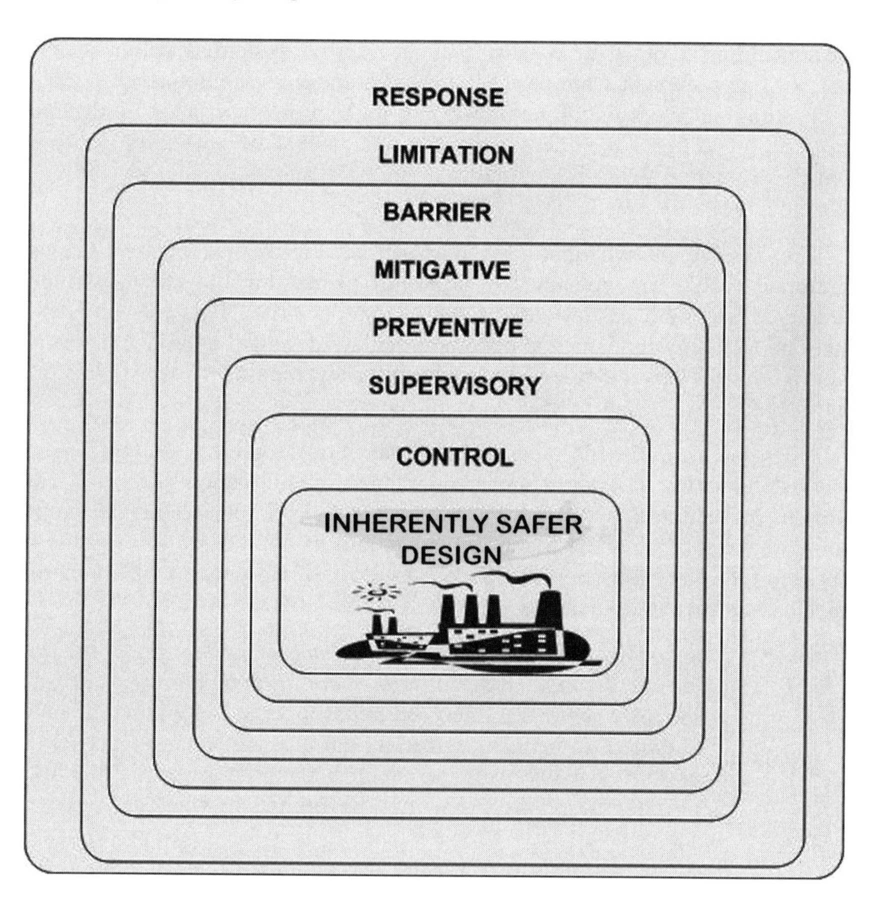

Figure 2.2 Layers of Protection

2.5 LEVELS OF INHERENT SAFETY

The steps of managing risk should ideally be done in a hierarchical manner and iteratively, as shown in Figure 2.3. The process risk management approach illustrated begins with the definition of goals for managing risk. Without a clear definition, the need for further investment in safety is unclear short of any regulatory requirements. This is important in determining how "inherently safe is safe enough."

Whether or not a particular safety feature in a chemical process is "inherent" can be debated. Such discussion may arise, in part, because different people have different perspectives on risk, or view the process at different levels of resolution, ranging from a global view of the entire process to a very detailed view of one specific feature. One could also be referring to addressing the hazards of the process vs. addressing other aspects of managing the risk of the process hazards.

In the strictest sense, or the **First order view of Inherent Safety,** one could argue that the definition of inherently safer applies only to the elimination of a hazard. Inherently safer strategies may absolutely eliminate a hazard, while hopefully not introducing another hazard of concern as a result.

Alternatively, inherently safer approaches could instead treat the hazard by making it less intense or less likely to occur. Such approaches can be labeled the **Second order view of Inherent Safety**. Such approaches are clearly in line with inherent safety philosophy, but may not be as powerful as a First order change. In the Second order of IS, the hazard is only reduced through the application of IS principles. It could be that Second order Inherently Safer design options result in a considerable reduction of hazard and, therefore, the risk is adequately addressed.

In the broadest sense, the overall hazard is not eliminated or reduced by way of Second order Inherently Safer strategies but, instead, sublevel hazards are minimized and the likelihood of the event occurring is reduced by adding layers of protection. The strength and reliability of a layer of protection can vary, with some layers designed to be more "robust" than others. This could mean it is more reliable, effective, simpler, or any other positive safety attributes achievable by comparison, but the fundamental hazard may still exist. This is the difference between the inherently safer design concepts being applied to the hazard, and layers of protection being applied to reduce the overall risk.

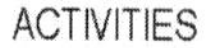

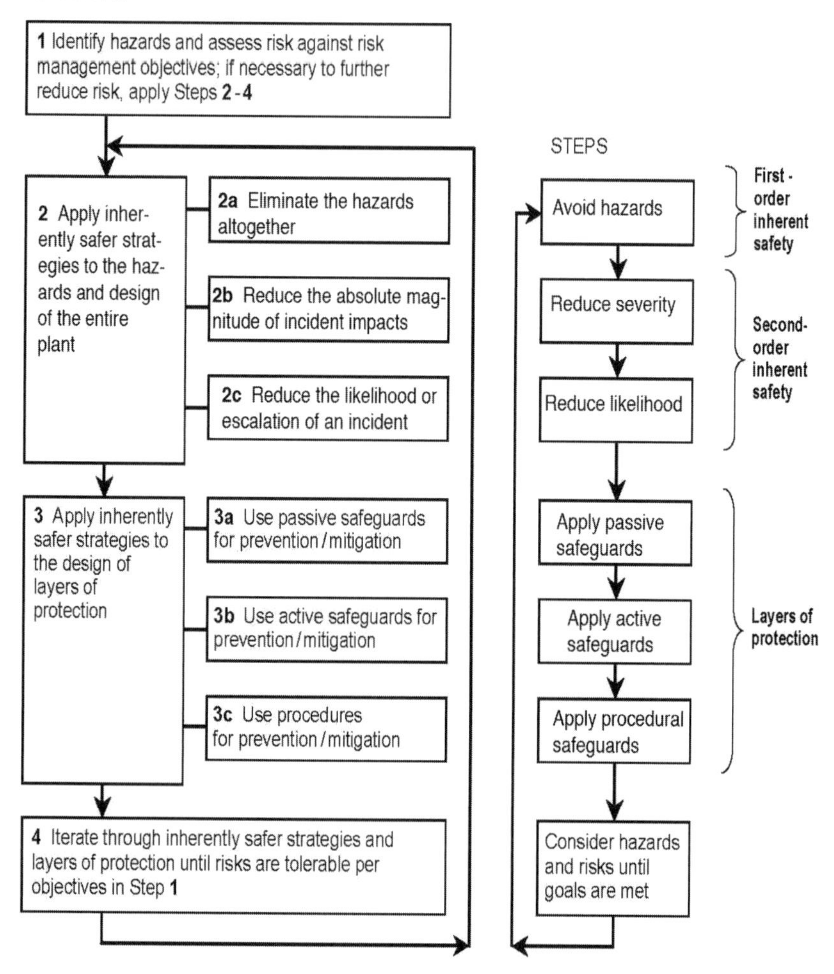

Figure 2.3 Inherent Safety Considerations in Process Risk Management (adapted from Amyotte, et al., 2006)

In considering process safety alternatives, it is important to consider the total life cycle costs and risks. There is an increasing interest in this concept in the environmental area, with recognition of the need to incorporate waste treatment, waste disposal, regulatory compliance, potential liability for environmental damage, and other long-term environmental costs into project economic evaluation. Similarly, we must consider life cycle *safety* costs. Some examples of the safety factors that should be considered when evaluating costs include:

- Capital cost of safety equipment, including active barriers and systems such as instrumented safety systems and mitigation systems.
- Capital cost of passive barriers, such as containment dikes, and vacant land to provide spacing, required by codes, regulations and insurers.
- Operating and maintenance costs for instrumented safety systems, fire protection systems, personal protective equipment, and other safety equipment.
- Increased maintenance cost for process equipment due to safety requirements, such as safety permits, cleaning and purging equipment, personal protective equipment, training, and restricted access to process areas.
- Operator safety training costs for hazardous materials or processes.
- Regulatory compliance costs.
- Insurance costs.
- Potential property damage, product loss, and business interruption costs if an incident occurs.
- Potential liability if an incident occurs.

In addition, the cost differential of the physical process technology, licensing, operating costs, and equipment must be considered in the trade-off analysis. This has to be weighed against the change in risk from adopting one alternative or the other.

2.6 WORKED EXAMPLE

Joe, a process control engineer, has the task to add layers of protection to reduce the risks from a process that uses flammable materials and operates at elevated pressure. One protection layer is a safety instrumented function (SIF) designed to prevent overpressure in the process vessel that could lead to catastrophic rupture and fire or explosion. He describes this particular SIF as "inherently safer" because it uses diverse multiple sensing elements, compared to an alternative design which uses multiple, but identical, sensors. Reliable diverse sensors are available and, unlike the identical sensing elements in this application, do not have the potential for common mode failure. [3]

A second proposed protection layer is a relief valve to a flare system. If the relief system is designed for the full heat input load to the vessel, then it would be considered as mitigation for the consequences of catastrophic vessel rupture and, if it is successful, the consequence would be less severe. The full-load relief system would be considered a passive SIF protection layer, but the flare might be considered an active system. Joe considers the SIF as inherently safer than the relief system to flare because the SIF stops the scenario without a release, while the relief system to flare only reduces the potential impact of the release. Depending on the risk reduction required, Joe may need to build both protection layers. The relief system to flare can be designed for the other high-pressure challenges that generate a smaller relief volume, such as the fire case. API Standard 521 recognizes this approach.

A third layer of protection might be a deluge system that protects against major damage to the facility, but does not protect against catastrophic rupture, flammable vapor release, or fire. If successful, the deluge system will stop a small fire from becoming a larger fire. Joe considers automatic activation of the deluge system by flammable vapor sensor to be inherently safer than activation by heat sensing detectors or nozzles since the sensors can start the water *before* a fire actually occurs. Likewise, Joe considers the activation by heat sensing detectors or nozzles to be inherently safer than manual activation of the deluge system because these sensors do not depend on an operator being present. Joe is looking at the process at a very detailed level, considering the characteristics of different layers of protection.

On the other hand, Mary, a research process development engineer, does not consider Joe's system to be inherently safer, because a truly inherently safer system would not require an SIF at all. The process uses flammable materials and operates at elevated pressure. Mary, looking at the entire process, would only consider it to be inherently safer if the flammable materials were eliminated or the process were operated at ambient pressure or the hazard is minimized in some other fundamental way. Mary is considering the inherent safety characteristics of the entire process, rather than one single element, such as an SIF.

Using the strict definition of "inherent safety" as only reducing hazard, Joe's design may have adequately reduced risk, but Mary's design made use of inherently safer design strategies. Joe's SIF is indeed more robust as a layer of protection than the alternative using identical sensors, but it is still part of a process that is inherently less safe than other feasible alternatives. Mary is

---

[33] Note that in many process systems, identical sensors may give better risk reduction than diverse sensors because there may be only one reliable sensor technology. The tradeoffs must be balanced, and the risk reduction for the SIF calculated as described in CCPS (2007).

searching for that inherently safer process, which may make Joe's layers of protection unnecessary. However, until she finds that alternative, it is shown to be feasible, and it is actually implemented, Joe's choices may be the best way to achieve the required risk reduction.

2.7 SUMMARY

Inherently safer design is a fundamentally different way of thinking about the design of chemical processes and plants. It focuses on the elimination or reduction of the hazards, rather than on risk management and control. This approach should result in safer and more robust processes, and it is likely that these inherently safer processes will also be more economical in the long run (Kletz, 1984, 1991b). It must be recognized, however, that an inherently safer plant is not necessarily the most efficient or economical, as the cost of changing an existing design to a more inherently safer technology may be unjustified or difficult to justify from an investment standpoint. For this reason, the options must be holistically weighed and the total Life Cycle costs and risks be analyzed for completeness.

Eliminating or reducing the hazard through the application of inherently safer design is the first element in the process risk management hierarchy. Other strategies—passive, active or procedural—constitute layers of protection around the hazard.

This book encourages engineers to make conservative choices that apply the principles and spirit of inherently safer design to the extent feasible. The discussion and examples range from basic process chemistry through details of the design of hardware and procedures. The authors recognize that some examples illustrate processes that would not be described as "inherently safe." However, inherently safer thinking can be applied at all levels of the process design, from the overall concept through the detailed equipment design and procedure development.

3

The Role of Inherently Safer Concepts in Process Risk Management

3.1 INTEGRATING INHERENT SAFETY IN PROCESS RISK MANAGEMENT SYSTEMS

How does inherent safety fit into an overall process risk management program? As discussed in the previous chapter, "risk" is defined as "the likelihood that a defined consequence will occur." Any effort to reduce the risk arising from the operation of a chemical processing facility can be directed toward reducing the likelihood of incidents (incident frequency), reducing the magnitude of the loss, injury or damage should an incident occur (incident consequences), or some combination of both.

A key engineering risk tool is a management system appropriate for the risks being addressed (i.e., health, occupational safety, process safety, equipment reliability, etc.). Process safety management (PSM) can be defined as the application of management principles to ensure the safety of chemical process facilities. Safety management systems are comprised of elements—management of change, process hazards analysis, mechanical integrity, and others. These systems are recognized and accepted worldwide as best-practice methods for managing risk. They typically consist of 10 to 20 program elements that must be effectively carried out to manage the risks in an acceptable way.

Inherent safety strategies can be applied to these process safety management elements. As an example, by including inherent safety guidewords in a management of change (MOC) program, the MOC protocol recognizes inherent safety as both a driving force for—and as an opportunity during—implementation. However, inherently safer design is a way of thinking involving a holistic approach, considering the process as a system with interacting concerns, such as toxicity, flammability, reactivity, stability, quality, capital, operating cost, transportation risks and site factors. These considerations make it important to understand how to manage tradeoffs and definition of acceptable risks—be they safety, environmental, or financial in nature. Containment of energy, pressure, temperature, or chemicals will be an important consideration in many inherently safer suggestions, but many constraints will be placed on the decision-making process due to economic limitations or perceived risks associated with new approaches. It must be remembered that there is no single answer to the question of what is the best way to build or operate a safe industrial facility that manufactures, stores, processes or uses chemicals. Inherently safer does not always equal "safer," but is one tool for reducing risk. Inherent safety must be used in conjunction with other safety methodologies.

Section 3.5 introduces four strategies for identifying and implementing inherently safer processes. These **inherent safety strategies** can be integrated into to the implementation of any of the four **process risk management strategies** - inherent, passive, active, and procedural (Table 3.1).

An inherently safer process can offer greater safety potential, often at a lower cost. However, selection of an inherently safer approach does not guarantee that the actual implementation of those approaches will result in a safer operation than an alternate process that is safer due to multiple layers of protection. The traditional strategy of providing layers of protection for a hazardous process can be quite effective, although the expenditure of resources to install and maintain the layers of protection may be very large. In some cases, the benefits of the inherently more hazardous technology will be sufficient to justify the costs needed to provide the layers of protection required to reduce its risk to a tolerable level.

Hendershot (1995b) compares the inherent safety characteristics of air and automobile transportation, and concludes that automobile transportation is inherently safer for reasons such as:

- The automobile, on the ground, will coast to a stop in case of engine failure, while the airplane will rapidly descend and may not be able to land safely.
- The automobile travels at a lower speed.
- The automobile contains a smaller inventory of passengers.
- The control of an automobile is simpler (in two dimensions) compared to the airplane, which must be controlled in three dimensions.

Table 3.1: Examples of Process Risk Management Strategies *(CCPS, 1993a)*

Risk Management Strategy	Example	Comments
1. Inherent	An atmospheric pressure reaction using non-volatile solvents that are incapable of generating any pressure in the event of a runaway reaction.	There is no potential for reactor overpressure because of the chemistry and physical properties of the materials.
2. Passive	A reaction done in a 250 psig reactor capable of generating 150 psig pressure in case of a runaway,	The reactor can contain the runaway reaction. However, if 150 psig pressure is generated, the reactor could fail due to a defect, corrosion, physical damage or other cause.
3. Active	A reaction capable of generating 150 psig pressure in case of a runaway, done in a 15 psig reactor with a 5 psig high pressure SIS to stop reactant feeds and a properly sized 15 psig rupture disk discharging to an effluent treatment system.	The SIS could fail to stop the reaction in time, and the rupture disk could be plugged or improperly installed, resulting in reactor failure in case of a runaway reaction. The effluent treatment system could fail to prevent a hazardous release.
4. Procedural	The same reactor described in Example 3 above, but without the 5 psig high pressure SIS. Instead, the operator is instructed to monitor the reactor pressure and stop the reactant feeds if the pressure exceeds 5 psig.	There is a potential for human error, the operator failing to monitor the reactor pressure, or failing to stop the reactant feeds in time to prevent a runaway reaction.

Note: These examples apply specifically to categorization of risk management strategies with respect to high-pressure hazards caused by a runaway reaction. The processes described may involve trade-offs with other risks arising from other hazards. For example, the non-volatile solvent in the first example may be extremely toxic, and the solvent in the remaining examples may be water. Decisions on process design must be based on a thorough evaluation of all of the hazards involved.

However, the benefits of air transportation, primarily speed, make it an attractive alternative for longer trips. These benefits have justified the expenditure of large amounts of money to provide extensive layers of protection to overcome the inherent hazards of air travel, including extensive redundancy in aircraft construction, management of airplane movement by the air traffic control system, rigorous maintenance of equipment, and many other systems. The result is that air travel, while inherently more hazardous, is, in fact, safer than automobile travel for long trips. Moreover, the rate of fatal accidents dropped from 1982-2005. Similar situations can be expected to occur in the chemical process industry (Hendershot, 2006b).

Even when we determine that the benefits of an inherently less safe technology justify its use, we should still continue to look for inherently safer alternatives. Technology continues to evolve and advance, and inherently safer alternatives that are not economically attractive today may be very attractive in the future. The development of these new technologies offers the promise of meeting process safety goals more reliably and economically.

Some other examples illustrate the concepts of inherent safety, and demonstrate that inherently safer design is applicable to many technologies, not just the chemical industry:

- A railroad crossing with gates and flashing lights (alarms, barriers, procedures) is not as inherently safe as an overpass or underpass;
- Traffic lights are not as inherently safe as grade-separated highway interchanges;
- Carpeting can be put on stairway steps to reduce the risk of slipping; a ramp replacing the steps would be an inherently safer approach.

3.2 TIMING FOR CONSIDERATION OF INHERENTLY SAFER OPTIONS

The search for inherently safer process options begins early and continues throughout the process life cycle. The greatest potential opportunities for impacting process safety occur early in process design and development. As stated by the National Research Council (Design, 1988), "Few basic decisions affect the hazard potential of a plant more than the initial choice of technology." Early in development, there is a great deal of freedom in the selection of chemistry, solvents, raw materials, process intermediates, unit operations, plant location, and other process parameters. As the process moves through its life cycle, it becomes more difficult and expensive to change the basic process.

Marshall (1990, 1992) discusses accident prevention, control of occupational disease, and environmental protection in terms of strategic and tactical approaches. Strategic approaches have a wide significance and represent "once and for all" decisions. The inherent and passive categories of risk management would usually be classified as strategic approaches. In general, strategic approaches are best implemented at an early stage in the process or plant design. Tactical approaches, which include active and procedural risk management categories, tend to be implemented much later in the plant design process, or even after the plant is operating, and often involve much repetition, increasing costs, as well as the potential for failure. However, it is never too late to consider inherently safer alternatives. Major enhancements to inherent safety have also been reported in plants that have been operating for many years (CCPS, 1993a; Wade, 1987; Carrithers, et al., 1996).

To be most effective, an inherent safety program should have as its objective creating an awareness of inherent safety in a broad range of chemists and engineers involved in the development of products and processes throughout an organization. The inherently safer way of doing things should become a way of thinking and working. Perhaps the critical moment in the life of any idea occurs immediately after the idea springs into a person's head. Does he/she pursue the idea, file it away for further thought, discuss it with a colleague, or just forget it as being of no further interest? If everyone in the organization understands that inherently safer products and processes are valued and desired, it is much more likely that ideas with the potential to develop into inherently safer systems will survive this critical moment, and will grow and mature.

3.3 INHERENT SAFETY CONSTRAINTS

Inherent safety is not a panacea and does not result in a zero risk situation. There will always be residual risks. It is rare that a process alternative will be inherently safer with respect to every hazard. Furthermore, with new technology, the hazards may not be understood as well, and, in fact, some hazards may not be known at all. The conflicts and constraints of inherently safer design are explored in Chapter 9.

3.4 RESOLVING INHERENT SAFETY ISSUES

Deciding among a number of process options that have both inherent safety advantages and disadvantages with respect to different hazards can be quite difficult. The first step is to *identify and understand all hazards* associated with the process options. Process hazard analysis and evaluation techniques are appropriate tools (CCPS, 1992). These include:

- past history and experience
- interaction matrices
- hazard evaluation techniques such as checklists, what-if analysis and HAZOP analysis

The hazard identification step is perhaps the most important, because any hazard not identified will not be considered in the decision process. For example, the impact of chlorofluorocarbons on atmospheric ozone was unknown for much of the time these chemicals were used. The potential hazard of chlorofluorocarbons was not considered until recent years.

Once the hazards have been identified, the process options can be ranked in terms of inherent safety with respect to all identified hazards. This ranking can be qualitative, with hazards placed into consequence and likelihood categories based on experience and engineering judgment (CCPS, 1992). Quantitative rankings for more specific types of hazards can also be used. Two examples of the latter are the *Dow Fire and Explosion Index* (Dow, 1994b; Gowland, 1996a) and the *Dow Chemical Exposure Index* (Dow, 1994a). A number of other quantitative tools for evaluating inherent safety have also been developed(Khan and Amyotte, 2005; Heikkilä, 1999; Edwards, et al., 1996). Unfortunately, none of these indices considers the full range of hazards. To get an overall assessment of process options, it is necessary to use a variety of indices and qualitative techniques, and then combine the results.

Sometimes the consequences of all hazardous incidents can be expressed by a single common measure, such as the dollar value of property damage, total economic loss, risk of immediate fatality due to fire, explosion or toxic material exposure. If all consequences can be measured on a common scale, the techniques of quantitative risk analysis (CCPS, 1989a, 1995b) may be useful in assessing the relative magnitude of various hazards, and in understanding and ranking the total risk of process options.

But, in many cases, it is not readily apparent how the potential impacts from different hazards can be translated into some common scale or measure. For example, how do you compare long term environmental damage and health risks from use of CFC refrigerants to the immediate risk of fatality from the fire, explosion, and toxicity hazards associated with many alternative refrigerants? This question does not have a "right" answer. It is not really a scientific question, but instead it is a question of societal risk. Individuals, companies, and society must determine how to value different kinds of risks relative to each other, and base decisions on this evaluation.[1]

3.5 INHERENTLY SAFER STRATEGIES

Approaches to the design of inherently safer processes and plants have been grouped into four major strategies by IChemE and IPSG (1995) and Kletz (1984, 1991b):

---

[1] See Chapter 9 of this book for a discussion of decision-making tools that can introduce additional rigor, consistency, and logic into the decision process.

Minimize Use smaller quantities of hazardous substances (also called *Intensification*).

Substitute Replace a material with a less hazardous substance.

Moderate Use less hazardous conditions, a less hazardous form of a material, or facilities that minimize the impact of a release of hazardous material or energy (also called *Attenuation* and *Limitation of Effects*).

Simplify Design facilities which eliminate unnecessary complexity and make operating errors less likely, and which are forgiving of errors that are made (also called *Error Tolerance*).

These inherently safer design strategies are discussed in more detail in Chapter 4. Examples can also be found in Chapter 5, which discusses inherently safer design opportunities through the life cycle of a chemical process.

During the development of the first edition of this concept book, CCPS consolidated all the strategies found in these publications into the four major categories noted above above. Table 3.2 presents a mapping of the original concepts to the CCPS ISD strategies as restated by Kletz in 1998.

3.6 SUMMARY

Inherently safer design represents a fundamentally different approach to chemical process safety. Rather than accepting the hazards in a process, and then adding on safety systems and layers of protection to control those hazards, the process designer is challenged to reconsider the process and eliminate the hazards. If the designer cannot eliminate the hazards, the challenge becomes to minimize or reduce them as much as possible by modifying the process, rather than by adding external layers of protection.

Process risk management strategies can be categorized as inherent, passive, active, and procedural. Inherently safer design strategies can be applied to any of these risk management strategies, as illustrated in Figure 2.3. It is possible, for example, to describe one procedure as inherently safer than another, perhaps because the “inherently safer” procedure is simpler. However, both procedures would still represent valid approaches to process risk management.

Inherent safety is one tool available to improve process safety, but is not the only tool and is most often used in conjunction with other safety practices, equipment, and methodologies. While inherently safer designs can most easily be developed and implemented in the earliest phases of process development and design, opportunities to apply inherent safety strategies may exist throughout a project’s life cycle.

Finally, in considering inherently safer design alternatives, it is essential to remember that there are often, and perhaps always, conflicting benefits and liabilities associated with the different options. Chemical processes usually have many potential hazards, and a change that reduces one hazard may create a new one or increase the magnitude of another existing one. It is essential that the process designer retain a broad overview of the process when considering alternatives that he/she remains aware of **all** hazards associated with each process option, and that appropriate tools are applied to choose the overall best option

Table 3.2: Strategies for Inherently Safer Design

Kletz (1998)	**CCPS (2007)**
Inherent Safety Strategies	**Inherent Safety Strategies**
• Intensification	• Minimize
• Substitution	• Substitute
• Attenuation	• Moderate
• Limitation of Effects	• Moderate and Simplify
"Friendly Plant Design" Strategies regarded as "add on" features	**Mapping of Kletz's "Friendly Plant Design" Strategies to CCPS Inherent Safety Strategies**
• Simplification	• Simplify
• Avoiding knock-on effects	• Moderate
• Making incorrect assembly impossible	• Simplify
• Making status clear	• Simplify
• Tolerance of misuse (error tolerance)	• Moderate
• Ease of control	• Moderate
• Understandable software for computer control	• Simplify
• Instructions and procedures	• Simplify
• Life cycle friendliness (construction and demolition)	• Consider all strategies with regard to all phases of a plant life cycle
• Passive safety	• Moderate

4

Inherently Safer Strategies

4.1 DEFINITION OF INHERENTLY SAFER STRATEGIES

Chapter 3 introduced four strategies for inherently safer design:

Minimize	Moderate
Substitute	Simplify

These four strategies form a protocol by which the risks associated with the loss of containment of hazardous materials or energy can be significantly reduced, and in some cases eliminated. The elimination of risk due to loss of containment is very difficult, if not impossible to achieve using other risk reduction measures, i.e., active or passive safeguards. These measures, while effective if installed and maintained properly, generally reduce the likelihood of release, and sometimes will mitigate the consequences of a release. However, they cannot reduce the risk to zero. Truly, Trevor Kletz's statement "What you don't have can't leak" embodies the ultimate goal (First order view) of inherently safer strategies, and describes the elimination of the risk of hazardous materials releases. However, while they are highly effective techniques, it is usually not possible to eliminate all process-related risks using inherently safer strategies.

As previously noted, some researchers and practitioners include other strategies within the inherent safety definition. In general, these may be considered subsets of the four core principles. However, in the literature, they are sometimes described as separate strategies (Amyotte, 2003; Khan, 2003; Overton and King, 2006; Lutz, 1995a). These inherent safety strategies include:

- Limitation of effects
- Knock-on effects
- Avoid incorrect assembly
- Making status clear
- Inherent robustness

This chapter provides examples of these five and other strategies. How IS concepts can be applied to the various stages of the process life cycle is discussed in Chapter 5, while additional examples of IS concepts are presented in Chapters 5, 7, 9, and 11.

While inherently safer strategies offer risk reduction measures that are not possible with active or passive safeguards, the application of inherently safer designs and measures is not without its own problems. This is because hazards are interrelated in a process, and the elimination or reduction of one risk may introduce new risks or cause existing ones to become worse. Both components of risk—both frequency of occurrence, and its consequences—must be evaluated and compared for each risk affected by the implementation of the IS measure, and a decision must be made as to which path forward is appropriate. It may be that the risk without implementing certain IS measure(s) is actually *lower* than with it implemented.

4.2 MINIMIZE

In the context of IS, *minimize* means to reduce the quantity of material or energy contained in a manufacturing process or plant. Process minimization is often thought of as resulting from the application of innovative new technology to a chemical process, for example, tubular reactors with static mixing elements, centrifugal distillation techniques, or innovative, high surface area heat exchangers. These types of minimization strategies are also discussed in this chapter. However, much can be accomplished in process inventory reduction simply by applying good engineering design principles to more conventional technology.

Often, the inventory of hazardous materials on-site is driven by operational and business considerations, particularly in the number of transportation containers that are stored or used on-site at any given time. Railroad dispatching schedules, trucking schedules, and other transportation related issues, most of which are established independently of safety considerations, often influence the amount of hazardous materials present on-site. Sometimes inventories are determined by on-site purchasing considerations, such as timing of incoming or outgoing shipments related to price. Careful coordination with shippers and carriers is required in order to minimize inventories related to transportation scheduling.

Changing operational and maintenance practices can also help minimize inventories. For example, the application of reliability-centered maintenance techniques can also increase the inherent safety of a plant by reducing plant downtime, thus reducing the need for intermediate inventory and storage. This in-process storage or surge capacity may be required to allow portions of the plant to continue to operate while other parts are shut down for equipment maintenance. Improving the reliability of critical pieces of equipment may eliminate or significantly reduce the need for in-process storage of hazardous chemical intermediates.

When designing a process facility or unit, the dimensions of every item of process equipment should be specified as large enough to accomplish its intended purpose, and no larger. Required surge capacities, either for normal operations or for emergency situations, sometimes demand larger equipment. They are part of the intended purpose of a process design, and must be maintained. But, this extra space must be kept empty and unused, and the process should not be modified in the future to accommodate additional process capacity. Raw material and in-process intermediate storage tanks should be minimized, if feasible. One should question the need for all in-process inventories, particularly of hazardous materials.

Minimizing the size of equipment not only enhances inherent process safety, but it can often save money. If equipment can be eliminated from a manufacturing process, it may eliminate the need for designing, engineering, purchasing, operating, or maintenance costs. Equipment which is eliminated also cannot leak or release hazardous material or energy into the surrounding environment. The true engineering art is to determine how to accomplish a given task with a minimum of equipment, and with the required equipment of the smallest size. Siirola (1995) discusses process synthesis strategies that are helpful in designing and optimizing a process route to minimize the required equipment and operations.

The term "process intensification" is used synonymously with "minimization," though the former is often used more specifically to describe new technologies that reduce the size of unit operations equipment, particularly reactors. In recent years, innovative process intensification techniques have received, and continue to receive, more attention. An international conference on process intensification (Akay and Azzopardi, 1995) held several years ago presented several interesting possibilities for a range of unit operations, including reaction, gas-liquid contacting, liquid-liquid separation, heat exchange, distillation, and separation. While the focus in process intensification in general, and at this conference in particular, is on improving process efficiencies and economics, many of the technologies described can also improve the inherent safety of processes by reducing in-process inventories resulting from their application.

Process intensification includes the following novel techniques and designs to minimize the size, inventory, and energy consumption of process equipment (Stankiewicz, 2004):

- Equipment:
  - Reactors: spinning disk reactors, static mixer reactors, microreactors.
  - Non-reaction equipment: static mixers, compact heat exchangers, packed bed contactors, centrifugal absorbers.

- Methods:
  - Multifunctional reactors: heat-integrated reactors, reactive separations, reactive extrusion, fuel cells.
  - Hybrid separations: membrane adsorption, membrane distillation, adsorptive distillation
  - Alternative energy sources: centrifugal fluids, ultrasound, solar energy, microwave, electric fields, plasma technology.
  - Other methods: supercritical fluids, process synthesis.

A few examples of process minimization will be presented here. See Kletz (1984, 1991b), Englund (1990; 1991a,b; 1993), IChemE and IPSG (1995), Lutz (1995a, b) and CCPS (1993a), and Stankiewicz (2004) for more examples.

4.2.1 Reactors

Reactors can represent a large portion of the risk in a chemical process. A complete understanding of reaction mechanism and kinetics is essential to the optimal design of a reactor system. This includes the chemical reactions and mechanisms, as well as physical factors, such as mass transfer, heat transfer, and mixing. A reactor may be large because the chemical reaction is slow. However, in many cases, the chemical reaction actually occurs very quickly, but it appears to be slow due to inadequate mixing and contacting of the reactants. Innovative reactor designs that improve mixing may result in much smaller reactors. Such designs are usually cheaper to build and operate, as well as being safer due to smaller inventory. In many cases, improved product quality and yield also result from better and more uniform contacting of reactants. With a thorough understanding of the reaction, the designer can identify reactor configurations that maximize yield and minimize size, resulting in a more economical process that generates fewer by-products and waste, and increases inherent safety by reducing the reactor size and inventories of all materials.

A relatively new development in reactor design is the spinning disk reactor. In this novel design, the reactions take place in an imposed acceleration field, in this case, centrifugal motion. The fluid acceleration greatly enhances both the mass and heat transfer processes, thereby allowing the same reaction rates to occur in a much smaller volume. The rotating surface of revolution (i.e., the spinning disc) creates an ideal environment for the rapid transmission of mass, heat, and momentum because the thin liquid films generated on the disc are highly sheared. This facilitates rapid physical or chemical processes involving liquids (even viscous liquids), such as polymerization, precipitation, and rapid exothermic organic reactions. For example, the manufacture of a particular pharmaceutical product may require a 2000 liter conventional stirred batch reactor vessel. Using a 30 cm disc reactor can produce 1000 tons/yr of the same pharmaceutical product at a continuous rate of 30 gram/sec (Stankiewicz 2004).

Another recent development in reaction technology is microreaction. The miniaturization of plant mixing and heat transfer equipment can generate very high mass and heat transfer rates. Firstly, the gradients driving mass and heat transfer, i.e., concentration and temperature, will be increased when the device dimensions are reduced, although viscous losses will increase. Secondly, the surface area-to-volume ratio of the system increases as dimensions decrease, thereby increasing the interface area per unit mass or volume. This increases the mass and heat transfer rates. Both factors together can create extremely efficient mixing devices and heat exchangers. Since volume is a three-dimensional property, the volume of the micro-device, and hence the amount of material in it, is reduced by the inverse of the third power of its characteristic dimensions. In doing so, material inventory is greatly reduced, as are concerns about large temperature and concentration differences because of the greatly reduced response time of the micro-device (Stankiewicz 2004).

4.2.2 Continuous Stirred Tank Reactors

A continuous stirred tank reactor (CSTR) is usually much smaller than a batch reactor for a specific production rate. In addition to reduced inventory, using a CSTR usually results in other benefits that enhance safety, reduce costs, and improve the product quality. For example:

- Mixing in the smaller reactor is generally better. Improved mixing may improve product uniformity and reduce by-product formation.
- Controlling temperature is easier and the risk of thermal runaway is reduced. A smaller reactor provides greater heat transfer surface per unit of reactor volume.
- Containing a runaway reaction is more practical by building a smaller but stronger reactor rated for higher pressure.

In considering the relative safety of batch and continuous processing, it is important to fully understand any differences in the chemistry and processing conditions that may outweigh the benefits of a continuous reactor's reduced size. Englund (1991b) describes continuous latex processes that have enough unreacted monomer in the continuous reactor to be less safe than a well-designed batch process.

4.2.3 Tubular Reactors

Tubular reactors often offer the greatest potential among reaction devices for inventory reduction. They are usually extremely simple in design, containing no moving parts and a minimum number of joints and connections. A relatively slow reaction can be completed in a long tubular reactor if mixing is adequate. There are many devices available for providing mixing in tubular reactors, including jet mixers, eductors, and static mixers.

It is generally desirable to minimize the diameter of a tubular reactor because the leak rate in case of a tube failure is proportional to its cross-sectional area. For exothermic reactions, heat transfer will also be more efficient with a smaller tubular reactor. However, these advantages must be balanced against the higher-pressure drop due to flow through smaller reactor tubes.

4.2.4 Loop Reactors

A loop reactor is a continuous tube or pipe that connects the outlet of a circulation pump to its inlet (Figure 4.1). Reactants are fed into the loop, where the reaction occurs, and product is then withdrawn. Loop reactors have been used in place of batch stirred tank reactors in a variety of applications, including chlorination, ethoxylation, hydrogenation, and polymerization. A loop reactor is typically much smaller than a batch reactor producing the same amount of product. Wilkinson and Geddes (1993) describe a 50 liter loop reactor for a polymerization process that has a capacity equal to that of a 5000 liter batch reactor. Mass transfer is often the rate limiting step in gas-liquid reactions, and a loop reactor design increases mass transfer, while reducing reactor size and improving process yields. As an example, an organic material was originally chlorinated in a glass-lined batch stirred tank reactor, with chlorine fed through a dip pipe. Replacing the stirred tank reactor with a loop reactor that fed chlorine to the recirculating liquid stream through an eductor reduced reactor size, increased productivity, and reduced chlorine usage. These results are summarized in Table 4.1 (CCPS, 1993a).

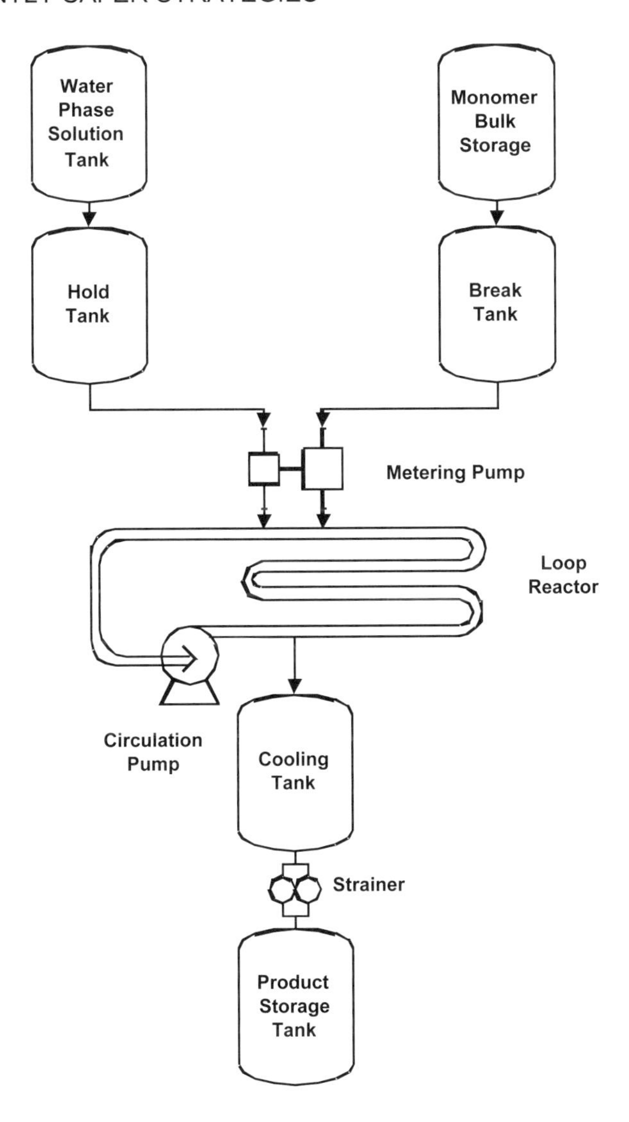

Figure 4.1: A Loop Reactor Production System

Table 4.1: Effect of Reactor Design on Size and Productivity for a Gas-Liquid Reaction (CCPS, 1993a)

Reactor Type	Batch Stirred Tank Reactor	Loop Reactor
Reactor Size (l)	8000	2500
Chlorination Time (hr)	16	4
Productivity (kg/hr)	370	530
Chlorine Usage (kg/100 kg product)	33	22
Caustic Usage in Vent Scrubber (kg/100 kg product)	31	5

4.2.5 Reactive Distillation

The combination of several unit operations into a single piece of equipment can eliminate equipment, minimize the inventory of reactants, and simplify a process. However, there may be inherent safety conflicts resulting from this strategy (see Chapter 9). Combining a number of operations into a single device increases the complexity of that device, but it also reduces the number of vessels or other pieces of equipment required for the process. Careful evaluation of the options with respect to all hazards is necessary to select the inherently safer overall option.

Reactive distillation is a technique for combining a number of process operations in a single device. One company has developed a reactive distillation process for the manufacture of methyl acetate that reduces the number of distillation columns from eight to three, while also eliminating an extraction column and a separate reactor (Agreda, et al, 1990; Doherty and Buzad, 1992; Siirola, 1995). Inventory is reduced and auxiliary equipment, such as reboilers, condensers, pumps, and heat exchangers are eliminated. Figure 4.2 shows the conventional design, and Figure 4.3 shows the reactive distillation design. For this process, Siirola (1995) reports significant reductions in both capital investment and operating cost for the reactive distillation process.

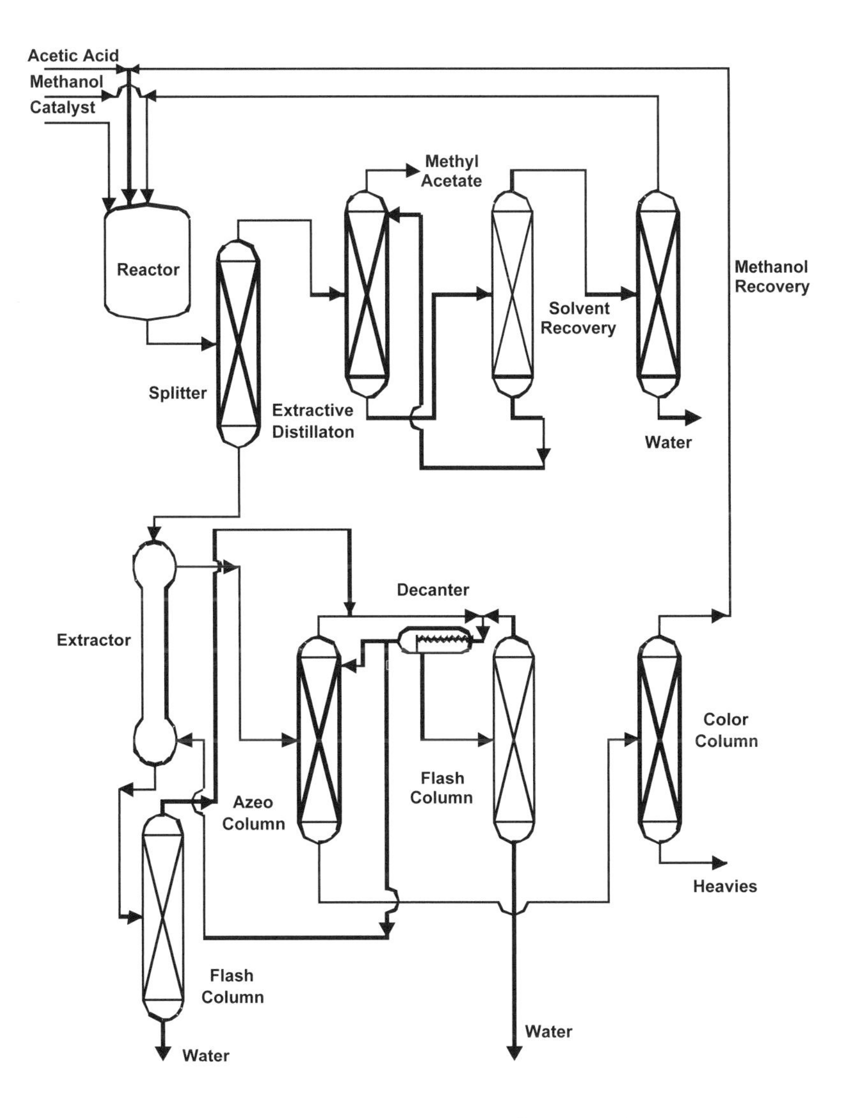

Figure 4.2: Conventional process for methyl acetate
(based on Siirola 1995)

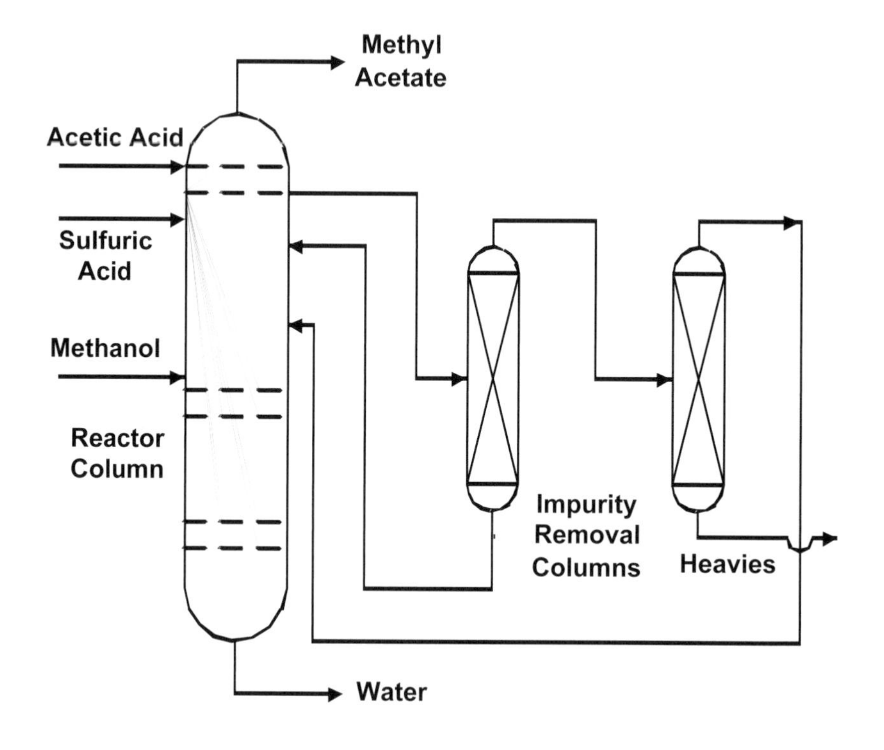

Figure 4.3: Reactive distillation methyl acetate process
(based on Agreda et al 1990)

Englund (1990) and Stankiewicz (2004) describe several other types of minimization:

- Reducing the inventory by changing the process chemistry. The production of latex using styrene and butadiene can be accomplished by the continuous addition of the reactants so that the reaction occurs as the batch reactor is being filled. The maximum inventory of unreacted monomer is greatly reduced and the temperature control is made easier. Scheffler (1996) also addresses the application of IS principles to the production of latex.

- Reducing the inventory by changing the mixing intensity. Some reactions that occur very quickly can be accomplished in piping systems using inline mixers or nozzles. This reduces the holdup volume of the reactants significantly. Static mixer reactors that incorporate heat transfer also reduce the holdup volume of reactants and solve the heat transfer problems associated with many static mixers. For gas-liquid reactions, which are an integral part of many processes, such as chlorinations, oxidations, sulfonations, nitrations, and hydrogenations, high intensity motionless static mixers can increase the reaction rate, particularly when the gaseous reactant must be transferred to the liquid phase (Stankiewicz, 2004). For example, sodium hypochlorite (bleach), an inherently safer substitute for chlorine in the treatment of potable and wastewater, is produced by reacting gaseous chlorine with liquid sodium hydroxide in a small static mixer pipe reactor.

4.2.6 Storage of Hazardous Materials

Raw material and in-process storage tanks and piping/pipelines often represent a significant portion of the risk of a processing facility. Attention to the design of storage and transfer equipment can reduce hazardous material inventory.

Storage tanks for raw materials and intermediates are often much larger than really necessary, usually because this makes it "easier" to operate the plant. The operating staff can pay less attention to ordering raw materials on time, or can accept downtime in a downstream processing unit because upstream production can be kept in storage until the downstream unit is back on line. This convenience in operation can come at a significant cost in the risk of loss of containment of the hazardous materials being stored. The process design engineers and operating staff must jointly determine the need for all intermediate hazardous material storage, and minimize quantities where appropriate. Hendershot (2006a), et al. describes a detailed risk analysis that demonstrated that changing from the bulk storage of bromine to cylinders reduced the calculated off-site consequences by three orders of magnitude. This minimization change also had the advantage of relieving the facility of having to comply with EPA's Risk Management Program (40 CFR Part 68). In this case, the usage rate of bromine was low, thereby not significantly increasing the number of deliveries. The existing bulk storage tank and its supports were also in need of replacement due to corrosion. Therefore, the proposed inherently safer modification was economically viable as well.

Similarly, hazardous raw material storage should also be minimized, with greater attention being given to "just in time" supply. Inventory reduction lowers inventory costs, while increasing inherent safety. However, in determining appropriate raw material inventories, the entire supply chain should be considered, as follows:

- Will the originating facility for the materials, distribution facility(ies) in the value chain, or both have to increase inventories to provide "just-in-time" service, and will this represent a greater risk than a larger inventory at the end-user facility?
- How much additional burden will "just-in-time" delivery place on operating staff? Will increased number of "just-in-time" deliveries increase the potential for human errors in loading/offloading operations?

- Will the additional time working in transient operating modes, due to unplanned shutdowns and the resulting restarts caused by running out of raw materials, increase the risks?
- Will transportation and temporary storage of raw material, either in parked railroad cars, tank trucks, barges, or other transportation containers along transportation routes or in transportation facilities, present perhaps an even greater risk than on-site storage in a well designed end-user facility?
- Does the increased number of deliveries increase the risk in the mobile portions of the value chain? Typically, increased rail, truck, or barge shipments increases exposure of populations, property, and the environment along the transportation routes to potential loss of containment events along these routes. Chapter 5 discusses inherently safer options in hazardous materials transportation in more detail.

The reduction in inventory resulting from greater attention to plant operations and design of unit interactions can be substantial. Wade (1987) gives several excellent examples:

- An acrylonitrile plant eliminated 500,000 pounds of in-process storage of hydrogen cyanide by shutting down an entire unit when the product purification area shut down. This forced the plant staff to solve the problems that caused the purification area shutdowns.
- Another acrylonitrile plant that supplied by-product hydrogen cyanide to various other units eliminated an inventory of 350,000 pounds by having the other units draw directly from the plant. This required considerable work to resolve many issues related to acrylonitrile purity and unit scheduling.
- A central bulk chlorine system with large storage tanks and extensive piping was replaced with a number of small cylinder facilities local to the individual chlorine users. The total inventory of chlorine was reduced by more than 100,000 pounds. This is another example of conflicting inherent safety strategies. The use of a central bulk chlorine system reduces the need for operators to frequently connect and disconnect chlorine cylinders, a step where increased rates of human error and equipment wear and failure are possible. But, the disadvantage is a large inventory that could be released if a leak occurs. The use of a number of local cylinder facilities results in a greater likelihood of a leak because of the necessity to connect and disconnect the cylinders more frequently, but the maximum size of the leak will be limited to the inventory of one cylinder. In such cases the total risk must be analyzed and considered; the increased frequency of failure, combined with the reduced consequences per event when using small chlorine cylinders, must be compared to the

combined frequency and consequences of using large bulk chlorine storage and delivery systems.

4.2.7 Process Piping

All process equipment and units must be connected by piping systems, making the layout of piping within a plant a significant influence on the inventories of materials on-site. In turn, the layout and placement of units and equipment will influence the length of inter-unit and intra-unit process piping and transfer piping systems which link these units. Since the volume of piping is a function of the square of the piping diameter, each additional linear foot of piping length can represent a measurable increase in site material inventories. Each 100 feet of piping, with an inner diameter of 2 inches, adds approximately 16 gallons of liquid to site inventory, and each 100 feet of 6 inch piping adds approximately 150 gallons. Although the length of piping systems is generally fixed during initial plant design, the decommissioning of units and equipment during the life of the facility will render some piping obsolete. This piping should be physically removed from the system. However, if this is not possible, then decommissioned piping should never be allowed to contain any hazardous materials, or to receive flow from "live" process equipment.

4.2.8 Process Equipment

Process systems contain a variety of equipment that perform many necessary functions, including heat exchangers, filters, strainers, dryers, heaters, rotating equipment, and valves. This equipment must be sized to adequately perform its required function, but should not have unnecessary additional volume. Inventory reductions in any one device may not have a substantial effect on the total amount of materials on-site, however, the cumulative effect of minimizing all of these components during an initial design or re-design (i.e., a de-bottlenecking project) will result in significant inventory reductions. Innovative designs that minimize the inventory of materials in the device at any given time, particularly for heat exchangers, can substantially reduce the volume of materials present in a process. Heat exchangers involve several variables whose optimization offer the opportunity to reduce the materials inventory, including:

- Heating/cooling medium
- Heating/cooling medium and process material flow characteristics
- Heating/cooling temperature
- Heat transfer surface area
- Heat transfer geometry
- Location of hazardous material in the heat exchanger (i.e., shell or tube side)

Section 5.3.2 describes different types of compact heat exchangers that achieve a significant degree of process minimization/intensification. These novel heat exchanger designs are characterized by high compactness ratios (heat transfer area-to-volume) and small hydraulic diameters.

In addition to reactors, the use of high gravity or centrifugal forces has also been developed for packed bed applications. A possible equivalent to a large packed-bed column to perform liquid/liquid extractions, gas/liquid interactions, and other similar operations, is a compact rotating packed bed contactor. The heavier component, in this case, the heavier liquid, is introduced at the eye of the packed rotating bed and moves outward, while the lighter component, such as a lighter liquid or gas, is introduced at the periphery and moves inward. The use of an accelerated fluid greatly reduces the size of the packed bed (Stankiewicz, 2004).

Another development is the potential for desktop manufacturing. Where annual production rates are relatively small, such as for certain pharmaceuticals, replacement of a large batch process that operates infrequently to satisfy desired production volume with a much smaller continuously operating lab or pilot scale process that operates at a very low rate results in a large degree of process minimization. For example, an annual production amount of 500 tons corresponds to a continuous rate of 70 mL/sec. This demand can be met with a desktop process. Scale-up design problems are minimized, and process loads, such as power demand and heat load, are distributed over much wider times, resulting in much smaller equipment (Stankiewicz, 2004).

Section 5.3.2 on Unit Operations – Specific includes additional examples of opportunities for process minimization or intensification for other unit operations and types of equipment.

4.3 SUBSTITUTE

In the context of IS, "substitution" means the replacement of a hazardous material or process with an alternative that reduces or eliminates the hazard. Process designers, line managers, and plant technical staff should continually inquire if less hazardous alternatives can be effectively substituted for all hazardous materials used in a manufacturing process. However, the substitution concept of inherent safety is best applied during the initial design of a process. Substituting raw materials and intermediates after the process has been built, while possible in some cases, is usually very difficult.

Examples of substitution in two categories—reaction chemistry and solvent usage—are discussed herein. However, there are many other areas where opportunities to substitute less hazardous materials can be found, including materials of construction, heat transfer media, insulation, and shipping containers.

4.3.1 Reaction Chemistry

Basic process chemistry that uses less hazardous materials and chemical reactions offers the greatest potential for improving inherent safety in the chemical/processing industry. Alternate chemistry may use less hazardous raw materials or intermediates, or result in reduced inventories of hazardous materials (minimization), or less severe processing conditions (moderation). Identifying catalysts that can enhance reaction selectivity, or allow desired reactions to be carried out at a lower temperature or pressure, is often the key to developing inherently safer chemical synthesis routes. Some specific examples of innovations in process chemistry that result in inherently safer processes include:

- The insecticide carbaryl can be produced by several routes, some of which do not use methyl isocyanate (MIC), or which generate only small quantities of this toxic material as an in-process intermediate (Kletz, 1991b). One company has developed a proprietary process for manufacturing carbamate insecticides that generates methyl isocyanate as an in-situ intermediate. Total methyl isocyanate inventory in the process is no more than 10 kilograms (Kharbanda and Stallworthy, 1988; Manzer, 1994). An Israeli company has changed the order of the three materials (α-naphthol, methylamine, and phosgene) used to produce carbaryl so that no MIC is generated. Rather than react α-naphthol with methylamine to produce MIC, α-naphthol and phosgene are reacted to produce a chloroformate ester that is then reacted with methylamine to produce carbaryl. No MIC is produced. Neither of these reactions is ideal because they involve the use of phosgene, which is a more toxic material than chlorine (Kletz 1998). However, process designs to produce the phosgene in situ and in "just-in-time" amounts, with little or no interim storage, are common in many applications where phosgene is a required material.

- Acrylonitrile can be manufactured by reacting acetylene with hydrogen cyanide:

$$CH \equiv CH + HCN \rightarrow CH_2 = CHCN$$

Acetylene   Hydrogen Cyanide   Acrylonitrile

A new ammoxidation process uses propylene and ammonia, which are less hazardous raw materials (Dale, 1987; Puranik, et al, 1990).

$$CH_2 = CHCH_3 + NH_3 + \frac{3}{2}O_2 \rightarrow CH_2 = CHCN + 3H_2O$$

Propylene   Ammonia   Acrylonitrile

This process does produce HCN as a by-product in small quantities. Puranik, et al. (1990) report on work to develop an improved, more selective catalyst, and to couple the ammoxidation process with a second reactor in which a subsequent oxycyanation reaction would convert the by-product HCN to acrylonitrile.

- The Reppe process for manufacturing acrylic esters reacts an alcohol with acetylene and carbon monoxide, using a nickel carbonyl catalyst having both high acute and long term toxicity, to make the corresponding acrylic ester:

$$CH \equiv CH + CO = ROH \xrightarrow[HCl]{Ni(CO)_4} CH_2 + CHCO_2R$$

Acetylene Alcohol Acrylic ester

The alternative propylene oxidation process uses less hazardous materials to manufacture acrylic acid, followed by esterification with the appropriate alcohol (Hochheiser, 1986).

$$CH_2{=}CHCH_3 + \frac{3}{2}O_2 \xrightarrow{catalyst} CH_2{=}CHCO_2H + H_2O$$

Propylene Acrylic acid

$$CH_2{=}CHCO_2H + ROH \xrightarrow{H+} CH_2{=}CHCO_2R + H_2O$$

Acrylic Acid Alcohol Acrylic ester

- Polymer-supported reagents, catalysts, protecting groups, and mediators can be used in place of the corresponding small molecule materials (Sherrington, 1991; Sundell and Nasman, 1993). The reactive species is tightly bound to a macromolecular support that immobilizes it. This generally makes toxic, noxious, or corrosive materials much safer. Although it is being phased out as a motor fuel additive for environmental reasons, methyl tert-butyl ether (MTBE) is produced commercially by the use of a polystyrene sulfonic acid catalyst with methanol and isobutene (isobutylene).

$$CH_3OH + CH_2{=}C(CH_3)_2 \xrightarrow{polystyrene\ sulfonic\ acid} CH_3{=}OC(CH_3)_3$$

Methanol Isobutane MTBE

- The chemistry of side reactions and by-products may also offer opportunities for increasing the inherent safety of a process. For example, a process involving a caustic hydrolysis step uses ethylene dichloride (EDC; 1,2-dichloroethane) as a solvent. Under reaction conditions, a side reaction between sodium hydroxide and EDC produces small, but hazardous, quantities of vinyl chloride:

$$C_2H_4Cl_2 + NaOH \rightarrow C_2H_3Cl + NaCl + H_2O$$

Ethylene Dichloride	Vinyl chloride

An alternative non-reactive solvent has been identified which eliminates this hazard (Hendershot, 1987).

- Hydrofluoric acid (HF) has a number of industrial uses, including petroleum refining, detergent component manufacturing, metals processing, glass etching, organofluoride manufacture, semiconductor manufacturing, and uranium enrichment. Two recent process developments have resulted in possible substitutes for liquid HF:
  - Research continues on the possible substitution of existing liquid HF alkylation processes with those that employ solid HF catalysts. Alkylation units in petroleum refineries are used to produce high octane gasoline blending components that have no sulfur or aromatic content. One such process, currently under development, uses a liquid phase riser reactor with a solid catalyst similar in concept to a Fluidized Catalytic Cracking Unit (FCCU). The reactor operates at about 350 psia and refrigerated temperatures of 50-100°F. These operating conditions are similar to existing liquid HF Alkylation units, as are the feed input rations (isobutane-to-olefin). Catalyst reactivation and regeneration are key steps in these processes (McCarthy, 1996).
  - Solid bed HF catalysts are used in the production of linear alkylbenzenes, a material used in the generation of sulphonation compounds. These are components used in the manufacture of bio-degradable detergents (Chemical Engineering, 1991).
- Phase transfer catalysis processes (Starks and Liotta, 1978; Starks, 1987) for the synthesis of many organic materials use less, or sometimes no, organic solvent. They may use a less toxic solvent, or less hazardous raw materials, such as aqueous HCl instead of anhydrous HCl), or they may operate at milder conditions. Some types of reactions where phase transfer catalysis has been applied include:
  - esterification
  - nucleophilic aromatic substitution

- dehydrohalogenation
- oxidations
- alkylation
- aldol condensations

Rogers and Hallam (1991) provide other examples of chemical approaches to inherent safety, involving synthesis routes, reagents, catalysts and solvents.

Innovative chemical synthesis procedures have been proposed as offering economical and environmentally-friendly routes to manufacturing a variety of chemicals. These novel chemical reactions may also potentially offer increased inherent process safety by eliminating hazardous materials or chemical intermediates, or by allowing less severe operating conditions. Some examples of interesting and potentially inherently safer chemistries include:

- Electrochemical techniques, proposed for the synthesis of naphthaquinone, anisaldehyde, and benzaldehyde (Walsh and Mills, 1993).
- Extremozymes, or enzymes which can tolerate relatively harsh conditions, suggested as catalysts for complex organic synthesis of fine chemicals and pharmaceuticals (Govardhan and Margolin, 1995).
- Domino reactions, in which a series of carefully planned reactions occur in a single vessel, used to prepare complex biologically active organic compounds (Hall, 1994; Tietze, 1995).
- Laser light "micromanaged" reactions, directed at the production of desired products (Flam, 1994).
- Supercritical processing, which allows the use in chemical reactions of less hazardous solvents like carbon dioxide or water. This benefit must be balanced against the high temperatures and pressures required for handling supercritical fluids. Johnston (1994), DeSimone, et al. (1994), and Savage, et al. (1995) review some potential applications of supercritical processing.
- The use of glucose in lieu of benzene (a toxic and flammable hydrocarbon) for the production of adipic acid. It may be possible to produce glucose from biological residue materials, such as plant husks and straw (Kletz, 1998).
- The substitution of toxic/flammable gases, such as phosphine, diborane, and silane, in the manufacture of semiconductors with less hazardous liquids, such as trimethyl phosphite, trimethyl borate, and tetraethyl-*o*-silicate (Kletz, 1998).
- Phosphorous oxychloride can be used in place of phosgene, and 1,1,1-trichloroethane can be used in place of hydrochloric acid gas

(Kletz, 1998). There are some tradeoffs associated with these substitutions. Trimethyl phosphite and phosphorous oxychloride are on the Department of Commerce's list of Chemical Weapons Convention treaty materials. This puts them under export controls because of their possible use as chemical weapons precursors. Phosphorous oxychloride is also a toxic material from an airborne release viewpoint, although it is less toxic than phosgene. In addition, 1,1,1-trichloroethane is a suspected human carcinogen and an ozone depleting chemical.

Many of these alternative chemistry options are still undergoing research, and there are few, if any, current commercial applications of these techniques. However, the potential environmental and safety benefits of these and other innovative chemical synthesis techniques will encourage further research and development.

A United States Environmental Protection Agency (EPA) report (Lin, et al, 1994) contains an extensive review of inherently safer process chemistry options that have been discussed in the literature. This report includes chemistry options that have been investigated in the laboratory, as well as some that have advanced to pilot plant and even to production scale.

- More than a decade ago, EPA established the Green Chemistry Program with the goal of promoting innovative chemical technologies that reduce or eliminate the use or generation of hazardous substances in the design, manufacture, and use of chemical products. Green chemistry is the use of chemistry for pollution prevention. More specifically, green chemistry is the design of chemical products and processes that reduce or eliminate the use and generation of hazardous substances. Green chemistry is a highly effective approach to pollution prevention because it applies innovative scientific solutions to real-world environmental situations. For these reasons, the Green Chemistry Program covered activities that are broader in scope than the IS concepts described in this book. Hendershot (2006b) provides a clear description of the relationship between inherently safer technologies and green chemistry. However, among the twelve principles of Green Chemistry, as published by EPA, are:
    - to design chemical products to be fully effective, yet have little or no toxicity, and
    - to design syntheses to use and generate substances with little or no toxicity to humans and the environment.

These principles, as well as the others that define the Green Chemistry Program are compatible with the goals and objectives of IS. Two examples of awards given to companies that have redesigned their processes under the Green Chemistry Program include:

- Disodium iminodiacetate (DSIDA) is a key intermediate in the production of Monsanto's Roundup®, an environmentally-friendly, nonselective herbicide. Traditionally, Monsanto and others have manufactured DSIDA using the Strecker process, which requires ammonia, formaldehyde, hydrochloric acid, and hydrogen cyanide. Furthermore, the chemistry involves the exothermic generation of potentially unstable intermediates, and special care must be taken to preclude the possibility of a runaway reaction. The overall process also generates up to one pound of waste for every seven pounds of product, and this waste must be treated prior to safe disposal. Monsanto has developed and implemented an alternative DSIDA process that relies on the copper-catalyzed dehydrogenation of diethanolamine. The raw materials have low volatility and are less toxic, and the dehydrogenation reaction is endothermic, thereby eliminating the danger of a runaway reaction. Moreover, this zero-waste route to DSIDA produces a product stream that, after filtration of the catalyst, is of such high quality that no purification or waste cut is necessary for subsequent use in the manufacture of Roundup®. This new technology represents an example of the substitution strategy of IS because it avoids the use of cyanide and formaldehyde. It also illustrates moderation, because it operates endothermically rather than exothermically, and simplification, because it has fewer process steps (*http://www.epa.gov/gcc/pubs/pgcc/winners/gspa96.html*).

- The use of transition metals for catalyzing reactions is of growing importance in modern organic chemistry. These catalysts are widely used in the synthesis of pharmaceuticals, fine chemicals, petrochemicals, agricultural chemicals, polymers, and plastics. Of particular importance is the formation of C–C, C–O, C–N, and C–H bonds. Traditionally, the use of an inert gas atmosphere and the exclusion of moisture have been essential in both organometallic chemistry and transition-metal catalysis. The research of Professor Chao-Jun Li has focused on the development of numerous transition-metal-catalyzed reactions, both in air and water. Specifically, Professor Li has developed a novel [3+2] cycloaddition reaction to generate 5-membered carbocycles in water; a synthesis of beta-hydroxyl esters in water; a chemoselective alkylation and pinacol coupling reaction mediated by manganese in water; and a novel alkylation of 1,3-dicarbonyl-type compounds in water. His work has enabled rhodium-catalyzed carbonyl addition and conjugate addition reactions to be carried out in air and water. A highly efficient, zinc-mediated Ullman-type coupling reaction, catalyzed by palladium in water, has also been designed. This reaction is conducted at room temperature under an atmosphere of air. In addition, a number of Barbier-Grignard-type reactions in water have been developed. These novel synthetic methodologies are applicable to the synthesis of a variety of useful chemicals and compounds. Transition-metal catalyzed reactions in water and air offer many advantages. Water is readily available and inexpensive, and is not flammable, explosive, or toxic. Using water as a reaction solvent can save synthesis steps by avoiding protection and deprotection processes that affect overall synthesis efficiency and contribute to solvent emission (i.e., simplification). Product isolation may be facilitated by simple phase separation, rather than energy-intensive and organic-emitting processes involving distillation of organic solvent. The temperature of reactions performed in aqueous media is also easier to control since water has such a high heat capacity *(http://www.epa.gov/gcc/pubs/pgcc/winners/aa01.html).*

4.3.2 Solvents

Replacement of volatile organic solvents with aqueous systems or less hazardous organic materials improves the safety of many processing operations and final products. In evaluating the hazards of a solvent, or any other process chemical, it is essential to consider the properties of the material at the processing conditions. For example, a combustible solvent is a major fire hazard if handled above its flash point or boiling point.

Some examples of solvent substitutions include:

- Water-based paints and adhesives, replacing solvent-based products.
- Less volatile solvents with a higher flash point, used for agricultural formulations (Catanach and Hampton, 1992). In many cases, aqueous or dry flowable formulations for agricultural chemicals can be used instead of organic formulations.
- Aqueous and semi-aqueous cleaning systems, used for printed circuit boards and other industrial degreasing operations (Mandich and Krulik, 1992; Davis, et al, 1994).
- Abrasive media cleaning systems, replacing hazardous organic solvents for paint stripping (Davis, et al., 1994).
- N-Methyl pyrrolidone, dibasic ethers, and organic esters, substituting for more hazardous paint removers (Consumer Reports, 1991; Davis, et al., 1994).

The effort to substitute inherently safer and more environmentally-friendly solvents in many industries continues to be active. Scores of solvent substitutions have been identified in a variety of industries, including food processing, textile, wood and furniture, printing and casting/metals processing. The United States Environmental Protection Agency developed an expert system to aid in solvent substitution for the printing industry, the Integrated Solvent Substitution Data System (ISSDS) (http://es.epa.gov/issds/). This online resource includes links to a number of websites and guidance documents that provide assistance on substituting materials for hazardous and polluting solvents. These include resources established by EPA as well as DOD.

4.3.3 Refrigerants and Firefighting Agents

Refrigeration systems require the use of a material that has a high vapor pressure, low flash point, and high heat capacity in order to function successfully as a cooling medium. Unfortunately, most of the materials that have desirable properties as refrigerants also have other properties that introduce undesirable risks. The three major refrigerants used in industrial refrigeration systems are light hydrocarbons (i.e., propane, propylene, etc), anhydrous ammonia, and chlorofluorocarbons (CFC). Each of these materials presents individual hazards that may preclude their use in refrigeration systems. Hydrocarbons, such as propane and propylene, are highly flammable and represent significant fire and explosion hazards if released due to leakage. Anhydrous ammonia is a toxic inhalation hazard, and CFCs released to the atmosphere deplete the ozone layer. The choice of refrigerant will depend on availability of the material on-site, ease of maintenance of the refrigeration system, and amount of refrigeration tonnage required for the application. Due to the international banning of CFC usage, many facilities have changed back to the use of hydrocarbon refrigerants. If a hydrocarbon-based refrigeration system only incrementally increases the total amount of flammable hydrocarbons on-site, it may be the best choice. Anhydrous ammonia refrigeration systems have been used extensively in the food processing industry because ammonia is an environmentally-compatible refrigerant with no ozone depletion potential (ODP). Ammonia also has advantageous thermodynamic properties at the temperatures and pressures common to refrigeration applications, resulting in smaller, more compact systems and less energy consumption than other refrigerants when used in large industrial systems.

There has been research in recent years on alternative refrigeration materials that have lower ODP levels, or lower flammability, such as hydrofluorocarbons, hydrochlorofluorocarbons (HCFC), and hydrofluoroethers. These materials can be used on their own or as carriers for flammable materials, resulting in a refrigerant that is less flammable than a pure hydrocarbon, such as propane. HCFCs are considered an interim solution to ozone layer damage and are scheduled for phase-out in the first third of the 21st century. Liquid nitrogen and liquid carbon dioxide also offer possibilities as refrigerants. These materials do not have the same safety or environment hazards as the materials discussed above. However, they are not risk free, either. Both of these materials are colorless and odorless, unlike ammonia, which has a very low odor threshold in humans and, if released in an enclosed space, would represent a significant asphyxiation hazard (Kletz, 1998).

Some of the same issues associated with CFC refrigerants affect some traditional firefighting agents. Bromofluorochlorocarbons (BFC), or Halon have been used extensively as firefighting agents, particularly in spaces containing electronic or electrical equipment where other fire suppression agents (i.e., water) would cause serious equipment damage. BFCs have the same adverse effects on the ozone layer as CFCs, and their use is being phased out accordingly. Newer halogenated hydrocarbon materials have been developed, however, they are not as efficient firefighting agents as BFCs, and larger systems, with larger inventories, are required to extinguish the same size fire. Carbon dioxide is a good firefighting alternative, however, the concentration needed to extinguish most fires is high enough to cause asphyxiation if someone is in the space and not properly equipped with emergency breathing capability. Halon and newer alternatives are used in concentrations that will not cause asphyxiation (Kletz, 1998).

4.4 MODERATE

In the context of inherent safety, *moderate*, also called attenuation, means using materials under less hazardous conditions. Moderation of conditions can be accomplished by strategies that are either physical (i.e., lower temperatures, dilution) or chemical (i.e., development of a reaction chemistry which operates at less severe conditions).

4.4.1 Dilution

Dilution reduces the hazards associated with the storage and use of a low boiling hazardous material in two ways:

1. By reducing the storage pressure.
2. By reducing the initial atmospheric concentration if a release occurs.

Materials that boil below normal ambient temperature are often stored in pressurized systems under their vapor pressure at the ambient temperature. The pressure in such a storage system can be lowered by diluting the material with a higher boiling solvent. This reduces the static pressure imposed on the storage container, as well as the pressure difference between the storage system and the outside environment, thereby reducing the rate of release in case of a leak in the system. If there is a loss of containment incident, the atmospheric concentration of the hazardous material at the spill location and the downwind atmospheric concentration and hazard zone are reduced.

Some materials can be handled in a dilute form to reduce the risk of handling and storage:

- Aqueous ammonia or methylamine in place of the anhydrous material.
- Hydrochloric acid in place of anhydrous HCl.
- Dilute nitric acid or sulfuric acid in place of concentrated fuming nitric acid or oleum (SO_3 solution in sulfuric acid).

If a chemical process requires the concentrated form of the material, it may be feasible to store a more dilute form, and to concentrate the material by distillation or some other technique in the plant prior to introduction to the process. The inventory of material with greater intrinsic hazard (i.e., undiluted) is reduced to the minimum amount required to operate the process, but the distillation adds a new hazardous process.

Chemical reactions are sometimes conducted in a dilute solution to moderate reaction rates, to provide a heat sink for an exothermic reaction, or to limit maximum reaction temperature by tempering the reaction. In this example, there are conflicting inherent safety goals - the solvent moderates the chemical reaction, but the dilute system will be significantly larger for a given production volume. Careful evaluation of all of the process risks is required to select the best overall system.

4.4.2 Refrigeration

Many hazardous materials, such as ammonia and chlorine, can be stored at or below their atmospheric boiling points with refrigeration. Refrigerated storage reduces the magnitude of the consequences of a release from a hazardous material storage facility in three ways:

1. By reducing the storage pressure
2. By reducing the immediate vaporization of leaking material and the subsequent evolution of vapors from the spilled pool of liquid
3. By reducing or eliminating liquid aerosol formation from a leak

Refrigeration, like dilution, reduces the vapor pressure of the material being stored, which, in turn, reduces the driving force (pressure differential) for a leak to the outside environment. If possible, the hazardous material should be cooled to or below its atmospheric pressure boiling point. At this temperature, the rate of flow of a liquid leak will depend only on liquid head or pressure, with no contribution from the vapor pressure of the material. The flow through any hole in the vapor space will be small and limited to breathing and diffusion.

Material stored at or below its atmospheric pressure boiling point has no superheat. Therefore, there will be no initial flash of liquid to vapor in case of a leak. Vaporization will be controlled by the evaporation rate from the pool formed by the leak. This rate can be minimized by the design of the secondary containment, for example, by minimizing the surface area of the liquid spilled into the dike area, or by using insulating concrete containment sides and floors. Because the spilled material is cold, vaporization from the pool will be further reduced. However, materials that have molecular weights that are less than air will, when released at cold temperatures, exhibit the dispersion characteristics of denser-than-air gases until they absorb enough heat and become neutrally or positively buoyant. This will inhibit dispersion of the material and increase its ground level concentrations, and hence any human exposure. If the material is flammable, the released vapor cloud will retain an explosive concentration for a longer period of time and greater distance.

Many materials, when released from storage in a liquefied state under pressure, form a jet containing an extremely fine liquid aerosol of small droplets, which are possibly micron and sub-micron sized. The fine aerosol droplets formed may not rain out onto the ground, but instead may be carried downwind as a dense cloud. The amount of material contained in the cloud may be significantly higher than would be predicted based on an equilibrium flash calculation assuming that all of the liquid phase rains out. This phenomenon has been observed experimentally for many materials, including propane, ammonia, hydrogen fluoride, and monomethylamine. Refrigeration of a liquefied gas to a temperature near its atmospheric pressure boiling point eliminates the two-phase flashing jet, and the liquid released will rain out onto the ground. Containment and remediation measures such as spill collection, secondary containment, neutralization, and absorption may then be effective in preventing further vaporization of the spilled liquid (CCPS, 1993a).

Figure 4.4 shows an example of a refrigerated storage facility for chlorine. This facility includes a spill collection sump that is covered to reduce evaporation to the atmosphere, both by containing the evaporating vapors and by reducing heat transfer from the surrounding atmosphere. The spill collection sump is vented to a scrubber that collects evaporated chlorine.

Marshall et al. (1995) provide a series of case studies that evaluate the benefits of refrigerated storage for six materials – ammonia, butadiene, chlorine, ethylene oxide, propylene oxide, and vinyl chloride. They conclude that, "refrigerated storage is generally safer than pressurized storage" for all of the chemicals studied, except ammonia. Ammonia was reported to be an exception "due to a density shift with temperature making it heavier than the surrounding air." Other materials may exhibit similar results, and it is essential that the designer fully understand the consequences of potential incidents.

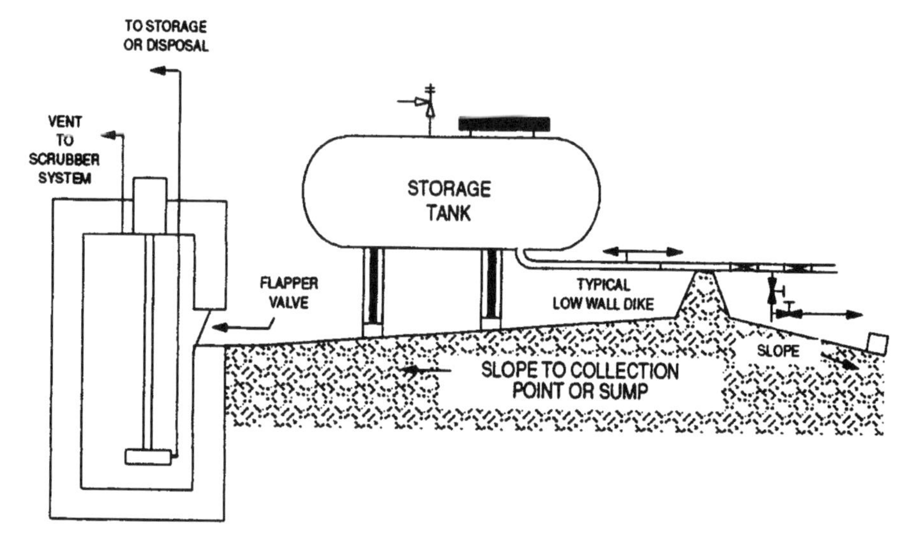

Figure 4.4: A chlorine storage system with collection sump with vapor containment *(Puglionesi and Craig, 1991)*

4.4.3 Less Severe Process Conditions

Operating under less severe conditions, as close to ambient temperature and pressure as possible, increases the inherent safety of a chemical process. Loss of containment events occurring at lower process temperatures and pressures are less energetic and result in releases that are slower and of less total material. Related to temperature is the phase of the material. Although the temperature of a fluid must be increased to vaporize a material, a loss of containment event in a portion of a process that contains a material in its vapor state will result in less overall material being released. For example, a piping rupture downstream of a chlorine vaporizer will result in a release of chlorine that is less than a corresponding rupture of liquid chlorine piping over the same time period.

Some examples of process improvements that have resulted in the moderation of process conditions include:

- Improvements in ammonia manufacturing processes have reduced operating pressures. In the 1930s, ammonia plants operated at pressures as high as 600 bar. In the 1950s, process improvements had reduced operating pressures to 300-350 bar, and by the 1980s, processes operating in the 100-150 bar range were being built. Besides being safer, the lower pressure plants are also cheaper and more efficient (Kharbanda and Stallworthy, 1988).

- Improvements in the production of phenol from cumene have reduced operating temperatures so that the process is not within 10° C of the temperature where a runaway reaction takes place. This eliminates the need for an emergency dump system and its associated active controls.
- The use of continuous loop reactors for the nitration of hydrocarbons. Nitration processes have historically been among the most hazardous in the chemical/processing industry. The use of a loop reactor with sulfuric acid as a diluent lowers the total inventory in the process, accomplishing minimization at the same time. The reaction is limited to a few seconds in a specific part of the reactor, thus limiting the contact time between the nitric acid and the hydrocarbon. The dilution ratio of approximately 30:1 precludes concentration of reactants that can support a runaway reaction, and the maximum temperature of the process is 15 °C. Sulfuric acid is a hazardous material, but it has a low vapor pressure so, if spilt or brought directly in contact with people, it does not represent a toxic inhalation hazard (Kletz 1998).
- Catalyst improvements allow methanol plants and plants using the Oxo process for aldehyde production to operate at lower pressures. The process also has a higher yield and produces a better quality product (Dale, 1987).
- Improvements in polyolefin manufacturing technology have resulted in lower operating pressures (Althaus and Mahalingam, 1992; Dale, 1987).
- Use of a higher boiling solvent may reduce the normal operating pressure of a process, and will also reduce the maximum pressure resulting from an uncontrolled or runaway reaction (Wilday, 1991).
- Semi-batch or gradual addition batch processes limit the supply of one or more reactants, and increase safety when compared to batch processes in which all reactants are included in the initial batch charges. For an exothermic reaction, the total energy of reaction available in the reactor at any time is minimized. However, the inherent safety benefits of semi-batch operation are realized only if the limiting reactant is actually consumed as it is fed, and there is no buildup of unreacted material. A number of process upsets, such as contamination with a reaction inhibitor, operating at too low a temperature or forgetting to charge a catalyst to the reactor or to start the agitator, could result in a buildup of unreacted material. If any upset causing loss of reaction can occur, it is important to ensure that the reactants are indeed being consumed as they are fed in order to realize the inherent safety benefits of a semi-batch process. The reactor could be monitored to provide confirmation that the limiting

reactant is being consumed, by on-line analysis or by monitoring some physical property of the batch that is reliably correlated to reaction progress (CCPS, 1993a).

- Advances in catalysis will result in the development of high yield, low waste manufacturing processes. Catalysts frequently allow the use of less reactive raw materials and intermediates, and less severe processing conditions. High yields and improved selectivity reduce the size of the reactor for a specified production volume (i.e., minimization). High selectivity for the desired product also reduces the size and complexity of the product purification equipment. It may be possible to develop a catalyst sufficiently selective to make purifying the product at all unnecessary, as in a process for HCFC-141b (CH_3CFCl_2) described by Manzer (1993). Allen (1992), Manzer (1993, 1994), and Dartt and Davis (1994) describe a number of catalytic processes, which are potentially environmentally-friendly and safer.

- Atorvastatin calcium is the active ingredient of Lipitor®, a drug that lowers cholesterol by blocking its synthesis in the liver. The key chiral building block in the synthesis of atorvastatin is ethyl (*R*)-4-cyano-3-hydroxybutyrate, known as hydroxynitrile (HN). Traditional commercial processes for HN require a resolution step with 50 % maximum yield or syntheses from chiral pool precursors. They also require hydrogen bromide to generate a bromohydrin for cyanation. All previous commercial HN processes ultimately substitute cyanide for the halide under heated alkaline conditions, forming extensive byproducts. These processes also require a difficult high-vacuum fractional distillation to purify the final product, which decreases the yield even further. Codexis has designed an alternative HN process around the exquisite selectivity of enzymes and their ability to catalyze reactions under mild, neutral conditions. The evolved enzymes are so active and stable that Codexis can recover high-quality product by extracting the reaction mixture. The process involves fewer unit operations than earlier processes, most notably obviating the fractional distillation of the product. The formation of byproducts and the generation of waste is reduced, avoiding hydrogen gas, and reducing the need for solvents and purification equipment. This process also uses less hazardous materials and moderate conditions (aqueous, pH ~7, 25–40 °C, atmospheric pressure) *(http://www.epa.gov/gcc).*

4.4.4 Secondary Containment - Dikes and Containment Buildings

Secondary containment systems are considered passive protective systems. They do not eliminate or prevent a spill or leak, but they can significantly moderate the impact without the need for any active device. Containment systems can actually be defeated by manual or active design features. For example, a dike may have a drain valve to remove rainwater, and the valve could leak or be left open. Another example is a door in a containment building that could be left open.

Harris (1987) provides an excellent set of guidelines for the design of storage facilities for liquefied gases that can minimize the potential for vapor clouds:

- Minimize substrate surface wetted area.
- Minimize pool surface open to atmosphere.
- Reduce heat capacity and/or thermal conductivity of substrate.
- Prevent "slosh over" of containment walls and dikes.
- Avoid rainwater accumulation.
- Keep liquid spills out of sewers.
- Shield the pool surface from the wind.
- Provide vapor removal system to a scrubber or other emission control device.
- Provide liquid recovery system to storage where possible.
- Avoid direct sunshine on containment surfaces in hot climates.
- Direct spills of flammable materials away from pressurized storage vessels to reduce the risk of a Boiling Liquid Expanding Vapor Explosion (BLEVE).
- Provide below-grade collection sumps for flammable or combustible materials released from tanks or vessels that will allow the released materials to burn harmlessly without any effects on other equipment or people. Figure 4.5 shows an example of a collection sump system, with a fire pit, for flammable/combustible liquid releases (CCPS, 1993a).
- Provide sealed below-grade collection sumps directly below or adjacent to tanks or vessels having volatile toxic materials that will rapidly collect and contain liquids and vapors that are released. Figure 4.4 shows an example of a collection sump system for chlorine releases (CCPS, 1993a).

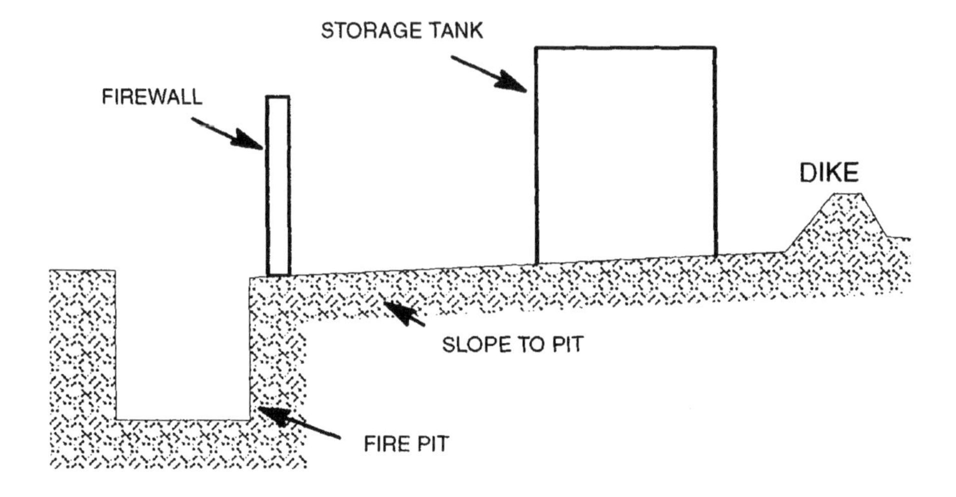

Figure 4.5: A diking design for a flammable liquid *(Englund 1991a).*

- Figure 4.6 shows a liquefied gas storage facility that incorporates many of the above principles.
- Containment buildings have been used to limit the impact of loss of containment incidents for many toxic materials, including chlorine and phosgene (CCPS, 1993a). These buildings can range from a simple, light structure to reduce evaporation of a relatively non-volatile toxic material spill, to a very strong pressure vessel designed to withstand an internal explosion. Englund (1991a) describes the evolution in the design of a phosgene handling facility from an open air plant through various stages of increasing containment, culminating in the design shown in Figure 4.7. The process is totally enclosed in a large pressure vessel capable of withstanding the overpressure of a flammable vapor deflagration.

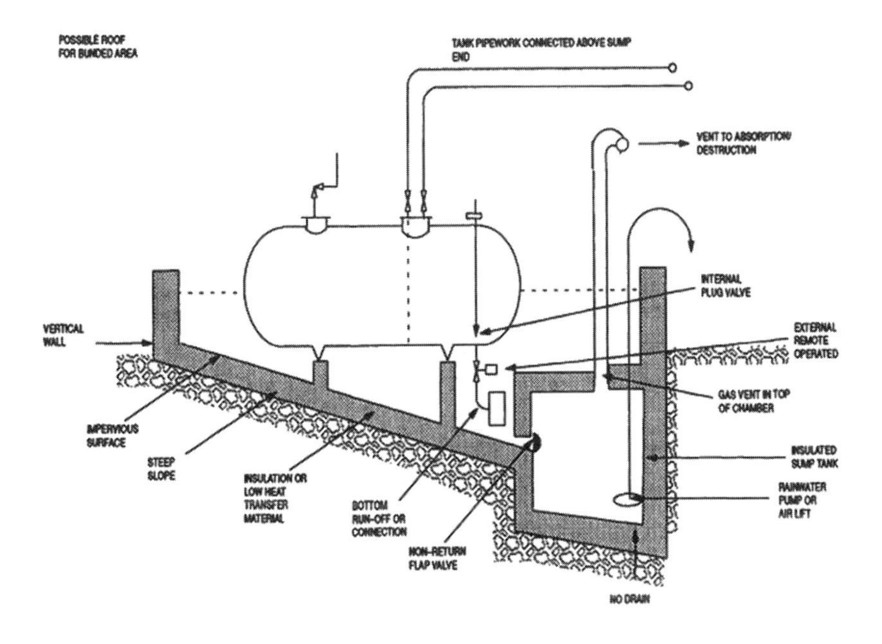

Figure 4.6: A liquefied gas storage facility (Harris, 1987)

Containment buildings offer a good example of inherent safety conflicts and tradeoffs. A containment building provides protection outside the building, but can also trap and concentrate material from small leaks inside the building, increasing the risk to personnel. Provisions must be made to ensure worker protection for a process located in a containment building, such as monitoring the atmosphere in the containment structure for hazardous vapors, remotely controlling operations from outside the containment structures, restricting access, and requiring proper personal protective equipment when entry into the containment structure becomes necessary.

In particular, great care must be taken when evaluating tradeoffs for a containment building for a flammable and toxic material, such as hydrogen cyanide. A leak or fire inside the building could cause a confined vapor cloud explosion, destroying the building, and possibly increasing the total risk.

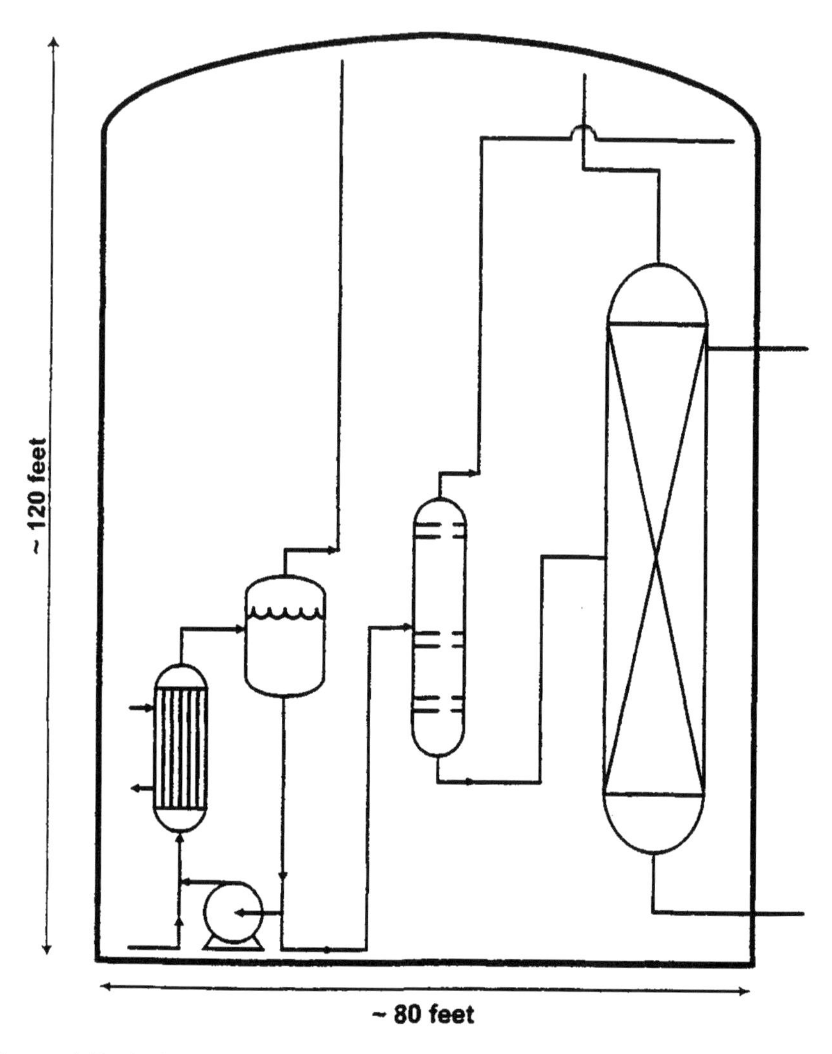

Figure 4.7: A chemical process totally contained in a large pressure vessel *(based on Englund, 1991a)*

Frank (1995) and Purdy and Wasilewski (1995) have published quantitative risk studies that evaluate the benefits of containment buildings for chlorine handling facilities.

4.5 SIMPLIFY

In the context of IS, *simplify* means designing the process to eliminate unnecessary complexity, thereby reducing the opportunities for error and misoperation. A simpler process is generally safer and more cost-effective than a complex one. For example, it is often cheaper to spend a relatively small amount of money to build a higher-pressure reactor, rather than spending a large amount of money for an elaborate system to collect and treat the discharge from the emergency relief system of a reactor designed for a lower maximum pressure. Kletz (1998) offered a few reasons why process designs are unnecessarily complex:

- *The need to control hazards*. Instead of avoiding hazard using inherently safer design principles, most designers choose to control them actively using controls, alarms, and safety instrumented systems.
- The *desire for technical elegance.* To some designers, simple equates to crude or primitive, whereas, if carefully designed, a simple process can achieve what it needs to do without excess equipment. A simple process design that contains only the essential elements to safely carry out its intended task(s) is actually more elegant than a complicated process that does the same thing.
- *The failure to carry out hazard analyses until late in the design.* PHAs and similar studies performed late in the design usually result in more active controls and equipment rather than more inherently safer solutions.
- *Following standards and specifications that are no longer appropriate or not completely applicable.* Active solutions to potential hazards that are sometimes contained in design/engineering standards and specifications can accumulate in a design and create an over-complicated process.
- *Flexibility and redundancy.* While some level of redundancy is necessary and desirable with basic process equipment, particularly where the failure of the component will have serious effects, this should be limited to what carefully performed PHAs and other studies reveal as the correct level. For every extra pump, heat exchanger, or other basic component, additional controls, utility requirements, piping/valves and other mechanical equipment will follow, thereby greatly expanding the complexity of the process. Additionally, not every risk can or should be solved by specifying some piece of equipment to deal with it. Only those risks that have been identified in the PHA process that exceed a pre-determined value should be addressed using active equipment, or where law or regulation specifies such a solution.

Kletz (1998) also offers some suggestions on use of simple technologies in lieu of high or more recent technologies to solve certain types of problems:

- Fire protection systems should be as simple as possible. For example, new technology for detecting fires by using sophisticated detectors to measure heat, ultraviolet radiation for fires, light absorption by smoke, detection of smoke using ionizing radiation, etc. is often less reliable than a simple detector that allows the fire to burn through a filament and break an electrical circuit. This melts a plastic tube that contains air under pressure and the low pressure, then activates an alarm. Fire suppression systems that involve complex piping systems with many sprinkler nozzles are also prone to plugging and difficult to maintain. A simpler, single-flow, manually-operated monitor device that can be directed to drench an area is a simpler solution, though it usually requires a larger supply of water. Fire insulation is even simpler, is completely passive if installed properly, and not prone to water supply failures or human error.
- Buildings containing compressors that process flammable gases should be open to natural ventilation. This usually means that the walls are partially enclosed and high points are vented properly. Completely enclosing a flammable gas compressor and relying on a series of active controls to detect and trip the machine when leaks occur is not as reliable.
- Oxidation processes usually require measurement of process flow stream oxygen levels. For mixed gas streams, where other vapors or liquids are present, this measurement is difficult. Complicated systems that scrub these other materials out of the stream to be measured generally don't work well, so oxygen monitors provide inaccurate measurements. Sometimes a simple solution, like an uninsulated vertical length of pipe that allows liquids to condense and flow back into a vessel or pipe, solves the problem without the complexity of a scrubbing or separation system.
- Flare systems should be kept as simple as possible, and not be equipped with other appurtenances, such as flame arrestors, water seals, filters, etc. These components are prone to plugging and reducing flare capacity.

These suggestions describe a design philosophy where simple—and sometimes old—technologies work just as well as newer, more sophisticated ones. Such a philosophy should be employed wherever possible before resorting to complex solutions.

A few examples of simplification and error tolerance are discussed in the following sections. Additional examples can be found in Kletz (1991b, 1998) and in Chapter 5 of this book.

4.5.1 Inherently Robust Process Equipment

In many cases, it is possible to design process equipment that is strong enough to contain the maximum positive or negative pressure (i.e., maximum overpressure or maximum vacuum) resulting from the worst case process incident(s), including an external fire that overheats the vessel (CCPS, 1993a). This is a form of passive safeguard design. Containment within the process vessel simplifies the design by eliminating elaborate active safeguards related to pressure detection, control, and relief. Emergency relief devices, such as rupture disks or relief valves, may still be required by regulations and codes, but their size and relief capacity, as well as hazards associated with the opening of relief devices, may be reduced or eliminated. It may also be possible to eliminate catch tanks, quench systems, scrubbers, flare stacks, or other devices designed to safely dispose of the effluent from emergency relief systems.

If external fire is a consideration, the design temperature of equipment must be high enough to withstand the temperature and design pressure generated by the fire in its material of construction. In general, the temperatures generated by fires will exceed the design temperatures of most materials used in the fabrication of process equipment, and therefore it is very difficult to design process equipment to be inherently robust with respect to a fire.

This same concept also applies to internal fires in equipment. For example, if it is possible to structurally design a column containing reactive or pyrophoric packing materials to withstand the temperatures from an internal fire, then the active safeguards required to maintain an oxygen-free environment would not be as critical. In this case, it would still not be advisable to allow an internal fire in a column to develop because of the other personnel and equipment hazards associated with fires. However, if the high temperatures of a fire would not affect the basic integrity of the column itself, the severity of the fire would be less, and the criticality of the safeguards would be lower.

Process equipment containing liquid levels should also be designed to withstand the maximum hydrostatic load that could be imposed on the equipment if it were completely filled. This is particularly true of tall equipment, such as columns, towers, large reactors, etc. Designing the structure of a column at a high level hydrostatic load without completely filling it requires that active safeguards be relied upon to protect the column against structural failure.

The concept of inherent robustness also applies to designing equipment to be impervious to the corrosion mechanisms that are present given the materials of construction and within the process, and the operating conditions (i.e., temperature, pH, concentration, viscosity, etc.). The use of certain alloys will eliminate certain types of corrosive attack. Recently, fiberglass reinforced plastic (FRP) has increasingly been used in applications where corrosion is particularly severe. However, the use of specialty alloys may increase the susceptibility of the equipment to a different corrosion mechanism, with a possible loss of pressure or temperature rating. The use of FRP always results in a loss of robustness with respect to temperature and pressure resistance. Therefore, these possibly competing inherent robustness goals (strength and corrosion resistance) must be carefully analyzed and balanced.

Although robust equipment design may be considered to be a passive safeguard rather than an inherently safer design, it considerably simplifies the remainder of the process design. Therefore, it fits within the definition of simplification. It is also highly effective in eliminating the possibility of an uncontrolled loss of containment. In a general sense, the removal of this possibility from a process design must be considered to be inherently safer.

The maximum pressure resulting from a deflagration of a combustible dust or flammable vapor in air initially at atmospheric pressure is often less than 10 bar. It may be feasible to build process equipment and structures that are strong enough to contain this type of event. When designing a system for combustion containment, factors such as highly reactive materials, oxygen or other oxidant enriched atmospheres, and congested geometry inside vessels or pipelines that could result in transition to detonation must be considered. These factors can significantly increase the maximum pressure of a combustion reaction.

4.5.2 Vacuum

Designing vessels for full vacuum eliminates the risk of vessel collapse. Many storage and transport containers have been imploded by pumping material out with the vents closed.

4.5.3 Runaway Reactions

Choosing a reactor design pressure sufficiently high to contain the maximum pressure resulting from a worst case runaway reaction eliminates the need for a large emergency relief system. It is essential that the reaction mechanisms, thermodynamics, and kinetics under *runaway conditions* be thoroughly understood to be confident that the design pressure is sufficiently high for all credible reaction scenarios. All causes of a runaway reaction must be understood, and any side reactions, decompositions, and shifts in reaction paths at the elevated temperatures and pressures experienced under runaway conditions must be evaluated. Many laboratory test devices and procedures are available for evaluating the consequences of runaway reactions (CCPS, 1995d; 1995e). Several of these reaction hazard testing methods are summarized in Table 4.8 (CCPS, 2003b).

Table 4.8 Summary of Reaction Testing Methods *(Leggett 2002)*

Hazards Test Stage	Method	Typical Information	Comments
Hazard Screening	Desk Calculation	Reaction enthalpy, ΔH_{RXN}	Need formation energy data or derive it Must know precise stoichiometry Known reactions only, no rate information
Hazard Screening	Mixing Calorimetry	Instantaneous heat of mixing, ΔH_{MIXING} Gas generation rates	Isothermal, from ambient to 150 °C Cannot test multi-phase systems
Hazard Screening	DSC/DTA	Reaction enthalpy, ΔH Reaction 'onset' temp, T_{ONSET}	Very quick (~2 hours), needs little sample No mixing, no pressure data, no multi-phase, although some systems mix the samples by rotating the sample container Difficult to get representative mixture
Hazard Screening	Adiabatic Screening	$\Delta H_{UNDESIRED}$, T_{ONSET}, ΔT_{ADIAB} P, T, t, dP/dT, dT/dt, Simple kinetics, E_A, A	Sample ~ a few grams Reasonably quick to test (~ 1/2 day) Poor/moderate sample agitation Not reliable for scale-up (high ϕ-factor)
Develop Desired Reaction	Reaction Calorimetry	$H_{DESIRED}$ Power output, Q_{RXN} Heat transfer rate Accumulation, X_{AC}	Normally 0.1 to 2 liter scale Mimics normal operation Essential information for safe scale-up Very useful for process development

Table 4.8 Summary of Reaction Testing Methods *(Continued)*

Hazards Test Stage	Method	Typical Information	Comments
Detailed Hazard Assessment	Low thermal inertia (ϕ–factor) Adiabatic Calorimeter	$\Delta H_{UNDESIRED}$, T_{ONSET}, ΔT_{ADIAB} dP/dT; dT/dt; T_{SADT}, T_{NR}, t_{MR} estimates Vent sizing data	Sample size ~ 100ml to 1 liter Safe for general laboratory work Good mimic of large scale runaway Ideal for 'what-if' scenario study
Special Studies	High Sensitivity Calorimetry	$\Delta H_{DESIRED}$ $\Delta H_{UNDESIRED}$, dT/dt; ΔT_{ADIAB}, Kinetics, E_A	Sample size 1-50 ml, μW/g sensitivity Shelf life studies by accelerated aging Combine with low adiabatic to confirm solids low self heating rate studies

Runaway reactions in batch reactors can sometimes be avoided by using separate reactor vessels for different stages of the process. In Figure 4.8, four reactants are added to a reactor to make a product. If materials C or D are added to the first stage (when A and B are added), or A or B are added to the second stage (when C and D are added), then a runaway reaction may occur. This situation can be rendered moot by adding materials A and B in one reactor, then piping the resulting materials to a separate reactor vessel where materials C and D are added. A runaway reaction is not possible in the two-reactor design (Kletz, 1998).

4.5.4 Containment Vessels

In many cases, if it is not feasible to contain a runaway reaction within the reactor, it may be possible to pipe the emergency device effluent to a separate pressure vessel for containment and subsequent treatment. Quench drums, blowdown drums, and other similar devices can be used to contain the effluent from exothermic/runaway reactions (CCPS, 1993a).

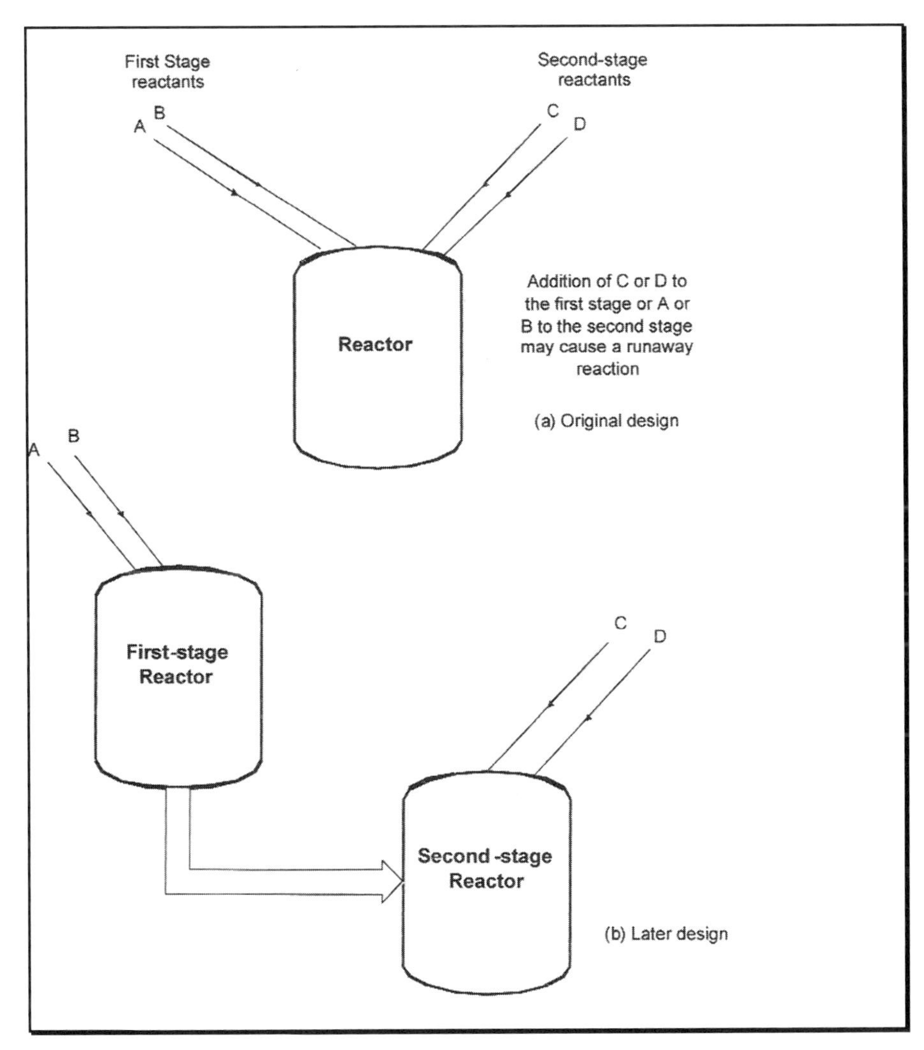

Figure 4.8: Old (a) and new (b) designs for a two-batch reaction system
(Kletz, 1998)

4.5.5 Heat Transfer

The shell and tube sides of liquid-to-liquid heat exchangers, and the tubes of air-cooled heat exchangers can be designed to contain the maximum attainable pressure on either side. Such a design eliminates reliance on pressure relief devices or other active safeguards.

Where possible, cooling systems should be designed so that they can provide adequate heat removal via natural convective cooling. This requires a thorough understanding of the hydraulic conditions created by elevation and temperature differences, and what types of mass and heat transfer mechanisms are established by these differences. Nuclear submarine reactor fluid designs allow an assumption of a significant fraction of the ship's propulsion and electrical load without the use of forced convection in the reactor. Reactor coolant pumps are only required at higher loads. In submarine operations, this saves acoustical emissions underwater, a prime operational consideration. However, it also saves significant energy, which would be a prime consideration in commercial applications. Military and commercial reactors also utilize elevation and temperature differences to drive emergency cooling systems, which will keep a shutdown reactor from overheating due to decay heat. The same principles can also be applied in chemical/process system designs.

4.5.6 Liquid Transfer

Liquid transfer systems can be designed to minimize leakage potential. For example, transfer systems that use gravity or pressure/vacuum differentials that occur in the process or are created using inert gases require no moving parts or seals. Examples include:

- Use of elevated feed tanks for processes. This is an effective method of simplification; however, only if the feed materials are not hazardous, because the pumps required to fill the elevated feed tanks take the place of feed pumps.
- The use of nitrogen as a motive force to unload transportation containers. This is effective only if the maximum nitrogen pressure cannot overpressurize the transportation container, not counting relief valves or pressure active reducing equipment.

If a pump is needed, centrifugal pumps with double mechanical seals, diaphragm pumps, jet pumps, eductors, and various types of seal-less pumps may be good choices. Seal-less pumps greatly reduce the risk of a process fluid leak, but they also introduce new hazards and concerns, such as overheating and internal leakage, which may be very rapid.

4.5.7 Reactor Geometry

Continuing research in the nuclear power industry has identified systems which utilize natural convection to provide emergency core cooling, rather than relying on pumped cooling water circulation, as well as improved reactor and core geometry, and novel materials of construction. These research efforts are to prevent core overheating more reliably (Forsberg, et al, 1989). Similar approaches are applicable in the chemical/processing industry. For example, maleic anhydride is manufactured by partial oxidation of benzene in a fixed catalyst bed tubular reactor. There is a potential for extremely high temperatures to develop due to thermal runaway if feed ratios are not maintained within safe limits. Catalyst geometry, heat capacity, and partial catalyst deactivation have been used to create a self-regulating mechanism to prevent excessive temperature (Raghaven, 1992).

4.5.8 Distributed Control Systems

A distributed control system (DCS) normally uses input and output modules that contain eight, sixteen, or more inputs or outputs. Failure of the module will simultaneously disable a large number of control loops. Attention to the assignment of input/output points to the modules makes the plant more tolerant of a failure in one of them (CCPS, 1993a). Distributed control systems also represent a source of common cause failure because so many of the controls-related safeguards pass through, or are part of, the DCS. Therefore, judicious choice of control signal groupings for I/O module assignment will reduce the possibility of common cause failures and result in inherently safer processes. Kletz (1998) offers some suggestions for simplifying process computer systems:

- Software should be well documented so that process engineers and others who need to know how the process is controlled can understand what the control system is doing.
- Old software should not be re-used on new applications without very careful review. The European space rocket Arianne failed due to re-used software.
- Safety instrumented systems must be independent of basic process controls.
- The process must be designed to be tolerant of the failure of control system components or faulty software.

For a more detailed discussion of process control systems, see the process control part of Section 4.4, and Sections 6.4 and 6.5.

4.5.9 Separation of Process Steps

A multi-step batch process can be carried out in a single vessel, or in several vessels, each optimized for a single processing step. The complexity of the batch reactor in Figure 4.9, with many potential process fluid and utility interactions, can be greatly reduced by dividing the same process into three vessels as shown in Figure 4.10. Again, this is an example of an inherent safety conflict. The system in Figure 4.9 requires only one reactor, although it is extremely complex, and process intermediates never leave the reaction vessel. The system in Figure 4.10 uses three vessels, each of which can be optimally designed for a single task. Although each vessel is considerably simpler, it is necessary to transfer intermediate products from one vessel to another.

If one of those intermediate products is extremely toxic, it may be judged to be preferable to use the single reactor—a “one pot” process –to avoid transfer of the toxic intermediate. Additionally, this type of minimization and simplification can sometimes adversely affect the process dynamics, resulting in reduced controllability and larger disturbances from normal process operation due to changes in process feed rates or conditions, utility disturbances, or other external environmental factors. Because a "robust" process, one that can tolerate significant changes from its external environment, is also a strategy for inherently safer design, this represents an interesting conflict in designing inherently safer processes. These disturbances in process operation can result in increased safety risks, as well as poorer product quality (Luyben and Hendershot, 2004). As always, the inherent safety advantages and disadvantages of each system must be evaluated with careful consideration of all of the hazards of a particular chemical process, as discussed in Section 2.5

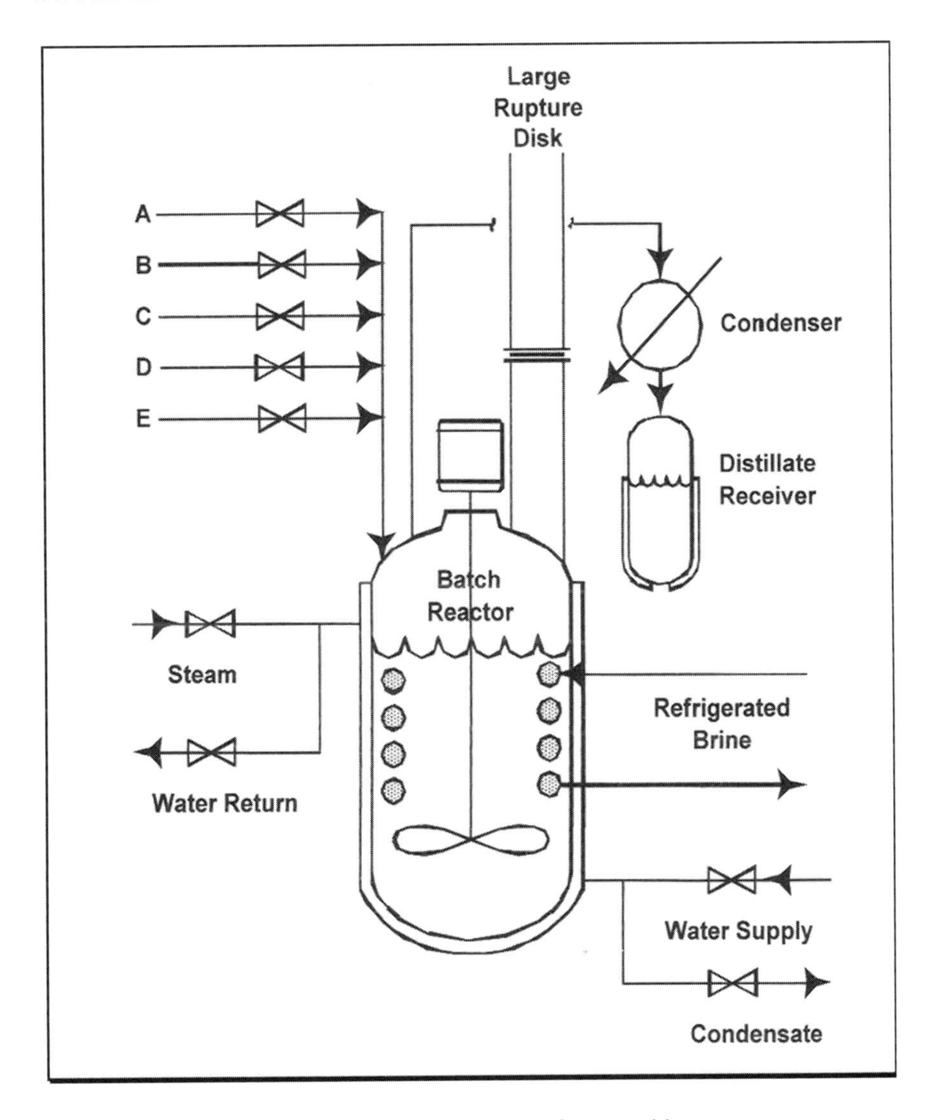

Figure 4.9: A complex batch reactor for a multistep process
(from Hendershot, 1987)

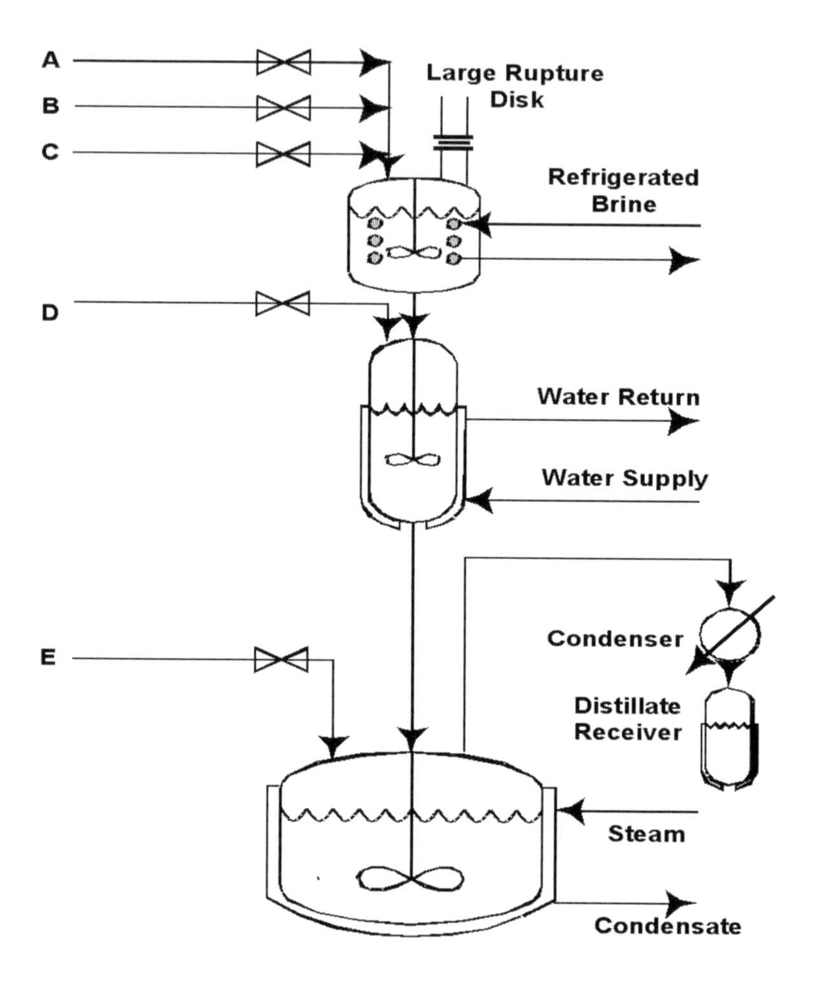

Figure 4.10: The same process as Figure 4.9 in a series of simpler reactors *(from Hendershot, 1987)*

4.5.10 Limitation of Available Energy

Adding energy to a chemical process is often a necessary component of product manufacture. However, in some cases, the energy addition method is not carefully matched to the amount of energy actually needed. Or, the method of energy addition used can result in excess energy being added because its design does not incorporate inherently safer design principles. Examples of proper matching of required energy are:

- Using immersion heaters that cannot add enough energy to cause a fire in the material being heated, or damage the container.

- Limiting process heating using steam, when possible, to the saturation temperature, which adds the needed amount of heat and no more. If the heating medium cannot be reduced, the heat transfer area should be adjusted to limit the energy transfer.
- Limiting pump discharge heads to less than the downstream relief valve setpoints.
- Ensuring that residual heat cannot be transferred inadvertently to a material via conduction or radiation, such as a hot vessel wall that transfers heat to a material that is sufficient to cause a runaway reaction (Kletz, 1998).

Some of the examples above also relate to designing equipment that is robust enough to withstand the maximum achievable temperature or pressure.

4.6 OTHER STRATEGIES

The four strategies described in this chapter represent the core concepts of inherent safety. There are several others identified by practitioners that are sub-sets of the four main strategies. These are:

- Limitation of effects
- Knock-on effects
- Avoid incorrect assembly
- Making status clear

4.6.1 Limitation of Effects

Within the context of inherent safety, "limitation of effects" (or limitation) means changing equipment design for less severe effects, and is a sub-category of moderation. Some examples of limitation include:

- The selection of some types of gaskets can reduce leak rates from equipment, hence limiting the hazard. For example, the use of spiral wound gaskets and flexible graphite type gaskets are preferred to fiber-based gaskets.
- Many runaway reactions can be prevented either by changing the order of addition, reducing the temperature or changing other parameters.
- Reduction of flow to the absolute minimum required to sustain operations and achieve desired unit/plant throughput. Limiting piping diameter to the maximum needed and using restriction orifices accomplishes this goal. Note that reduced piping diameter also

reduces the inventory of material and is an example of the minimization strategy of IS.

- Adequate secondary containment for tanks, vessels, or for entire process units that prevents the spread of liquid releases and the surface area for evaporation. Common containment structures that surround multiple tanks should be avoided or minimized if possible. The ongoing integrity of containment structures is also an important issue, particularly for earthen berms that can settle, erode, or otherwise weaken or lose their design capacity. Penetrations through secondary containment walls should be avoided or sealed properly.
- The use of blast walls and other barriers to absorb the energy from explosions and limit their radius of effect. These barriers are usually classified as passive safeguards, rather than as inherently safer devices.
- The application of facility siting concepts, that is, the siting of occupied on-site structures so as to limit the overpressure and radiant heat effects of flammable releases on the people working in these structures.
- Buffer zones inside or outside the plant fence that creates additional space between the location of the potential hazards and public receptors.

4.6.2 Global Hazards

Within the context of inherent safety, global hazards (sometimes referred to as knock-on or domino effects) means the cascading of effects from one area of the unit or plant to another, thereby increasing the consequences. Global hazards can place demands on multiple pieces of equipment and associated IPLs simultaneously. Considerations in avoiding global hazards include:

- Proper equipment spacing. Most of the industry standards for spacing have been established by the industrial insurance companies for the purpose of limiting the spread of fires.
- As discussed under Section 4.6.1 above for limitation of effects, adequate secondary containment for tanks, vessels, or for entire process units that prevents the spread of liquid releases, thereby also helping to prevent domino effects.
- The failure condition of equipment should be designed to limit the movement of materials and preserve the ability to remove energy. This concept applies to valves, agitators, pumps, heat exchangers, and control systems. For example, control system equipment should fail to a safe state of the process, which often activates shutdown condition on loss of power or instrument air.

- Avoid common cause failures where possible, so that the loss of one utility system or piece of equipment does not result in multiple failures throughout the plant.

4.6.3 Avoid Incorrect Assembly

Within the context of inherent safety, "avoid incorrect assembly" means designing equipment so that assembly, connections and other manual configurations of the equipment are difficult or impossible to perform in error. This concept is a sub-category of simplification. Some examples of avoid incorrect assembly include:

- Designing hose fittings so that a particular hose can only be connected to a fitting that is mated to receive it. These unique male-female fitting pairings eliminate the possibility of connecting the wrong hose to a manifold and inadvertently transferring a material into or out of a tank/vessel in error.
- Designing components with unique shapes so that they can be assembled in only one manner.

4.6.4 Making Status Clear

Within the context of inherent safety, "making status clear" means designing controls, indications, alarms and other man-machine interfaces so that there is no confusion, especially during accident conditions, of the status of plant equipment. This concept is a sub-set of simplification and is a staple of human factors engineering analysis of process equipment (see Chapter 6). Some examples of making status clear include:

- Designing distributed control system and other digital control system output terminals to be clear and unambiguous through the judicious use of colors, fonts, and other display characteristics.
- Avoiding information overload and difficult navigation in the design of digital control system output terminals.
- Arranging indications and control devices so that they are logical and avoid human error.
- Designing indications (gauges and meters) so that they match the ranges of the process parameters being monitored.
- Designing control systems that avoid alarm overload.
- Designing field conditions and equipment that allow operators to easily determine the condition of their equipment. This goal covers environmental conditions, such as lighting and ease of access, as well as intelligent equipment design, i.e., valve stems and operating mechanisms that quickly and easily indicate the position of the valves.

4.7 SUMMARY

This chapter describes the four main design strategies for development of inherently safer processes: minimize, substitute, moderate, and simplify. These strategies can be applied at any phase of the process life cycle, although the substitution and moderation strategies are best employed during initial process design. Examples of each strategy are given, drawn from all phases in the life cycle. In the next chapter, the focus will shift and the opportunities for application of the strategies described at specific stages in the overall life cycle of a chemical process will be discussed.

5

Life Cycle Stages

5.1 GENERAL PRINCIPLES ACROSS ALL LIFE CYCLE STAGES

As discussed previously, a process goes through various stages of evolution, including:

- Research
- Process development
- Detail design and construction
- Operations, maintenance and modification
- Decommissioning

This progression is typically referred to as the process life cycle.[1]

Throughout the life cycle of a process, opportunities will arise to apply the concepts and practices of inherently safer technologies and strategies. These opportunities should be evaluated to determine if the strategies can be applied, and whether the risks and costs are commensurate with the possible reductions.

This chapter demonstrates that applying inherently safer strategies can enhance process safety, while also improving economic and other objectives, such as quality, productivity, energy conservation, and pollution prevention. This application, utilizing formal review methods by trained individuals as identified in Chapter 7, will link the general principles of inherently safer concepts to all life cycle stages. It should be noted here that there is a tendency for some inherent risks to be volume or quantity dependent. A material that is considered "safe" at lab scale may carry high risk when managed at bulk manufacturing quantities. Eventual scale must be considered when evaluating risks even at the research and development phase. For example, doing a process hazards review of a pilot plant run is essential for safe pilot plant operations, but may not reveal inherent hazards

[1] See Figure 1.2 for a graphic illustration of this concept.

of bulk scale production. Virtual "scaling-up" must be part of the early life cycle inherent safety review. A similar issue may arise as a result of lab equipment being different from manufacturing equipment. A reaction that is safe in lab equipment may not be in the plant. These issues need to be taken into consideration in moving from lab scale to pilot plant to full-scale manufacturing.

5.2 RESEARCH

Research chemists, working with other research and development (R&D) professionals, play a fundamental role in applying inherent safety to process development. They are responsible for decreasing, to a practical minimum, the risks imposed by a future product, as well as the process used to make and distribute it. They have this important responsibility early in the development of a process and can significantly affect the outcome. Changes to the process design to incorporate inherently safer strategies during the research stage of the life cycle can be easy to accomplish and relatively inexpensive. As the life cycle proceeds to successive stages, the difficulty of implementation and the costs grow dramatically.

To apply inherent safety appropriately, research chemists must make an in-depth investigation into the process chemistry and into the entire process that may develop based on that chemistry. An adequate investigation necessitates input from a diverse team of people, including business, engineering, safety, environmental personnel, and research chemists. To choose the "inherently safest" chemistry, the team needs to take into account:

- the environment
- process hazards
- worker safety
- upstream, downstream, and ancillary unit operations (including waste disposal)
- inventory needs
- transportation of raw materials and final products

On the following page is a very simplified description of how IS can be included in the research function of the company.

Implementation of Inherent Safety in Research
in a Major Chemical Company

First, IS is included in the procedures that researchers must follow when developing a process chemistry and/or a process design.

The researchers need to review hazards, and document a process hazards analysis (PHA) for each experimental set-up and significant change in that set-up. A checklist is a required part of that PHA effort and IS questions have been included in the instructions for completing the checklist.

The format (template) for technical reports on product, chemistry, and process development includes a section on IS, as does the report format for applying for permission to file for a patent. The principal instruction on the template for each is similar to the following:

If the process or chemistry being addressed is aimed at the implementation of a new manufacturing process or to improvements/changes to an existing manufacturing process, a discussion of the anticipated hazard/risk level ***at the commercial scale*** must be included. This discussion should be from an IS perspective and include consideration of the quantities of hazardous materials involved and the severity of process conditions.

A standard "index" sheet, a form that dictates IS be considered, is also required. The index sheet is a chart that gives 5 levels of definition, from low to high, for toxicity, flammability, and reactivity (the material factor), quantity (the quantity factor), and reaction severity, pressure, temperature, corrosiveness/erosiveness, dust content, operability, and experience (the process factor). The researcher is asked to assign a "level of severity" to each factor, and then total all the figures.

From an ISP standpoint, the higher the resulting number, the higher the hazard of the chemistry or process. Several ***alternative chemistries*** and/or processes ***must be proposed*** and an index sheet completed for them as well. If competing chemistries and/or processes exhibit a lower total "level of severity," then the researcher is obligated to defend his choice. No chemistry that exhibits "severe" factors in all categories is accepted.

5.2.1 Inherently Safer Synthesis

Chemists and others directly involved in basic research have many opportunities to incorporate inherent safety in the development of process technology, including:

- synthesis routes
- catalysis leading to less severe operating conditions
- the use of a less reactive reagent, or enzyme-based chemistry and bio-synthesis

- reduction or elimination of hazardous solvents
- reduction of solvent hazards by tempering a reaction by using a more volatile solvent that will boil and more reliably remove the heat of reaction
- immobilization of hazardous reagents and catalysts by attaching active groups to polymeric or immobile backbones
- dilution of reagents
- reactions in water as opposed to those that proceed in a hazardous solvent
- elimination of hazardous unit operations
- near-critical and super-critical processing
- replacement of batch reaction processes with semi-batch or continuous processes reducing the quantity of reactant present
- use of processes that are less sensitive to critical operating parameter variations

The wide array of choices available necessitates a diligent search for hazards, and the selection of the inherently safest chemistry. To determine which type of processing offers the most inherently safe technology will require thorough literature searches, looking in particular for reports of incidents occurring in processes using the same or similar process chemistry being considered. Technologies that have contributed to the root cause(s) of dangerous incidents or near misses should be withdrawn from consideration.

5.2.2 Types of Hazards Associated With Research

Table 5.1 is a representative list of the types of hazards and hazardous events that researchers attempt to find and to minimize in searching for the best chemistry.

FIRE & EXPLOSION HAZARDS

Important flammability characteristics are the lower and upper flammability limits, the flash point, the minimum ignition energy, the minimum oxygen concentration, and the autoignition temperature. Values of these properties are available in many different publications and sources, including:

- **Material Safety Data Sheets (MSDS)**. The MSDS, which is required by OSHA's Hazard Communication Standard (29 CFR 1910.1200), is the primary reference for this information because manufacturers are required by regulation to create these documents and provide them to all users upon request. In recent years, a number of online resources have been developed providing easy access to MSDS information. A note of caution: Despite improvement in MSDS quality and completeness in recent years, they still sometimes

provide inconsistent information from manufacturer to manufacturer, and are sometimes still incomplete. However, for basic information such as flammability limits, they are the primary resource.

Table 5.1. Types of Hazards and Hazardous Events

Fires:
- Flash fires
- Pool (large, sustained) fires
- Jet fires

Explosions:
- Vapor clouds
- Confined deflagrations
- Detonations
- Pressure vessel ruptures:
  - *Exothermic runaway reactions*
  - *Physical overpressure of pressure vessels*
  - *Brittle fracture*
  - *Polymerizations*
  - *Decompositions*
    - *Undesired reactions catalyzed by materials of construction or by ancillary materials, such as pipe dope and lubricants*
    - *Boiling liquid, expanding vapor explosions (BLEVE)*

Toxicity-related hazards:
- Environmentally toxic to plant, animal or fish:
  - *Chronic or acute*
  - *Toxic to individual species or broadly hazardous*
    - *Pesticides, fungicides, herbicides, insecticides, fumigants*
- Toxic to humans:
  - *Chronic or acute*
  - *Reversible injury or irreversible injury or death*
  - *Carcinogens*
  - *Endocrine modifiers (estrogen mimics)*
  - *Persistent bioaccumulative toxins (PBT's)*
- Long-term environmental hazards:
  - *Greenhouse gases*
  - *Ozone depletors*
- Product hazards:
  - *Customer injury*
  - *Waste disposal environmental hazard*

- Several **CCPS publications** on fires and explosions provide flammability properties information for liquids and vapors (CCPS, 2003a; CCPS, 1994b) and for solid materials and dusts (CCPS, 1999; CCPS, 2003a).

- The former **U.S. Bureau of Mines (USBM)** has published a large amount of information regarding the flammability properties of liquids, gases, and dusts (Kutcha, 1985; Coward, 1952; Zabetakis 1965). Most of the USBM's safety, health, and materials programs were taken over by the Department of Energy.

- The **U.S. Coast Guard** has published a document, *Chemical Data Guide for Bulk Shipment by Water* (USCG, 1990) that provides the flammable, toxic, and reactivity properties of chemicals that are shipped via vessels.

- The **National Institute of Occupational Safety and Health (NIOSH),** which is part of the Centers for Disease Control and Prevention (CDCP), has published the *Pocket Guide to Chemical Hazards* (NIOSH, 2005) that provides the flammable, toxic, and reactivity properties of chemicals.

- Other open literature sources of information (Yaws, 1999).

These properties have typically been developed under standardized test conditions; therefore, process developers should be cautioned to evaluate the properties published in the open literature at the projected process conditions, as these conditions may influence the published values.

- The temperature and pressure caused by the ignition of a flammable vapor cloud are usually represented by static pressure, impulse (pressure integrated over time), and radiant heat flux result from the deflagration or detonation of a flammable vapor cloud. An ignited vapor cloud whose flame speed meets or exceeds the speed of sound (i.e., a detonation) can cause destructive overpressures, while ignited clouds whose flame speed is less than the speed of sound (i.e., a deflagration) can also cause damaging overpressures. The actual response of any structure to blast overpressures is a function of the pressure over time (i.e., the impulse). A vapor cloud that ignites but whose flame speed is not rapid enough to cause an overpressure (i.e., a fireball) will still generate significant radiant heat which will represent a hazard to both people and property (CCPS, 1994b).

- The ignition of a flammable material in tightly congested spaces (indoors or congested outdoor locations), or the interior of process equipment (vessels, as well as piping) can result in overpressures that are significantly greater than unconfined vapor cloud explosions (NFPA, 2002; Lewis and von Elbe, 1987).

- Pool fires are of much longer duration than vapor cloud ignition events and the thermal radiation intensity near the pool is usually high. In general, pool fires result in property damage, but do not usually result in significant numbers of casualties. However, one by-product of a pool fire is its possible effects on adjacent process equipment. The most severe effect from this external heat source is a possible Boiling Liquid Expanding Vapor Explosion (BLEVE). These events, which result in very large fireballs, occur when process vessels containing flammable materials with high vapor pressures (generally light hydrocarbons or chemicals with similar flammable properties, such as vinyl chloride monomer) are exposed to significant amounts of external heat, either from flame impingement or thermal radiation (CCPS, 1994b). In general, the exposed vessel must be very close to the pool fire location for this hazard to be realized.

- A runaway reaction in a vessel or a physical overpressurization of a vessel can cause it to lose its structural integrity. The reaction stability is a complex function of temperature, concentration, impurities, and degree of confinement. Knowledge of the reaction onset temperature, the rate of reaction as a function of temperature, and heat of reaction is necessary for analysis of a runaway reaction (CCPS, 1995e).

TOXIC HAZARDS

The dispersion and consequences resulting from the release of toxic materials require complex analyses that attempt to simulate many physical and chemical phenomena in nature, and model the movement and change in concentration of released materials in the atmosphere versus time (CCPS 1996a, 2000). Toxicological effects for humans are often expressed as a concentration (i.e., parts per million), and a number of resources are available to understand acute and chronic toxic effects. As with information regarding flammability properties, the MSDS is the primary reference for toxicological data. However, caution must be exercised in using toxicological data. Some data are intended to describe chronic exposures to workers, while others are intended to measure the short-term exposures associated with accident situations. Most of these data have been extrapolated from laboratory animal experiments and need to be corrected for the size and physiology of humans (Patty, 1991; Rand and Petrocelli, 1985; USDOE, 2005; ACGIH, 2000; NIOSH, 2005; CCPS, 2000).

PRODUCT HAZARDS

Parshall (1989) notes that product liability issues are complex and varied. An attorney knowledgeable in product liability issues may be an important contributor to process hazards discussions at the research stage.

5.2.3 Research Stage - Hazards Identification Methods

Process hazard analysis during the research stage includes identification of hazards that can be reduced or eliminated by inherently safer design, and those which are controlled by safety systems and administrative procedures. Several research methods can be used to identify and evaluate hazards. Several of these methods are described below.

Molecular Structure and Compounds. Certain molecular groupings are likely to introduce hazards into a process. The research chemist should identify these groupings. A search of the open literature will assist in identifying which types of compounds are likely to create potential hazards. Table 5.2 presents molecular structures and compound groupings associated with known hazards. The groupings in the table were developed from CCPS (1995e) and Medard (1989). The table is not all-inclusive.

The hazards of new compounds may not be known but the hazards of analogous compounds, or of those with the same or similar molecular groupings may be known. Testing may be necessary to determine the exact hazardous characteristics of any given compound.

Table 5.2: Representative Potentially Hazardous Molecular Groupings

The *primary* associated hazard is indicated to the right of the molecular grouping	
Ammonia	Toxicity and fire
Chlorinated hydrocarbons	Toxicity
Cyano compounds	Toxicity
Double and triple bonded hydrocarbons:	Fire and explosion
Epoxides:	Explosion
Hydrides and hydrogen	Explosion
Metal acetylides	Explosion
Nitrogen compounds	All explosion
Amides and imides and nitrides	
Azides	
Azo- and diazo- and diazeno- compounds	
Difluoro amino compounds	
Halogen-nitrogen bond containing compounds	
Hydrazine-derived nitrogen compounds	
Hydroxy ammonium salts	
Metal exolates	
Nitrates (including ammonium nitrate)	
Nitrites	
Nitroso compounds	
N-metal derivatives	
Polynitro alkyl and aryl compounds	
Sulfur-nitogen bond containing compounds	
Oxygenated compounds of halogens	Explosion
Oxygenated manganese compounds	Explosion
Peroxides (and peroxidizable compounds)	Fire and explosion
Polychlorinated biphenyls (PCBs)	Environmental
Poly cyclic aromatic hydrocarbons	Environmental

Reactivity of Types of Compounds. Many additional hazards derive from the hazardous reactivity of combinations of chemicals. The open literature contains numerous lists of the reactivity of different types of chemical combinations. Table 5.3 presents examples of combinations of compounds that are known to be reactive. More complete discussions and lists of highly energetic chemical interactions are found in CCPS (1995e), Yoshida (1987), Medard (1989), FEMA (1989), and Bretherick (1995). A complete review of the topic of chemical reactivity is also found in CCPS (1995e and 2003b).

The Interaction Matrix. The chemical interaction or reaction matrix is a recognized useful hazard identification tool. Figure 5.2 is a conceptual reactivity matrix. Matrices are referenced in CCPS (1992) and described in CCPS (2003b).

	Chemicals	Materials of Construction	Utilities	Lifeforms
Chemicals	X	X	X	X
Materials of Construction		X	X	X
Utilities			X	X
Lifeforms				X

NOTES:

- Chemicals include all raw materials, intermediate, product, and by-product chemicals, as well as any other chemicals used in the process (for example, catalysts).
- Materials of construction include ancillary materials like pipe dope
- Utilities include ambient air.
- Lifeforms include the operator, the plant neighbor, the general public, the ultimate consumer of the product. They also include the aquatic and terrestrial plant, fish and animal life.
- The X represents a reference to notes. These notes should be sufficiently complete to highlight the type of hazard and the degree (extent, severity) of the hazard.

Figure 5.2 A Conceptual Interaction Matrix

A general method for producing a plant/site–specific chemical compatibility matrix is given by the American Society for Testing and Materials (ASTM) International (ASTM 2006), and an example matrix for a process is presented in Gay and Leggett (1993). Necessary utilities, such as inerting nitrogen, and the materials of construction should be listed as components in a matrix, as should the operator and other populations impacted by the process. Tests or calculations may be appropriate if the effect of an interaction identified in a matrix is unknown as described below. The U.S. Coast Guard (USCG 2005) has published a comprehensive general chemical compatibility matrix for use in determining the possible reactions between different types of marine cargoes. This matrix stems from work performed by the USCG and the chemical industry.

Software and online resources are available to assist in making and maintaining reactivity/interaction matrices. Examples include:

- National Oceanic & Atmospheric Administration (NOAA)/Environmental Protection Agency (EPA) Chemical Reactivity Worksheet (NOAA, 2006).
- American Society for Testing and Materials (ASTM) International, CHETAH: Chemical Thermodynamic & Energy Release Evaluation.
- ColePalmer online chemical compatibility tool http://www.coleparmer.com/techinfo/chemcomp.asp)
- US Seal (http://www.usseal.com/jmchem.html).
- iProcessamart.com (http://www.iprocessmart.com/techsmart/compatibility.htm)
- Upchurch Scientific (http://www.upchurch.com/TechInfo/chemComp.asp)
- FLW (http://www.flw.com/material/index.html)

Reactivity Testing: There are a number of testing methods available to determine the thermal stability and the onset temperature of exothermic reactions, as well as the rate of reaction and heat generated per unit mass of the material(s) involved. These are summarized below, and described in full in CCPS (2003), Englund (1990), and Fauske (2006):

- Differential scanning calorimetry
- Differential thermal analysis
- Insulated exotherm test
- Decomposition pressure test
- Carius sealed tube test
- Mixing cell calorimetry
- Vent sizing package
- Accelerating Rate Calorimeter®
- Reactive System Screening Tool/Advanced Reactive System Screening Tool

Table 5.3: Reactive Combinations of Chemicals

Substances		Type of Hazard
A +	B →	Hazardous Event
Acids	Chlorates Chlorites and Hypochlorite Cyanides Fluorides Epoxides	Spontaneous ignition Spontaneous ignition Toxic and flammable gas generation Toxic gas generation Heat generation, polymerization
Combustibles	Oxidizing agents Anhydrous Chromic Acid Potassium permanganate Sodium peroxide	Explosion Spontaneous ignition Spontaneous ignition Spontaneous ignition
Alkali	Nitro compounds Nitroso compounds	Easy to ignite Easy to ignite
Ammonium Salts	Chlorates Nitrites	Explosive ammonium salts formed Explosive ammonium salts formed
Alkali Metals	Alcohols, Glycols Amides, Amines Azo- and diazo-compounds	Flammable gas and heat generation* Flammable gas and heat generation* Flammable gas and heat generation*
Inorganic Sulfide Metals	Water Explosives Poymerizable compounds	Toxic and flammable gas generation Heat generation and explosion Polymerization and heat generation

Additional testing methods exist to determine the shock sensitivity of materials, the flash point, autoignition temperature, and flammability limits of liquids and vapors, as well as the explosive properties of dusts, such as the minimum ignition energy (MIE), minimum explosive concentration (MEC), and rate of pressure rise (K_{st}).

Other Hazard Identification Tools: The What if, Checklists, and Hazard and Operability (HAZOP) methods are well-publicized and widely-used hazard identification tools. These methods are typically used to analyze the potential hazards associated with a preliminary or detailed process design. However, their basic brainstorming concept and format are also applicable to the process development/research stage of the process life cycle. In addition, the Preliminary Hazard Analysis (PrHA) technique was specifically designed to evaluate the potential hazards associated with the very early stages of a process design, while the basic processing technology and chemistry are being chosen. CCPS (1992) presents guidance on the use of these tools. Note that the application of these analytical methods will not only offer the opportunity to examine the inherently safer design associated with the basic chemistry of a process design, but also help identify active and passive safeguards that would be appropriate for any given process chemistry. Chapter 8 provides additional information and direction on the choice of process hazard analysis techniques when evaluating inherently safer strategies.

Alternative Chemistries and IS Measurement Methods: Alternative process chemistries will likely pose different hazards, as well as different degrees of the same hazards. The comparison of alternatives will inevitably lead to tradeoffs. The chosen candidate, though not devoid of hazards altogether, will likely be the one with the least risk. There is, at present, no universally accepted analytical method to conclusively judge among alternatives (see Sections 3.4 and 3.5). Edwards et al. (1996) proposed such a tool consisting of seventeen potential parameters, including pressure, temperature, yield, toxicity, flammability, corrosivity, etc. These parameters were spilt into ranges and a relative scoring index, consisting of an ordinal number (i.e., 1, 2, 3, etc.) was assigned to each range. Each step in the alternative process is analyzed, with the result being a summed relative score. The alternative process with the lowest score is the most inherently safe. Lawrence (1996) also proposed a similar relative IS scoring index consisting of a process index and a chemical index that were then combined. These original relative scoring methods have been modified and extended to add additional parameters, such as type of equipment, equipment layout, safety of process structures, and chemical interaction to the analysis (Heikkilä, 1999). A further refinement to the basic relative scoring system was added by (Gentile, 2004) to account for the discontinuities associated with using fixed ranges for the important parameters (i.e., the relative score for temperature changes from 2 to 3 for a 1° change from 199 °C to 200 °C). This refinement is accomplished using fuzzy logic, and moves the indexing scores in a continuous manner rather in discrete, sudden jumps.

Gupta (2003) modified the relative scoring approach, which is biased by the number of steps in the alternative being analyzed, and by the assumed equal weight of hazard assigned to each dimensionless value given to each range of the parameters of interest. For example, does a "3" assigned to a temperature range of 200-299 °C have the same level of hazard as a "3" assigned to a pressure range of 150-200 psig? To avoid these issues, the safety-related process parameters of interest (i.e., pressure, temperature, and a non-dimensional combined index that characterized flammability, explosiveness, and toxicity) for each step in the process were simply plotted and compared to each other. When this graphical method was used to analyze the alternatives for producing methyl methacrylate (MMA), compared to those generated by the Lawrence relative scoring method for the same alternatives, a different conclusion was reached. Of six possible routes for producing MMA, the acetone cyanohydrin alternative was the least inherently safe alternative using the relative scoring method, but the best alternative using the parameter graphical method. The differences involve the importance of the process parameters selected to characterize the risk. The graphical method displays the importance of these values in each step, whereas the relative scoring method does not highlight them as clearly. In choosing which type of alternative analysis is appropriate, the key is to determine which parameters are most important and how they are measured with respect to each other. In the example described herein the number of steps, and hence the number of hazardous chemicals, dominate the relative scoring method results, thereby emphasizing substitution and simplification to eliminate or replace the materials used and reduce the number of steps. The operating pressure dominates the graphical method results, thereby emphasizing moderation to operate at lower values of pressure.

Khan et al. (2003) compared several available IS index methods—the Lawrence IS Index; the *Dow Fire & Explosion Index* (Dow, 1994b) and *Dow Chemical Exposure Index* (Dow, 1994a); the Mond Fire, Explosion, and Toxicity Index (ICI, 1985); and the Safety Weighted Hazard Index (Khan 2001)—to an ethylene oxide production process. Several of these indices predate the formalization of the IS strategies described above. The conclusion was that none of the four indices evaluated were comprehensive methods that were easy, required little interpretation, and could be applied at all times during the life cycle of a process. The Lawrence IS Index allowed all of the IS strategies to be analyzed, but only at the process development stage. The other methods had variable success in capturing the IS strategies adequately in their respective analytical methodologies. In using these types of methods to analyze IS alternatives, it is important that the results be analyzed with a full understanding of the strengths and weaknesses of the method employed. And, perhaps using more than one method is appropriate, especially when the overall risks are high.

The INSIDE (INherent SHE In DEsign) project was a European government/industry project established by the Commission of the European Community in August 1994 to encourage and promote inherently safer chemical processes and plants. The goal of the *INSIDE* Project was to develop practical ways to encourage the use of inherent safety in process development and plant design. The result of this work has been a collection of tools and methods known as the *INSET* Toolkit (INSET, 2001) that is specifically designed for this purpose. The *INSET* Toolkit provides chemists and engineers with the tools and methods to systematically identify, evaluate, optimize and select inherently SHE processes and designs. Whether a project is completely new, or an existing process in a new plant, or whether a client is considering modifications to an existing plant and process, safety, health and environmental hazards are treated in an integrated way to ensure the conflicts and synergies between them are recognized and effectively managed. The Toolkit pays special attention to the key early stages of a project when almost all the main decisions that determine the SHE performance of the plant are taken. It should be noted that the tools focus on only the inherent safety aspects in the decision-making, not on the total safety, health, or environmental picture. Also, the INSET tools do not replace the need to apply conventional safety studies, hazard analyses, and risk assessment.

The INSET Toolkit is arranged to be used in four stages, as described in Table 5.4. Specific tools are used in each stage to achieve the analysis described in Table 5.4. INSET (2001) provides all of the detailed guidance on applying each tool.

Table 5.4 INSET Toolkit Stages (INSET 2001)

Stage I: Chemistry route selection
This stage is where potential chemistry routes to manufacture the product are sought; for some products there could be hundreds of routes available. Some simple screening is carried out on these to see which (say five) should be evaluated further.
Stage II: Chemistry route detailed evaluation
This stage involves taking a few potential chemistry routes, gathering the relevant chemical data, and assessing the routes in detail. A final selection of the best route, or perhaps two routes, to be further optimized/developed or to be used directly as the basis of the plant process should result. It is particularly applicable where the available options have many conflicting aspects and there is no immediately obvious route alternative.
Stage III: Process design optimization
The selected route(s) from *INSET* Stage II are assessed to optimize the conditions and take account of the practicalities of industrial scale processing and the implications of using particular processing equipment.
Stage IV: Process plant design
The initial process design is developed and "challenged" to identify further changes in sequencing, feed profiles, conditions, unit operations and equipment selection in order to improve the performance. The detailed aspects of equipment sizing and pipework fittings are subsequently evaluated to try to identify means of reducing the process inventories and eliminating complexity and hence the possible leak points.

Additionally, the Dow Fire and Explosion Index (FEI) (Dow, 1994b), the Dow Chemical Exposure Index (Dow, 1994a), the Mond Index (Lewis, 1979; ICI, 1985), and the Mond-like acute toxicity Hazard Index (Tyler, et al, 1996) are analytical methods that can assist in choosing the most appropriate process chemistry among alternatives. Guidance on alternative selection is recognized as a future need in Chapter 12.

Life Cycle Costs. The life cycle cost of a process is the net total of all expenses incurred over its entire lifetime. The choice of process chemistry can dramatically affect this life cycle cost. A quantitative estimate of life cycle cost cannot be estimated at the research stage with sufficient accuracy to be of practical value. However, making a qualitative attempt to estimate the life cycle costs of competing chemistries can be beneficial. Implicit in any estimate of life cycle cost is the estimate of risk. One alternative may seem more attractive than another until the risks associated with product liability issues, environmental concerns, and process hazards are given due consideration. Cost benefit analyses (CCPS, 1995a) are useful in predicting and comparing the life cycle costs of alternatives.

Examples of Inherently Safer Synthesis Routes. As summarized by Bodor (1995), the ethyl ester of DDT is highly effective as a pesticide and is not as toxic. The ester is hydrolytically sensitive and metabolizes to non-toxic products. The deliberate introduction of a structure into the molecule to facilitate its hydrolytic deactivation to a safer form can be a key in creating a chemical product with the desired pesticide effects, but without the undesired environmental effects. This technique is also being used extensively in the pharmaceutical industry, and is applicable to other chemical industry sectors as well.

Synthetic rubber latex is produced using a process with a large and hazardous inventory of butadiene and styrene. In a modified process, the reactor has an initial charge of water and emulsifier. Monomers are added to the reactor as one pre-mixed stream, and the emulsified aqueous sodium persulfate is added as the other stream. The improved scheme, discussed by Englund (1991a), contains less hazardous material and operates at a lower, more controllable temperature. It illustrates that large and established processes may be made safer by applying inherently safer technologies.

5.3 PROCESS DEVELOPMENT

Process chemists and process engineers have essential roles to play in applying inherently safer strategies during the process development stage. At this point, the chemistry has already been chosen, thus defining the basic hazards of the materials. Process development personnel need to focus primarily on process synthesis, unit operations, and the type of equipment required for an inherently safer process. A thorough understanding of the necessary operational steps, as well as the alternate operational steps, is essential to develop an efficient and safe process. Many of the methods discussed in Section 5.2 for the research stage of the process life cycle are useful in the process development stage as well. The Dow FEI (Dow, 1994b), Dow CEI (Dow, 1994a), and Mond Index (ICI, 1985; Tyler, 1985), SWeHI (Khan 2001), IS Index (Lawrence 1996), graphical IS comparisons (Gupta 2003), and INSET (2001) Toolkit are examples of these methods, as well as the variety of available PHA techniques (CCPS 1992). It is also appropriate to revisit the basic chemistry to study alternate options that may be available, given the unit operations that are anticipated to be used in the process design.

5.3.1 Unit Operations - General

There are a variety of ways of accomplishing a particular unit operation. Alternative types of process equipment have different inherently safer characteristics, such as inventory, operating conditions, operating techniques, mechanical complexity, and self-regulation (i.e., the process/unit operation is inclined to move itself toward a safe region, rather than unsafe). For example, to complete a reaction step, the designer could select a continuous stirred tank reactor (CSTR), a small tubular reactor, or a distillation tower to process the reaction.

Before studying alternative types of equipment, the process requirements must be understood. For example:

- Is a solvent necessary?
- Must the products or by-products be removed to complete the reaction?
- What mixing and/or time requirements are necessary?
- What sequencing is necessary for material additions?
- Is the reaction exothermic, endothermic, or adiabatic?

These and other relevant questions must be answered before alternate reaction schemes can be evaluated.

Similarly, different unit operations are available to accomplish the same processing objective. For example,

- Should a filter, a centrifuge, or a decanter be used to separate a solid from a liquid?
- Should crystallization or distillation be used for a purification step?

It is inherently safer to develop processes with wide safe operating limits that are less sensitive to variations in the operating parameters, as shown in Figure 5.3. Sometimes this type of process is referred to as a "forgiving" or "robust" process. If a process must be controlled within a very small temperature band in order to avoid hazardous conditions, that process would have narrow safe operating limits; a process with a larger temperature band will have wider safe operating limits, also known as a "safe operating envelope" or a "safe operating window." For some reactions, using an excess of one reactant can enlarge the safe operating limits. Determination of the size of the safe operating limits requires a complete understanding of the processes under consideration.

5.3.2 Unit Operations - Specific

Some examples and considerations for specific common unit operations are described as follows.

Reaction. Reactor design is particularly critical because reactors involve chemical transformations, and often potentially significant energy releases. Evaluation of the safety characteristics for a given reactor design requires an understanding of what physical or chemical processes control the rate of reaction (catalysis, mass transfer, heat transfer, etc.), as well as the total potential energy consumption or generation involved in the reaction. Energy generating pressure and/or undesired side reactions should also be evaluated. This information is usually necessary for evaluating the suitability of various reactor types (CSTR, batch, tubular, various novel designs such as eductors in loop reactors, static mixers, extruders) for the desired reaction. Mixing and mass transfer are often the critical elements in reactor design as chemicals often react quickly once the molecules are brought together.

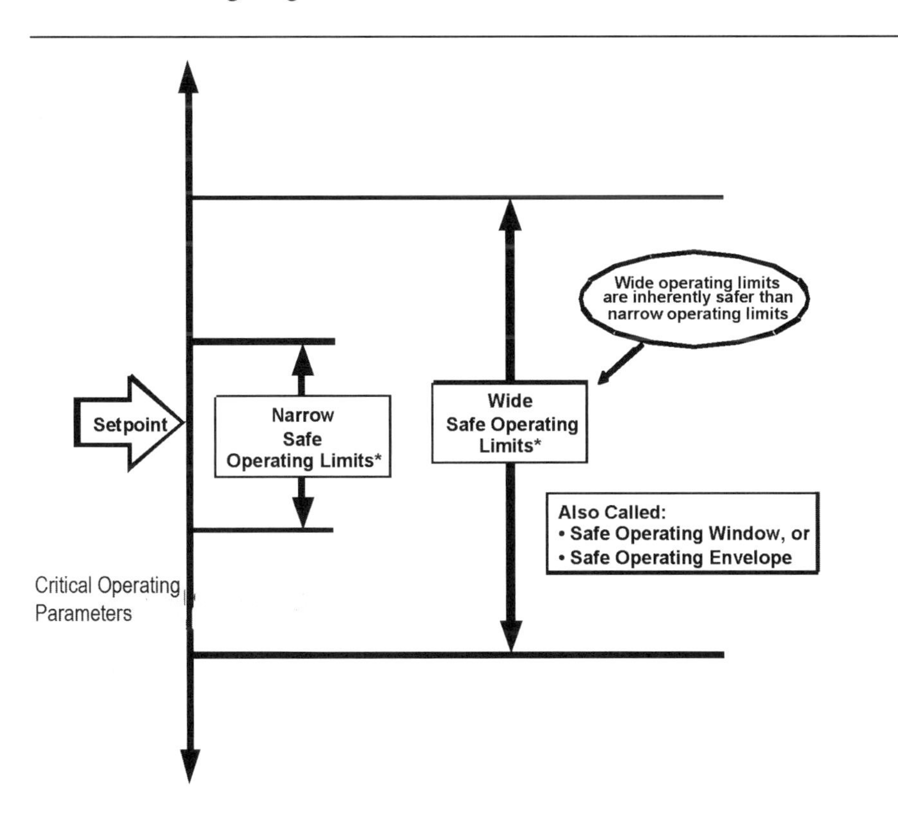

Figure 5.3. Process Design Safe Operating Limits

Not all reactions take place in a designated reactor. Some occur in piping systems, a heat exchanger, a distillation column, or a tank. It is vitally important to understand the reaction mechanisms and know where the reactions will occur before selecting the final design.

Some batch reactions have the potential for generating very high energy levels. If all of the reactants (and catalysts, if applicable) are charged into a reactor before the reaction is initiated, and two or more of the materials in the reactor react exothermically, a runaway reaction may result. The use of continuous or "semi-batch" reactors to limit the energy present and to reduce the risk of a runaway reaction should be considered. The term "semi-batch" refers to a system where one reactant and, if necessary, a catalyst is initially charged to a batch reactor. A second reactant is subsequently fed to the reactor under conditions such that an upset in the reacting conditions can be detected and the flow of the second reactant stopped, thus limiting the total amount of potential energy generated in the reactor.

Additional discussion regarding reactor design strategies is covered in Section 4.2 on minimization (as an inherently safer design strategy), and in Section 5.4 on design and construction.

Distillation. There are options to minimize the hazards when distilling materials that may be thermally unstable or have a tendency to react with other chemicals present. Some options include:

- Trays without outlet weirs
- Proprietary designs and sieve trays
- Wiped film evaporators
- An internal baffle in the base section to minimize hold-up
- Reduced base diameter (Kletz, 1991b)
- Vacuum distillation to lower temperatures
- Smaller reflux accumulators and reboilers (Dale, 1987)
- Internal reflux condensers and reboilers where practical (Dale, 1987)
- Column internals that minimize holdup without sacrificing operational efficiency (Dale, 1987)

Another option is to remove toxic, corrosive or otherwise hazardous materials early in a distillation sequence, reducing the spread of such materials throughout a process (Wells and Rose, 1986).

Low-inventory distillation equipment, such as the thin film evaporator, is also available and should be considered for hazardous materials. This equipment offers the additional advantage of short residence time and is particularly useful for reactive or unstable materials.

Solids Handling. The handling of solid materials frequently has the potential for producing dust clouds that can lead to potential health and explosion hazards. Handling solids in the form of larger particle size granules or pellets, rather than a fine powder, reduces the potential for worker exposure and also increases the minimum explosive concentration, thus reducing the hazard of dust explosions. Worker exposure and dust explosion hazards are also reduced by formulating dyes as liquids or wet pastes rather than dry solids or powders (Burch, 1986).

If the solid is unoxidized and hence combustible, the dust explosion hazard can be greatly reduced or even eliminated by using a larger particle size material. However, it is important to remember that particle attrition can occur during handling and processing, resulting in the generation of small particles that can become suspended, increasing dust explosion hazards. The sequence of size reduction steps or even the required particle size must be studied to minimize the number of processing steps that involve very small particles. Another option would be to change the form to a shape less prone to dust generation, such as pellets, beads, prills, etc. Handling of solids as a wet paste or slurry can also reduce hazards. For example, using wet benzoyl peroxide instead of the dry form reduces the hazards of this extremely reactive material (Yoshida, et al., 1991).

It may even be possible to eliminate solids handling by processing in a solution. However, this may require an assessment of the hazards of a toxic or flammable solvent in a process compared to the hazards of the solvent-free process. Additionally, hybrid mixtures of solid material dusts with flammable gases and liquids are generally more hazardous than the dust by itself (CCPS 2003a).

Inherently safer approaches to dust explosion hazard control include building equipment and structures strong enough to contain an explosion. A more thorough discussion of dust explosion hazards and their control may be found in CCPS (1999, 2003a), Eckhoff (1997), NFPA (2004, 2006).

Heat Transfer. Some processes have large heat transfer requirements. This may result in large inventories of material within the heat transfer equipment. If the material is thermally unstable, it would be inherently safer to reduce the residence time in the heat exchanger. Options to minimize heat exchanger inventory include the use of different types of heat exchanger designs. As noted in Section 4, inventories in shell and tube heat exchangers can be reduced by the use of 'turbulators" in the tubes to enhance heat transfer coefficients, and by placing the more hazardous material on the tube side.

Heat transfer equipment has a great variation in heat transfer area per unit of material volume. Process inventory can be minimized by using heat exchangers with the minimum volume of hazardous process fluid for the heat transfer area required. Table 5.5 compares the surface compactness of a variety of heat exchanger types. Falling film evaporators and wiped film heat exchangers also reduce the inventory of material on the tube side.

In the past several decades, a number of technology advances in compact heat exchanger technology have occurred. However, due to lack of awareness, use only in non-process plant applications, absence of reliable design methods, and lack of testing and investigation in actual process operating situations, these fairly advanced heat exchange technologies have not seen widespread use in the process industries. Compact heat exchangers are generally defined as having a compactness ratio of 700 m^2/m^3 or greater, and a hydraulic diameter of 1 mm to 10 mm (Stankiewicz, 2004). By contrast, shell and tube heat exchangers have much lower compactness ratios and larger hydraulic diameters. Therefore, their efficiencies are much lower and their footprint is much larger. The following is a summary of the current types of compact heat exchangers and their applications (Stankiewicz, 2004):

- *Plate heat exchangers (PHE):* Originally developed for use in the food industry (milk pasteurization), PHEs are made of pressed corrugated plates with heat transfer surface areas of 0.02-3.0 m^2 per plate and hydraulic diameters of 2-10 mm. Compactness ratios can be as high as 1000 (see Table 5.5). PHEs have seen use in chemical, district heating, and power applications, but mainly for single-phase duty. However, dual phase use of PHEs has been applied in the refrigeration industry. In automotive applications as heaters and evaporators, PHEs have been used to achieve a reduction factor of volume-to-heat capacity of 2.
- *Spiral heat exchangers:* Spiral heat exchangers consist of plates that are welded together and rolled to create a spiral passage. Heat transfer surface areas vary widely (0.05-500 m^2). Spiral heat exchangers have been used in phase change applications.
- *Plate and shell heat exchangers:* This type of heat exchanger consists of a bundle of plates inserted into a shell. In the process industry, these plate and shell heat exchangers have been used as boilers because the shell side can be designed to easily accommodate a higher pressure. The variety of geometries that can be used make it possible to achieve very high compactness ratios (see Table 5.5). Alternative materials of construction other than aluminum have been used to accommodate high operating temperatures and pressures and/or high corrosivity environments (copper, stainless steel, titanium).
- *Plate-fin heat exchangers (PFHE):* Aluminum PFHEs were first developed for the aircraft industry because of the need for a high compactness and light weight. Today, PFHE technology is used for air separation, hydrocarbon separation, and the liquefaction of natural gas (LNG).
- *Flat tube-and-fin heat exchangers:* This type of heat exchanger has been developed primarily in the automotive industry for use in air conditioners and engine cooling. These are primarily gas-liquid, with a few liquid-liquid heat exchange situations. The gas side is the finned side, and the liquid side is the small diameter channel side. Mechanically, these heat exchangers can accommodate high pressures (20-140 bar).

- *Microchannel heat exchangers:* This type of heat exchanger is characterized by hydraulic diameters of less than 1 mm. To manufacture these devices chemical etching, or micromachining, technology is required. Microchannel heat exchangers have been used in severe applications such as high temperature nuclear reactors, and offshore platforms. The most common type is the printed circuit heat exchanger in which a channel is chemically etched into a plate. Hydraulic diameters in the 50-200μm range are possible. The plates are then stacked and bonded together. Very high temperature (900 °C) and pressure (500-1000 bar) applications are possible with this type of heat exchangers. The main limitation of a microchannel heat exchanger is high pressure drops across the channel.

Additional work has been performed in recent years to develop multi-function heat exchangers that combine heat exchange and reaction unit operations in one device (Stankiewicz, 2004). Several such devices already exist, including catalytic plate reactors, where a plate-type heat exchanger is coated with a reaction catalyst. A heat exchange reactor has to meet several design objectives:

- The residence time in the device must be sufficient to complete the desired reaction.
- The fluid temperature must be controlled, implying high heat transfer coefficients.
- If the feed and reactant are not pre-mixed well, the channel geometry must create turbulence that is sufficient to accomplish adequate mixing.
- The pressure drop across the device must be acceptable.

Some compact heat exchanger designs meet these characteristics and have high heat and mass transfer coefficients even at low flows, and have flows that are turbulent enough ($Re > 300$) to ensure adequate mixing. However, additional design work needs to be done before such devices are ready for widespread application in the process industries.

Piping. Inventory in piping systems can represent a major risk. For example, a quantitative risk analysis of a chlorine storage and supply system identified the pipeline from the storage area to the manufacturing area as the most important contributor to total risk (Hendershot, 1991b). To minimize the risk associated with transfer lines, their lengths should be minimized by careful attention to unit location and pipe routing. Pipe size should be sufficient to convey the required amount of material and no larger. However, small bore piping is less robust and less tolerant of physical abuse when compared to large diameter piping, and additional attention to proper support and installation is required. In some cases, such as chlorine for water treatment applications, it may be possible to transfer material as a gas rather than a liquid, greatly reducing inventory in the transfer line.

Piping systems should be designed to minimize the use of components which are likely to leak or fail. Sight glasses and flexible connectors such as hoses and bellows should be eliminated wherever possible. Where these devices must be used, they must be specified in detail so they are structurally robust, have the same temperature and pressure ratings as the fixed piping (or as close as possible), are compatible with process fluids, and are installed to minimize the risk of external damage or impact. In addition, these more fragile elements of the piping system should be inspected on a more frequent basis than the rest of the piping system.

Table 5.5: Surface Compactness of Heat Exchangers
Adapted from Kletz (1991b)

Type of Exchanger	Surface Compactness (m^2/m^3)
Shell and tube	70-500
Plate	120-225 up to 1,000
Spiral plate	Up to 185
Shell and finned tube	65-270 up to 3,300
Plate fin	150-450 up to 5,900
Printed circuit	1,000-5,000
Regenerative - rotary	Up to 6,600
Regenerative - fixed	Up to 15,000 *
Human lung	20,000

** Some types have a compactness as low as 25m^2/m^3.*

Where flanges are necessary, spiral wound gaskets and flexible graphite type gaskets are preferred. The construction of these gaskets makes them less likely to fail catastrophically resulting in a large leak. Proper installation of spiral wound gaskets, particularly the torquing of the flange bolts, is important in preventing leaks.

5.4. DETAILED DESIGN AND CONSTRUCTION

As the process moves from the process development stage to the detailed design and construction stage, the chemistry, unit operations, and type of equipment have been set. The design and construction stage focuses primarily on detailed equipment specifications, piping and instrumentation design, and installation details. Although the opportunity to incorporate inherently safer strategies still exists during detailed design and construction, many of the moderation and substitution opportunities are no longer possible because the process chemistry and the safe operating envelope will already have been specified. However, if the equipment/unit layout was not established in earlier stages, an opportunity to apply the minimization strategy may exist at this stage. And, many opportunities exist to apply the simplification strategy during detailed design.

Plant designs should be based on a risk assessment that considers the process and the site in detail, as well as all of the principles of inherently safer operation. Earlier decisions may limit the options during the detailed design stage, but inherently safer principles can still be applied. The detailed design step is the last step at which changes can be made at moderate cost because most of the equipment is purchased after this detailed design is approved. Once the equipment has been purchased and fabricated, and the facility is constructed, the cost of modification increases substantially.

5.4.1 Process Design Basis

To reduce the potential for large releases of hazardous materials:

- If not already accomplished, minimize or eliminate the in-process inventory of hazardous materials to the lowest amount necessary consistent with the minimum operational needs. This includes inventory in the process equipment, as well as in vessels and tanks. Elimination of intermediate storage tanks will likely require improvements in the reliability of the upstream and downstream equipment in order to preclude unit or plant shutdowns caused by running out of material.
- Review secondary containments, impoundments, and spacing for tanks storing flammable materials. A sump inside a dike facilitates the collection of small spills. Sump drains or pumps can direct material to a safe and environmentally acceptable place. See NFPA 30, the flammable liquids code (NFPA, 2003).
- Review the layout to minimize the length of piping containing hazardous materials.
- In batch operations, minimize pre-charging the most energetic chemical. Consider adding energetic material in a "semi-batch" mode. That is, add most of the ingredients initially, followed by the addition of the energetic material under flow control. An SIS is used to isolate the feed to the reactor when high temperature or high pressure indicates abnormal operation outside the safe operating limits. A passive safeguard (pipe size, orifice, limited pump

capacity) could be considered to limit available energy. Low temperature can be dangerous if the energetic material "pools" unreacted in the reactor and then the reaction initiates. The pooled material could have enough potential energy to result in catastrophic releases.

When dealing with flammable materials, selection from inherently safer design options may vary according to the site and process. For example:

- Use non-flammable materials, if this is still an option at this stage of the life cycle.
- Design the vessel to withstand the maximum pressure generated (i.e., inherent robustness).

Additionally, a number of active and passive safeguards should be provided as appropriate for the storage and handling of flammable materials, including using inerting vessels, and installing explosion suppression and relief directed to a safer location (NFPA, 2002a, 2002b; CCPS, 1994b, 2003a).

5.4.2 Equipment

Some engineers specify the maximum allowable working pressure (MAWP) for pressure vessels, considering the intended operating temperatures and pressures only. This can result in a vessel design that is not tolerant of process upsets resulting from control system, communications, or utilities failure. If these process upsets potentially result in process safety incidents, protection layers are required to reduce the risk. An inherently safer choice is to minimize the use of protection layers by designing robust equipment. For example, if the vessel MAWP is higher than the maximum expected pressure, protection layers, such an SIS to shutdown the pressure sources on high pressure, are not necessary. The proper specification of process equipment should consider potential overpressure scenarios, such as those listed in API 521. This requires knowledge of the chemistry outside the design conditions to evaluate effects of loss of utilities and control systems.

Vessels with passive design can fully withstand any overpressure without exceeding the yield stress of the materials. If the overpressure in a vessel remains in the elastic range, the metal returns to its normal crystalline state after stretching. Systems designed to "bend but not break" slightly exceed the plastic region of the metal and are deformed (hardened). The vessel is then actually made stronger by this process; however, the new hazard that is introduced is that the vessel will not stretch and will usually burst if the scenario is repeated. Thus, vessels subjected to plastic range stresses require more frequent inspections for deformation and integrity. A truly passive design is not only safer, it is more cost effective when the lifetime test and inspection requirements are considered.

A passive design must be extended to include all system hardware elements. Little is gained if containment is lost when pipes, joints, or instruments fail due to overpressure. In designing the process and equipment, the same engineering principles described in Section 5.3 should be used to minimize the accumulation of, and to contain, energy or materials:

- Specify design pressures high enough to contain pressures generated during exothermic reactions and avoid opening the relief valve and/or exceeding the maximum allowable working pressure of the vessel (the safe upper pressure of the safe operating envelope).
- Use physical limits of pipe size, restrictive orifices, and pump sizing to limit excessive flow rates.
- Use gravity flow in the equipment layout where feasible to minimize the need for pumps or solids handling equipment for hazardous materials.
- Review injection points and piping runs for erosion concerns. Design injection points, elbows, turns, and other erosion-prone areas for lower velocities.
- Use materials with low corrosion rates.
- Use materials that are applicable over the full range of operating conditions, such as normal, startup, routine shutdown, emergency shutdown, and system draining. For example, carbon steel may be acceptable for normal operating conditions, but may be subject to brittle fracture at low temperatures under abnormal conditions, as in the case of a liquefied gas. Cold water, less than 60 °F, during hydrotesting may cause brittle fracture of some carbon steels.
- Avoid materials that crack or pit; uniform corrosion is safer than non-uniform, localized corrosion patterns.
- Avoid incompatible materials that could come into contact under abnormal conditions.
- Do not use copper fittings in acetylene service, or titanium in dry chlorine service. These principles also apply to gaskets, lubricants and instruments.
- When reviewing the materials of construction consider external corrosion concerns. Insulation capturing chlorides or containing chloride can initiate stress cracking of austenitic stainless steel. Stainless steel surfaces should be primed, and a weather barrier is needed.
- If possible, eliminate inherently weak equipment such as sight glasses, hoses, rotameters, bellows, expansion joints, and most plastic equipment.
- Minimize possible contamination via fewer cross-connections and fewer hose stations. Minimize the number of hoses required in loading/unloading facilities. Cross-contamination, sometimes even

from catalytic amounts of material, can result in undesired hazardous reactions. To prevent contamination due to rainwater and spills, consider storing a material that can react vigorously with water indoors. Do not store water-reactive materials below grade.

- Flexible connections should never be used as a cure for improper piping alignment and piping support concerns. Figure 5.4 illustrates both good and poor piping alignment. Where expansion joints are required in piping systems containing toxic materials, consider using double-walled expansion joints with pressure indication between the two walls for leak detection.
- All welded pipe is preferable to flanged piping, and threaded piping should be avoided for flammable and toxic materials.
- Pinch valves that have no packing to leak can leak if the tube breaks.
- There are trade-offs for magnetic drive and canned pumps versus centrifugal pumps with double-mechanical seals. The former have no seals to leak, but need active SISs to prevent high temperature when handling temperature-sensitive materials. Similarly, diaphragm pumps, which have no shaft seals to leak, have potential for process leaks out of the exhaust line and air leaks into the process on diaphragm failure. Double diaphragm pumps need two failures before material leaks out the exhaust line.
- Consider a weak roof seam for API tanks, if the tank could split under internal pressure. The roof seam should fail first, not the bottom seam. The weak roof seam must be specifically included in the specifications, and the mechanical design must address this issue. This emphasis is made because smaller tanks of less than 50 feet in diameter, manufactured under API RP 650, will not automatically have a weak roof seam. This standard is designed to provide the petroleum industry tanks of adequate safety and reasonable economy for use in the storage of petroleum, petroleum products, and other liquids commonly handled and stored by the various branches of the petroleum industry. It is intended to help purchasers and manufacturers in ordering, fabricating, and erecting tanks. API 650 (API, 1998) covers material, design, fabrication, erection, and testing requirements for vertical, cylindrical, aboveground, closed- and open-top, welded steel storage tanks, in various sizes and capacities for internal pressures approximating atmospheric pressure. But, a higher internal pressure is permitted when additional requirements are met. This standard applies only to tanks whose entire bottom is uniformly supported, and to tanks in non-refrigerated service, that have a maximum operating temperature of 90 °C (200 °F). This standard has been used extensively in the chemical and petrochemical industry, as well as in the petroleum industry.

Additional guidance for the design of storage tanks is provided in API 620 (API 2002) and UL-142 (UL 2002).

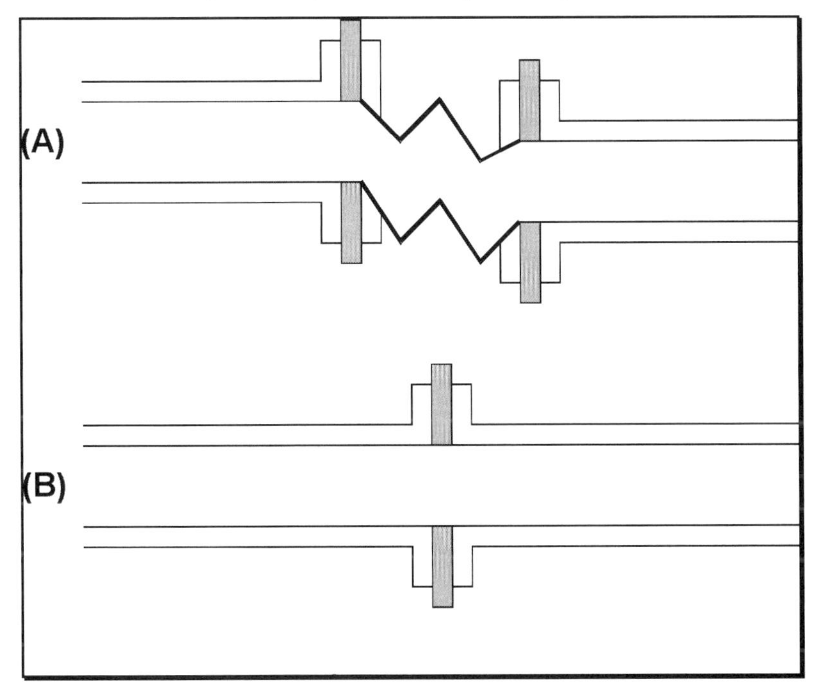

Figure 5.4: Proper and improper piping design. (A) shows improper use of a flexible connection to compensate for poor piping alignment, while (B) shows proper design and alignment

---

Users are cautioned that the details of the construction of the weak seam are extremely critical. The design may specify a steel of a certain grade, using the minimum strength of that grade's specification. The steel actually supplied may have a strength significantly higher than the minimum grade specification. For all the joints of the tank except the weak seam, higher strength is better and makes the tank inherently safer. However, for the weak seam, higher strength increases the burst pressure and negates its benefit. Likewise, the thickness of the steel plate may be greater than that specified, increasing the strength of the weak seam.

5.4.3 Process Controls

Many of the inherently safer design aspects discussed here appear in *Guidelines for Safe Automation of Chemical Processes* (CCPS, 1993b) and in *Guidelines for Safe and Reliable Instrumented Protective Systems* (CCPS, 2007).

The ultimate goal of inherently safer design is elimination of all hazards, with no need for active or other forms of risk controls. However, some control systems are always required. The design logic of a specific control system can make it inherently safer than other alternatives.

While it may be argued that the discussion below reads like a design guide, the point is that a management system that requires specific consideration of operating ranges and limits in the design of controls and instrumented protective systems can make the resulting systems inherently safer than those designed without such considerations. We urge organizations to apply the inherently safer principle of "simplify" to build management systems for the design process that considers the issues discussed below.

Developing inherently safer process chemistry and physical design features may be the most economical way to eliminate a hazard. A hazard avoided by inherently safer design does not require active controls. When designing a chemical or physical process, keeping the process within a safe operating envelope requires identification of all the system phenomena. Inherently safer designs ensure that the normal or quality operating limits are well within the safe operating limits, that these limits are within the instrumentation range, and that the instrumentation range is within the equipment containment limits, as shown in Figures 5.5 and 6.3.

If active controls are needed to prevent the process operating parameter from reaching a hazardous condition, the control system design should specify the desired operating conditions to provide adequate time for controls to function before reaching the equipment limits. Calculating the adequate time requires knowing the speed of the transient, the response time of the device being controlled, and the lag time of the control sensing element and the final controlled element (i.e., a control valve). For example, an upset in feed to a tank could lead to an overflow on high level. An inherently safer design provides enough time for the tank level control to sense the upset and to take corrective action on the flow into or out of the tank before it overflows. For such a tank, the maximum setpoint for the level must be reduced to allow adequate response time.

For some processes, human intervention is included in the response to an upset. See Section 6.5 for a discussion of the human response time in process control.

Response time considerations lead to the establishment of "never exceed" limits and "never deviate" actions. The never exceed limit is that point at which unsafe consequences will occur. The mandatory action point is a value set low or high enough to allow time for the instrumentation or human controls to prevent the hazard. One example of response lag is a system's ability to maintain a heat balance. In reacting systems, control is lost when the system generates heat faster than it can be transferred to other systems or to the surroundings. This excess heat produces an immediate pressure rise in systems containing volatile materials. In other systems, the reactants decompose to gases resulting in a pressure rise. The mandatory action temperature limit is set low enough to allow time for the response to prevent loss of containment from high pressure.

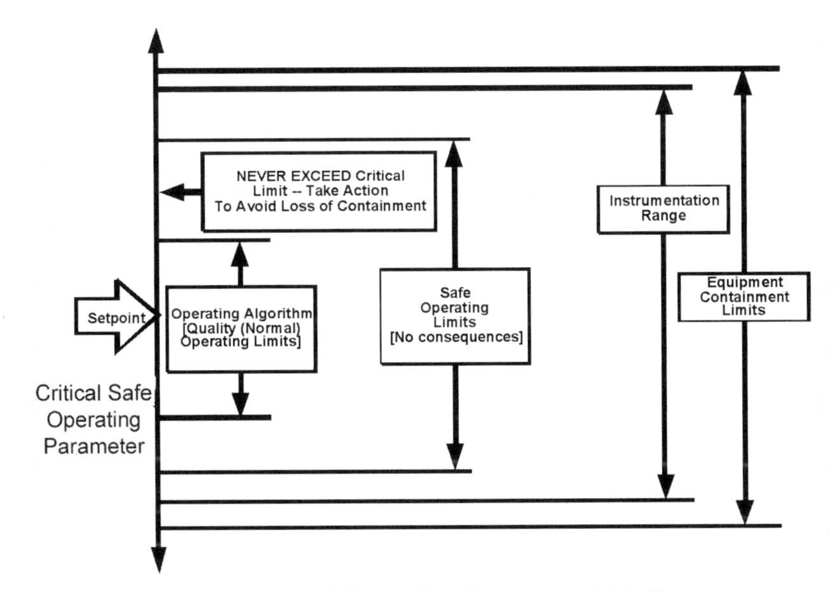

Figure 5.5 Operating Ranges and Limits

Basic Process Control System (BPCS) and Safety Instrumented System (SIS). There are few chemical plants that are so robust that an active control system is not required. Using both active and passive controls can assure product yield and quality, and maintain safe operating conditions. This type of control system is known as a basic process control system (BPCS). The BPCS acts to alarm and moderate a high or low operating condition within the never exceed limits. The SIS (CCPS, 2007) is provided to rapidly shutdown or otherwise place the process in a safe state if the BPCS fails to maintain safe operating conditions. A BPCS should not be used as the sole source of a process safety shutdown. Many of the following guidance items related to the design, operation, and testing of BPCSs and SISs are not inherently safer technology in a strict sense, because they relate to active safeguards. However, much of this guidance can also be considered part of the inherently safer strategy to simplify systems.

Inherently safer SIS should be fully independent of the process control system logic residing in the BPCS. Common-mode failure can result if the BPCS and SIS share components, including power supplies and any other utility system, such as instrument air. The SIS should normally be fail safe, designed to achieve or maintain the safe state of the process on loss of power. This de-energize to trip design is implemented using a solenoid operated valve to vent air from the valve actuator when power is lost. The valve actuator spring takes the valve to its shelf position, the so-called “fail-safe” position. This valve position should place the process in a safe state. The BPCS should also be programmed to take its outputs to the safe state if input or output signals are lost. Due to the high potential in most facilities for power or other utility loss, it can be very difficult to achieve adequate risk reduction with non-fail safe design. Fail-safe designs are significantly less complex, thereby helping to implement the IS simplification strategy.

Operators should always receive notification from the BPCS and the SIS as the mandatory action points are approached. An everyday example is a high temperature warning light on an automobile. It gives no indication of an abnormally high temperature until the temperature reaches the high alarm point. A temperature gauge gives advance warning before the alarm point. The gauge pointer provides feedback of its functionality by normal fluctuations.

When choosing SIS input variables, where possible, use direct readings of the process parameter being controlled, not an indirect reading. If pressure is being controlled, measure pressure directly rather than inferring it indirectly from temperature. This choice eliminates the lag time in processing that occurs when an indirect variable is chosen. Direct readings also eliminate potential errors in the inferred relationship between the variables.

Programmable Electronic Systems (PES) are designed to control the desired reactions or physical process using a BPCS, or produce a shutdown using a SIS when the never exceed limits are reached. CCPS (2000) deals extensively with redundant input, logic solver (PES), and output device functions. Signal gathering is also important in BPCS and SIS design. Routing all data through a single data link input/output (I/O) card can impair BPCS and SIS integrity by producing a common cause failure.

Startup, Shutdown, Tests, Inspections. Process control and safety shutdowns must be provided during all modes of operation, not only in the normal, steady-state operating mode. Hardwire devices, like timers or software logic, can be used to actuate the BPCS and SIS control functions that are pertinent to the operating mode.

Properly designed control systems may, where possible and if needed, include options for on-line functional tests of the entire control function, as well as for calibrations of individual devices, such as sensors. An SIS design must include provisions for testing at an appropriate frequency to maintain the specified integrity (or availability) for the SIFs. Inherently safer designs of SIS functions would provide means for testing in a manner that does not create an actual transient condition in an operating unit or plant. For example, a high level shutdown that can be tested without actually raising the level in a tank or vessel is inherently safer than one that requires raising the level. If this is not possible, then a procedure should be devised to raise the level safely. Remember that the fire and radiation release at Chernobyl occurred in part because safety systems were bypassed to allow testing another safety system with an actual challenge (Gruhn, 1999). Turbine tests are one exception to performing tests under actual operating conditions. Turbine mechanical overspeed trips must be tested by increasing the speed, since centrifugal force is needed to activate the trip.

In order to test the full control loop/function, the final controlled element(s) must also be tested. To do so without creating transient or frequent unit/plant shutdowns, it is often necessary to perform full SIS functional tests during shutdown periods. Depending on the nature of the plant response, it may be possible to initiate a planned shutdown by tripping the equipment during a test. This should only be attempted if the resulting shutdown will be orderly and stable and will not cause transients in other process parameters that are outside their normal limits.

Alarm Management Ergonomics. With the ability to make every signal into an alarm in a BPCS or SIS, operator information overload is a genuine safety concern. Proper alarm design will produce an alarm hierarchy. This design permits training operators to understand the importance of the never to exceed alarms. BPCS/SIS digital and analog alarm displays should be grouped to be readily identifiable by color, physical position, and distinctive sound annunciation. An on-screen list of potential action alternatives should be displayed. The navigation of digital BPCSs and SISs should be intuitive and user-friendly, particularly with respect to alarm screens. The ergonomics of control panels and other operator interfaces are further discussed in Chapter 6.

Further SIS analysis, design, testing, and other life cycle guidance is available in ANSI/ISA S84.01 (ANSI/ISA, 2004), and in CCPS (1993b, 2007) for both BPCSs and SISs.

5.4.4 Supporting Facilities

Factors to consider in regard to the support of utility systems include:

- Do not allow nitrogen or pressurized air supplies to overpressure tanks or vessels. These components should be designed to withstand the maximum pressure that can be generated by the pneumatic system, assuming that all pressure control equipment/regulators have failed. An alternative solution is to alter the maximum pressure that can be generated in the pneumatic system to be lower than the maximum allowable working pressure or design pressure of any equipment that can be exposed to the pneumatic pressure. If these inherently robust solutions are not possible due to operational requirements, then active controls, such as pressure relief valves will be required.
- Use water or steam as a heat transfer medium rather than high temperature flammable or combustible oils (Kharabanda and Stallworthy, 1980; Kletz, 1991b, 1998).
- Use heat transfer oils with high flash points and boiling points and operate them below their boiling point. Alternatively, use molten salt if water or steam is not feasible as a heat transfer medium (Dale, 1987; Kletz, 1991b, 1998).

- Seek alternatives to chlorine for water treatment and disinfecting applications. For example, sodium hypochlorite has been used successfully in both industrial and municipal water treatment applications (Somerville, 1990). Other alternatives for water treatment include calcium hypochlorite, ozone, ultraviolet radiation and heat treatment (Negron, 1994; Mizerek, 1996).
- Use magnesium hydroxide slurry to control pH, rather than concentrated sodium hydroxide (Englund, 1991a).

5.4.5 Batch Processes

Batch processes present some unique risks because of their characteristics and use. Some batch processes are dedicated to one product or one product family. However, many batch processes are multi-purpose, multi-product, and multi-stage. This provides the maximum flexibility to owners, particularly in the specialty chemicals sector, to produce a wide range of different products in the same unit/process. Pilot plants also exhibit the same multi-purpose/product/stage characteristics to allow for maximum research and experimentation flexibility. Operators often have a more direct role in the process operations and control than for continuous operations, and transient operating modes predominate in a batch or pilot plant. The entire batch run is usually an elongated transient operation with very little operation that can be characterized as steady-state. The status of plant equipment also changes with time, and the same equipment may be exposed to a wide variety of changing conditions during the batch. A non-hazardous deviation, such as a charge of an incorrect material, at one place and time, may represent a substantial risk in a different place and time. As a consequence of these characteristics, there are a number of risks that can be abated, or even eliminated, using inherently safer strategies. Examples are:

- The design of vessels, piping, and other piping system components should be chosen such that the pressure and temperature ratings are higher than the maximum anticipated conditions for any anticipated product. For pilot plants, this is difficult to predict when the plant is initially being designed. However, the nature of the products being tested should provide an idea of the pressure and temperature ranges that the equipment will be exposed to over its life cycle. Unanticipated process conditions that arise during the life of the batch or pilot plant will have to be evaluated and accommodated using the management of change process.
- The design of vessels, piping, and other piping system components should be chosen such that the most severe corrosion, erosion, and other similar conditions are accommodated in both the materials of construction, as well as the thickness of the equipment.

- To the extent possible, the use of flexible hoses and other "weak" components should be minimized or eliminated. This is often very difficult to do in a multi-product facility.
- Owing to the multi-purpose and multi-product nature of batch and pilot plants, multiple connections to and from raw material storage vessels, the reactors/mixing vessels, and intermediate/final product storage vessels exist. Valve manifolds are used to select the proper routing of materials, and especially those with multiple rows or identical valve operators present an increased likelihood of human error. Careful and distinctive labeling, particularly using different colors, will simplify the operations of valve manifolds and reduce human errors. The same is true of other equipment where there are multiple choices and the operating features are identical. One example is pump controllers that indicate panels with lights. See Chapter 6 for further discussions of the human factors associated with the design of such equipment to make it simpler to operate, more error resistant, and more error tolerant.

5.4.6 Other Design Considerations

A design which has deficiencies invites change during construction. The changes made in the field may not be well designed due to perceived time pressures, nor do they typically receive the same level of engineering or hazard review. Moreover, change may not be recognized as significant. In the Kansas City Hyatt Regency Skywalks incident, the initial design of a hanging support was "not buildable" and the contractor improvised an alternate design. The second design was not strong enough to withstand the forces involved when a number of guests started dancing on it. The Skywalk fell, resulting in injuries and death to many of the guests (Petroski, 1985). Actually, there are two lessons to be learned from this incident:

- Be sure the design is buildable.
- Thoroughly review any field engineering changes made during construction.

Layout considerations include avoiding crane lifts over operating equipment, especially equipment that contains hazardous materials and, in particular, hazardous materials held under elevated pressure or temperature. Allow space for maintenance access that avoids damaging other equipment.

Additional references on proper design features for chemical/processing facilities include:

- *Guidelines for Safe Storage and Handling of High Toxic Hazard Materials* (CCPS, 1988a)
- *Guidelines for Vapor Release Mitigation* (CCPS, 1988b)

- *Guidelines for Evaluating Process Plant Buildings for External Explosions and Fires* (CCPS, 1996b)
- Chapter 2 of the CCPS publication *Guidelines for Engineering Design for Process Safety* (CCPS, 1993a) provides an Inherent Safety Checklist. This checklist has been adapted and is provided in Appendix A.

These references do not address inherently safer strategies explicitly. However, they do provide relevant guidance that fits within the definitions of the four strategies.

5.5 OPERATIONS, MAINTENANCE, AND MODIFICATIONS

The longest stage in the life cycle of a process is the operations, maintenance, and modifications stage. This phase will likely last for decades and will span many changes in personnel, operating and maintenance philosophy, and perhaps ownership. There are two issues that are important with respect to inherent safety that should be addressed during this phase:

- Preserving the inherent safety features and practices provided during the process development phase of process life.
- Seeking opportunities for continued improvement in inherent safety.

5.5.1 Preservation of Inherent Safety

A primary objective of any safety program is to maintain or reduce the level of risk in the process. The design basis of the process, especially those inherently safer features that are built into the design and installation of the process, must be clearly documented. This is necessary so that any anticipated change may be completely evaluated to determine whether it strengthens or weakens the inherent safety features or practices, or has a neutral effect. Without adequate documentation, it may not be possible to make this evaluation for features or practices that were implemented years before. Management of Change (MOC) programs must preserve and protect against elimination of inherently safer features. For example, debottlenecking projects are intended to increase throughput and process efficiency. However, to increase unit or plant throughput requires that equipment size be increased at various points in the process(es). Increasing the size of a valve, or installing a larger pump could result in high pressure in a vessel, thus increasing the risk of a release. Sanders (1993) presents a number of examples of changes affecting the safety of a plant. MOC procedure and forms should be modified to include appropriate reviews and verifications that confirm:

- that inherently safer strategies previously incorporated have not been compromised, and,
- that the design, installation, operation, and maintenance of the modified equipment or practice incorporates the four inherently safer strategies, where appropriate.

Sometimes an inherently safer solution will be identified and implemented on one project with the realization that it applies throughout the unit or plant. Depending on the importance of the issue, a special project can be established to implement the remainder of the solution, or it can be implemented as the opportunity presents itself over time while other modifications are made. For example, if a portion of a process has been experiencing severe corrosion due to a specific combination of materials and operating conditions, and a new and different alloy solves this problem, the replacement of the piping and other components subject to the corrosion mechanism may be performed in total or in stages. This decision depends on the remaining life of the equipment and the possible severity of the consequences associated with the release of the material. MOC programs should recognize that inherently safer solutions such as these must be tracked and managed when the entire implementation is deferred over time.

Plant modifications also provide an opportunity to incorporate inherently safer strategies that were overlooked or deferred during the initial design. For example, if a pressure vessel must be replaced because it is at or near retirement thickness, this is an opportunity to specify a replacement vessel that has a higher pressure rating that cannot be reached or exceeded by the worst-case credible transient event. Pressure vessel replacement is also an opportunity to upgrade the materials of construction, if this is desirable and possible. The MOC program should trigger review of such considerations during the change review and approval process.

Although it is usually difficult to modify the process chemistry and operating conditions during the operations and maintenance stage of process life, there may be opportunities to make partial changes. It may not be possible to substitute chemicals, but it may be possible to use them in more moderate conditions. Even a small adjustment in pH may have a significant effect on the rate of corrosion. Small adjustments in flow may reduce erosion problems in certain components.

A final check that inherent safety issues have been properly addressed on projects can also be added to pre-startup safety review (PSSR) procedures and checklists.

5.5.2 Inherent Safety - Continuous Improvement

At a minimum, a PSM or risk management program (RMP)-covered facility will perform process hazard analyses (PHA) or similar studies at least once every five years. Many companies and facilities also perform PHAs as part of projects. During these PHAs, evaluate each safety device or procedure to see if it can be eliminated or modified by applying inherently safer principles of simplification. These include the use of:

- Valve designs that offer a visual indication of actual position.
- Mechanical connections (or disconnections) for blanking, draining, cleaning and purging connections so that maintenance activities cannot be started without first disconnecting lines that might add hazardous materials to the equipment. One example is to route a nitrogen line across or through a manway so the vessel cannot be entered without disconnecting the nitrogen line.
- Accessible valves and piping to minimize errors.
- Adequate spacing to avoid crowded vessel access.
- For operability and maintainability, minimize the complexity of any protection layers, where possible (CCPS, 2007).
- Logical numbering of a group of equipment. Figure 5.6 is an example of poor assignment of equipment numbers for pumps.
- Logical control panel arrangements. Figure 5.7 is an illogical arrangement of burner controls on a kitchen stove.
- Organized and complete technical information for the process equipment that highlights potential trouble spots for maintenance materials, including materials of construction, lubricants, or packing.

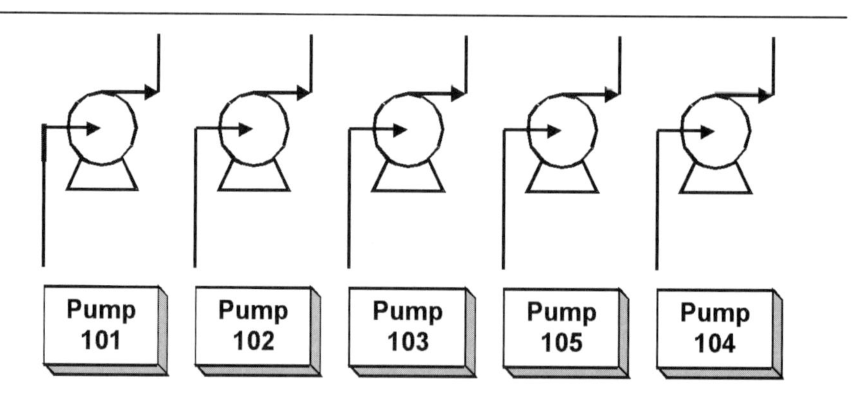

Figure 5.6: An example of poor assignment of equipment identification numbers

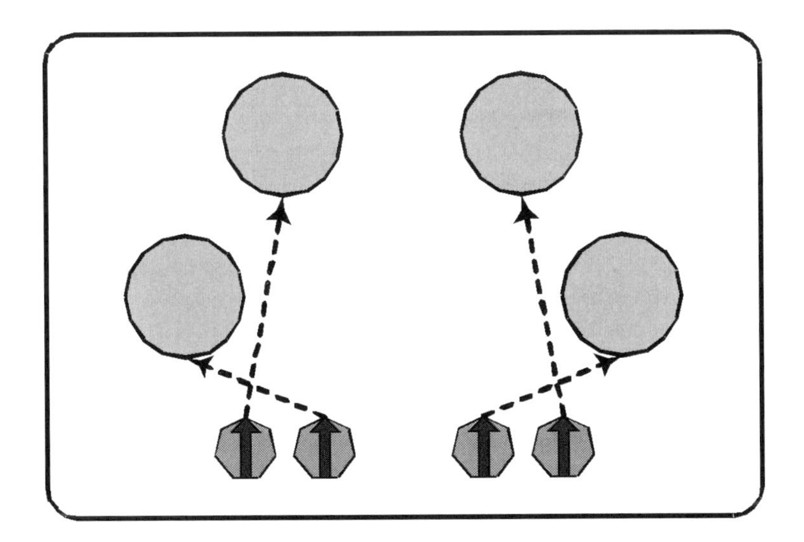

Figure 5.7: An illogical arrangement of burner controls for a kitchen stove
From Norman (1992). Note: The author who drew this figure had a difficult time getting it "right." His instinct was to connect the controls and burners in a logical fashion.

During PHAs and at other opportunities when operating procedures undergo review, they should be reviewed for the following inherently safer provisions:

- Using checklists in operator rounds.
- Using a second person to double-check certain log entries.
- Scheduling to minimize cross-contamination and clean-ups.
- Recording a value for a process variable rather than just a check mark.
- Reformatting of the procedures to make them more user-friendly. For example, placing an easily recognized tab at the page containing the emergency procedures. See Chapter 6 for further discussion of human factors.

In addition to operating procedures and practices, mechanical integrity program practices and procedures also offer an opportunity for the application of the inherently safer strategy of simplification:

- Maintenance procedures should be written using the same philosophy described above for operating procedures. Data sheets and other forms/media for recording inspection, testing, and preventive maintenance results should be complete and user-friendly.
- Spare parts and materials must be correctly specified and installed during maintenance. Sometimes people use what they have, not what

was specified. At a minimum, repair and spare parts management procedures should be modified so that work orders for maintenance always include the part number of the replacement materials, and this same number is used to label the part and/or its storage location in the warehouse.

- Spare parts management procedures should be modified to track the expiration dates of any material in stock that has a shelf life. This can be accomplished using commonly used computerized inventory systems, and also by simply placing distinctive labels on the warehouse storage locations.
- Plant equipment should be designed (or redesigned) for ease and reduced risk in performing inspections, tests, and preventive maintenance tasks. For example, if the pipe thickness can be measured from a platform, it is more likely to be done (and done on time) than if a crane is required.

5.6 DECOMMISSIONING

The design and implementation of inherently safer chemical processes includes not only consideration during the stages of the process life cycle when the unit/plant is being designed and is operating, but also when it is no longer a productive unit/plant and is decommissioned. The application of inherently safer strategies during the decommissioning phase of the life cycle is just as important as its design or operations phases. This is because the equipment is not under the daily attention and care of personnel who are responsible for its safe continued operation. Decommissioned equipment may be "abandoned," with respect to operating or maintenance personnel, for years before some action is taken to recommission or dismantle it. There may be no one who can remember the state in which the equipment was left. In particular, the inherently safer strategies of moderation and substitution are important while in a decommissioned state. The key actions necessary to ensure that a decommissioned unit/plant or individual piece of equipment remains in a safe and stable condition during the decommissioning process are:

- The decommissioned equipment must be completely and reliably isolated mechanically from other portions of the plant that are still productive so it does not represent a leakage path of hazardous materials to the environment. Maintaining the decommissioned equipment in a vented (or purged) state that is completely isolated from any other equipment is an application of the moderation strategy (i.e., keeping the conditions in the decommissioned equipment at low or even ambient temperatures, pressures, and zero flows).

- The decommissioned equipment must be completely isolated electrically from the productive portions of the plant. This includes power supplies, as well as control signals which will prevent any equipment in the decommissioned portion of the plant from operating or changing state.
- The decommissioned equipment must be completely free of inventories of hazardous materials, including residual amounts. In this context the inherently safer strategy of substitution is used to replace the inventory of hazardous materials with ambient air, or an inert purge liquid or gas (i.e., nitrogen) that will preserve a safe atmosphere inside the decommissioned equipment. However, the equipment necessary to maintain a purged condition would probably be considered an active safeguard.
- The exact state of the decommissioned equipment must be clearly documented, using the management of change process or an equivalent, so that at some point in the future, possibly years downstream, any actions taken to recommission, modify, or dismantle the equipment can be done safely. This is a form of the inherently safer strategy of simplify.

Example 5.1

A 50-gallon stirred pot reactor was used for the production of sodium aluminum hydride, which reacts exothermally with water with enough heat to cause the hydrogen that is released to explode. The reactor was emptied, cleaned thoroughly (by report), and then placed in an outdoor "surplus equipment yard" with the nozzles open to "weather." About one year later, a maintenance man was ordered to clean up the reactor in preparation for reuse. He was told to put on full protective fire gear before opening the vessel. He did not put on the fire gear and proceeded to open the vessel and hose it out with a fire hose. An explosion resulted when water dislodged crusted-over sodium aluminum hydride trapped in a nozzle. The worker was burned, requiring a two-week hospital stay and several months of recuperation.

Attention must be given to the long-term protection of people or the environment from the hazards of abandoned equipment. Equipment that meets the criteria for disposal in a landfill, i.e., it has been properly cleaned, may not be suitable for other uses. Problems such as the one related in the following example can be avoided by making the abandoned equipment unusable if it cannot be cleaned adequately.

Example 5.2

Reactors equipped with heavy agitators used for tetraethyl lead manufacture during World War II were disinterred from bomb rubble and were found by the people who dug them up to be ideal for processing fish paste for human consumption. The reactors were washed, but this did not prevent poisoning a number of people.

Example 5.3

Heavy-walled drums, once used for lead anti-knock chemicals, have been used for water storage or as barbecue pits, with subsequent risk to the user from residual toxic material. Equipment from the industry cited has for many years been cleaned, cut up, and sent under supervision to steel mills for recycle to eliminate the possible misuse of scrapped containers.

There is a temptation for managers to delay the cleanup of decommissioned or abandoned plants as long as possible, usually due to the costs of dismantlement. However, experience teaches that there will never be a time for chemical plant closure activities that will be less expensive or less hazardous than immediately after the plant is closed, because it is the time when:

- the most people who know how to handle the materials and where the materials are located are available.
- units are still intact, decontamination can be most easily performed, the procedures are well-known, and the equipment is available.
- waste disposal contracts that cover the materials in the plant are still open or can be re-opened easily.
- design documents, waste manifests, maintenance records, and other files are most likely to be readily available.
- equipment has probably not corroded to the point that it can't be handled safely—valves will open, nuts are not frozen, and instruments are in working condition.
- you are least likely to encounter tanks, drums, etc., with contents nobody can identify without an expensive analytical investigation.

Generally, if there are contaminants in the groundwater, the longer remediation is delayed, the larger will be the area to be cleaned up.

5.7 TRANSPORTATION

Addressing transportation risk at various life cycle stages can increase the inherent safety of the overall operation. This may mean minimizing the number of railcars sitting on track within or outside the plant, the selection of suppliers close to the plant, or other options. This chapter provides an overview of inherently safer concepts and their application to potential transportation risks.

In choosing the location of a new plant, or in assessing risks related to an existing plant, transportation risk must be considered. The design of new chemical processing units should include at the earliest opportunity a qualitative or quantitative risk assessment of the whole system, including production, use, and transportation of materials in order to minimize overall risk. Risk assessments of existing processes, including an assessment of transportation risk, may result in the conclusion that the process should be moved. Process risk assessment techniques are available in numerous references (CCPS, 2000) and the analysis of transportation risk is documented in *Guidelines for Chemical Transportation Risk Analysis* (CCPS, 1995b, 2008).

The assessment of transportation risk must include consideration of the capabilities, equipment, and practices of both raw materials suppliers and customers. Some apparently attractive options may need to be discarded because of the decisions of third parties. In some cases, suppliers' or customers' capabilities may need to be upgraded to acceptable standards.

The chemical/processing industry and carriers have historically worked together to improve both transportation safety and response to emergencies. The American Association of Railroads (AAR), among others, has worked with the American Chemistry Council (ACC) to recommend improvements in equipment, routing, and procedures to enhance safety. The Chlorine Institute has for years acted to provide inherently safer transport of chlorine. The ACC Responsible Care®, CHEMTREC®, and TRANSCAER® programs have resulted in significant improvement in transportation safety and emergency response.

There are many regulations governing the transportation of chemicals, and any evaluation of transportation risks and options must include consideration of those regulations. In addition, some companies have policies that require going beyond legal requirements for specific materials. With the addition of security concerns associated with the shipment of chemicals in the transportation sector in the past few years since the events of September 11, 2001, it is important that the implementation of inherently safer strategies take into account, at a minimum, the on-site transportation infrastructure.

5.7.1 Location Relative to Raw Materials

It may be possible to reduce or eliminate transportation risk by locating the plant where hazardous raw materials or intermediates are produced, if the risk from transporting the raw materials or intermediates outweighs the risk of transporting the final product(s). Locating the starting and ending points in the value chain of a given chemical product at the same site will probably provide additional opportunities for risk reduction by inventory reduction. Of course, increasing the number of facilities at a particular site may increase the overall risk at that site.

Example 5.4

A plant produced methyl methacrylate by reacting hydrogen cyanide with acetone to produce acetone cyanohydrin, followed by further processing to produce methyl methacrylate. The hydrogen cyanide was produced at another site and was transported to the methyl methacrylate plant by railcar. A hydrogen cyanide plant was subsequently installed at the methyl methacrylate plant site to eliminate the need for shipping hydrogen cyanide or acetone cyanohydrin.

Example 5.5

A company produced bromine in Arkansas and brominated compounds in New Jersey. A risk assessment resulted in a recommendation to consider the transfer of the bromination processes to the bromine production site in Arkansas. Economic considerations and the decrease in risk justified such a transfer, and it was implemented. Although safety was not the only consideration, it was an important factor in this decision.

5.7.2 Shipping Conditions

The physical condition and characteristics of the material shipped should be considered in transportation risk assessments on a case-by-case basis. There may be options available to reduce transportation risk by reducing the potential for releases or the severity of the effects of releases. A few possible ways of improving safety by modifying conditions are:

- Refrigerate and ship the material at atmospheric pressure or at reduced pressure.
- Ship materials in a concentrated state to reduce the number of containers, then dilute the concentrate at the user site.
- Ship and use the material or a substitute in diluted form, i.e., aqueous ammonia instead of anhydrous ammonia, or bleach instead of chlorine.
- Ship and use intermediates rather than raw materials.

5.7.3 Transportation Mode and Route Selection

Select a transportation mode to minimize risks to the extent practicable. Drums, ISO containers, tank trucks, rail tank cars, barges, and pipelines offer tradeoffs in inventory, container integrity, size of potential incidents, distance from supplier or customer, and the frequency of incidents.

Barges may have fewer accidents than tank trucks, but the consequences of a major release from a barge may be severe enough to make the tank truck shipments a more attractive choice.

The transportation mode used will affect the shipper's options with regard to the selection of the shipment routing. Using certified drivers to ship full truck loads of drums, ISO containers, and tank trailers, may allow the shipper to specify routes to avoid high risks. The time of day and duration of travel can also be specified.

Railroads choose the routing of rail tank cars and should be contacted to see if arrangements can be made to minimize risks. Improved tracking of rail shipments by the railroads should reduce hazards such as long-term storage of tank cars containing toxic or flammable materials on spurs adjacent to residential areas. The routing of barge shipments is essentially fixed by the location of the shipper and the receiver, and there is generally no choice of routing with a pipeline. Data on accident rates by mode and references are given by CCPS (1995b), and can be used to select the safest alternative.

Carriers and the chemical industry are working together to improve transportation safety. The American Association of Railroads has agreed to designate routes that handle 10,000 loads per year or more of chemicals as "Key Routes." Routes designated as "Key Routes" will receive upgraded track, enhanced equipment to detect flaws in equipment or trackage, and lower speed limits.

"Just-in-time" supply of materials may affect the mode of transportation and could actually increase associated risks. For example, if drums of a chemical are stockpiled near a user, the material is not under the same level of control that could be provided by either the supplier or user if the inventory maintained in a storage tank at one or the other facility. This type of risk should be included when contemplating "just-in-time" shipments.

5.7.4 Improved Transportation Containers

Transportation risk can be reduced by applying inherently safer design principles to transportation containers. Some examples of design improvements are as follows:

- The shipment of environmentally-sensitive materials in general purpose rail cars has been phased out, and DOT specification 105 pressure cars are used instead.
- Remote controlled shutoff valves can reduce the severity of incidents.
- Thermal insulation can be used to maintain lower temperatures in the containers and to provide improved protection from fire.

- Tank cars, trailers, and other containers can be specified without bottom outlets, or be provided with skid-protection for bottom outlets.
- All rail tank cars must be equipped with roller-bearing wheels.
- Using a container designed for the maximum pressure that the contents can generate given their vapor pressure at the maximum anticipated ambient conditions en route will eliminate the need to refrigerate the container for safety.
- Rail cars for chemicals have been designed with "shelf" couplers or double shelf couplers and reinforced tank ends to reduce releases from collision accidents.
- Some barges have double hulls.
- "Low-boy" trailers are used for truck movement of ISO containers. Low center of gravity trailers are also available for tanks. The lower center of gravity reduces the risk of a turnover.
- Baffles can be used in large containers to improve stability.
- Non-brittle containers can be used to improve resistance to impact or shock damage.

5.7.5 Administrative Controls

In addition to improving safety during transportation by optimizing the mode, route, physical conditions, and container design, the way the shipment is handled should be examined to see if safety can be improved. For example, one company performed testing to determine the speed required for the tines of the forklift trucks used at its terminal to penetrate its shipping containers. They installed governors on the forklift trucks to limit this speed below what was required for penetration. They also specified blunt tine ends be installed on their forklifts.

Another way of making transportation inherently safer, although by using procedural means, is a program to train drivers and other handlers in the safe handling of the products, to refresh that training regularly, and to use only certified safe drivers.

5.7.6 Management of Transportation Containers On-site

Transportation containers are delivered to the site to offload raw or intermediate materials, or arrive at the site to load intermediate materials or final products. These containers are also used as temporary storage vessels when they are connected directly to feed the processes. Chlorine is typically offloaded in this manner, as well, as certain industrial gases (i.e., boron trifluoride, hydrogen) which are offloaded from tube trailers connected directly to the process. The presence on-site of fully- or partially-loaded rail cars, tank truck trailers (bulk or tube trailers), or barges represents an inventory management issue. Often times, the number of these containers present on-site, either connected or staged/stored awaiting connection, represent the highest inventory of hazardous materials on-site. This is particularly true of rail cars, where the number present on-site is often a function of both the required production feed rates and the railroad's dispatching practices. This results, in many cases, in multiple fully loaded 90-ton chlorine rail cars being stored on-site.

The risk of on-site transportation containers loaded with hazardous materials can be reduced as follows:

- Reduce the number of loaded transportation containers to the absolute minimum, consistent with required operational needs. This will require negotiating with the railroad for modified switching schedules, which may be difficult to achieve.
- Do not store loaded railcars with raw materials, intermediate materials, or final products that are hazardous materials on sidings that are not within the site security fence, and controlled by site security personnel.
- Do not allow railcars on-site that do not have current and valid stencils or other records that indicate that the required pressure and relief device tests have been performed.
- Use gravity, or pneumatic pressure in conjunction with full-flow vents or vapor recovery systems to offload transportation containers and avoid overpressure or damaging vacuum conditions.

6

Human Factors

6.1 OVERVIEW

From a human factors perspective, the chemistry of a process can be made inherently safer by selecting materials that can better tolerate human error in handling, mixing, and charging. If a concentrated reagent is used in a titration, precision in reading the burette is important. However, if a dilute reagent is used, less precision is needed.

Likewise, the equipment can be made inherently safer relative to human factors by:

- making it easier to understand
- making it easier to do what is intended
- limiting what can be done to the desired actions

The "process" includes more than the equipment and the chemistry. It includes the systems of training, supervising, and providing tools to the people who operate and maintain the plant, including the design of operating and maintenance procedures and other management systems. If one designs certain features into the system of training and supervising people, they will operate and maintain the plant in a safer way. For example, avoiding fatigue through the use of optimum shift rotation schedules is inherent in the operating system. Such features, which help the operators and maintainers run the plant in a safer way, are inherently safer if they are incorporated into a *system*. One cannot depend on the operators and mechanics learning such things by chance.

As new equipment becomes available and new insights about ergonomics and human factors become available, new and existing facilities should be reviewed to optimize the person/process interface in order to increase the application of inherently safer human factors design.

One may argue that many techniques listed in this chapter are simply "safer," rather than "inherently safer." It could be that, while considering human factors, a design or operation could be made inherently safer to the 1st or 2nd order, or the actions may only increase layers of protection. A technique becomes inherently safer only if it is systematically designed into the process, equipment, and people systems in a unit, plant, or company. If only one or two people practice it on their own, it may be "safer," but it is not an inherently safer "system," since it has not been institutionalized into "the way things are done" at the company or facility in question.

New facilities should be reviewed for ergonomics and human factors issues during design, construction, and startup. Existing facilities should be reviewed periodically for opportunities to improve human factors in an inherently safer way, including through process hazards analyses. Such reviews are usually performed both periodically for an entire process, as well as for significant modifications. The use of human factors checklists, such as the one provided in the CCPS book, *Human Factors Methods for Improving Performance In the Process Industries, 2006,* can help improve the application of inherently safer design in existing and modified processes.

The CCPS Human Factors book describes the subject as the discipline of addressing interactions in the work environment between people, a facility, and its management systems. Reference is made to an International Association of Oil and Gas Producers (2005) model for Human Factors that is applicable to the process industries (Table 6.1). This model is based on three major areas, each with a number of sub-topics. Culture is a factor overriding all of these issues, as it defines the norms in which a system operates, both socially and technically.

Table 6.1. Culture and Working Environment *(OGP, 2005)*

Facilities & Equipment	***People***	***Management Systems***
Work space design	Human characteristics & behavior (physical and mental)	Management commitment
Maintenance	Fitness	Safety culture
Physical Characteristics	Stress	Procedures
Reliability	Fatigue	Training
		Hazard identification
		Risk assessment

Simply put, human factors involves working to make the environment function in a way that seems as natural as possible to people. The goal of human factors is to fit the task and environment to the person, rather than forcing the person to significantly adapt in order to perform the work. This reduces the potential for human error that can cause or contribute to process safety and other types of incidents.

Human factors has its origins in the Industrial Revolution and emerged as a full-fledged discipline during World War II when it was recognized that aircraft cockpit designs needed to consider the human interface for controls and displays to ensure safety and reliability of operations. Likewise, human factors have an essential role in the application of inherently safer design. A system or procedure that is designed with human factors as a core focus will be less prone to human error, resulting in reduced risk of safety, process safety, or environment-related incidents.

The subject of human factors in the process industries is treated in depth in CCPS (2006), which includes approaches for implementation of such strategies in the designs of plants and their management systems. A Human Factors Tool Kit is also provided.

Several types of human errors have been previously identified (Reason 1990):

- **Slips** are defined as errors in which the intention is correct, but failure occurs when carrying out the activity required (CCPS, 1994).
- **Lapses** are defined as an error in operator recall (Bea, Holdsworth, and Smith, 1996).
- **Mistakes** are defined as an error in establishing a course of action.
- **Violations** are defined as errors when an intended action is made that deliberately ignores known operations rules, restrictions, or procedures (excluding sabotage), and can be routine (short cuts), optimizing (violations for the thrill of it), or necessary (non-compliance is necessary to complete the job).

Human factors considerations are designed to minimize the potential for these types of errors, or to improve the potential for identification and corrective action in order to minimize the consequences of the error.

The guiding premises for making systems inherently safer against human error are:

- **Humans, and the systems designed and built by them, are susceptible to error.** Human factors design reviews of new and existing facilities and modifications, such as through PHAs or separate human factors evaluations, as well as reviews of human factors-related root causes or contributing factors in incident investigations (particularly near-misses), can help identify means to reduce the potential for human error.

- **Existing facilities can contain many traps to cause human error**. It is important to identify these potential traps based on operator input, as they alone may be aware of them. Input from both experienced and newer operators should be sought because newer operators may be more aware of the traps that more experienced operators have become used to, and found ways to routinely avoid. Elimination of such traps is inherently safer than training and expecting people to avoid them. Input from operators and maintenance personnel can also be valuable in identifying other human factors-related issues. Training for personnel in human factors often helps them identify issues that they may have previously recognized, but were unable to understand and express in terms of human factors and the potential for error that could lead to adverse safety consequences.

- **Designers can provide systems to facilitate error or deviation detection and enable recovery before the error or deviation becomes serious**. This is the essence of human factors-based design that will be addressed further in this section. Like other aspects of inherently safer design, addressing human factors issues in design of new and modified processes is generally easier and less costly than implementation after startup. More detailed descriptions of human factors designs, including a detailed human factors checklist can be found in several CCPS publications (1994a, 2006).

CCPS (1994a) and Lorenzo (1990) discuss human error in detail. They:

- offer guidance on the underlying theories of human error, and the role of performance-influencing factors,
- summarize techniques for human error analysis and quantification,
- suggest methods for collecting data on human error,
- provides sample case studies, and
- discuss a systematic approach to human error.

The tools in CCPS (1994a) can be used in each stage of the chemical process life cycle to help evaluate the tradeoffs involving human factors between various options. In many cases, low cost options in design can make the operations inherently safer from a human factors perspective. Additionally, the tools in CCPS (1994a) can be used to build inherently safer human systems, including:

- appropriate training,
- reviews,
- audits, and
- error correction cycles.

Human factors should also be applied to the employee hiring process to help ensure that new and transferred employees have the capabilities, both physical and mental, to perform the required job tasks (CCPS 2006).

Well-designed human systems can produce inherently safer plant designs and operating procedures. Plants and processes that are designed and constructed with careful attention to human factors are inherently safer than those that are not. If we understand how humans work and how human errors occur, we can design better systems for managing, supervising, designing, reviewing, training, auditing, and monitoring. Human factors consideration is an integral part of an inherent safety effort in a company.

The discussion in the following chapters is not comprehensive, but is only intended to provide examples. The inherently safer design strategy used in most of the examples is *Simplify* (see Chapter 4.5). Because human factors relates to the roles that people play in the overall process safety management system, it relates to the **layers of protection** discussed in Chapter 2 (see Figure 2.3). Attention to the principles of human factors does not eliminate or reduce the hazard, but reduces the overall risk of an incident by improving the level of robustness and reliability of human-based safeguards. However, even in the best of situations, these safeguards are not considered as reliable as properly designed, installed, and maintained passive or active safeguards. It is important to re-emphasize the point made earlier that improving the inherent safety and reliability of any of the process safety strategies improves the overall inherent safety level of a process.

6.2 OPERABILITY AND PERSONNEL SAFETY

There may be well-run facilities in which the operators are doing their best to "be careful" with facilities or systems that could be redesigned to be inherently safer. These facilities will be inherently safer if designed for operability. Note that inherently safer human factors features can reduce both the risk of injury to employees (improved personnel safety) as well as the risk to the process from the worker (improved process safety).

Ergonomics, which is one aspect of human factors, should be applied in the layout of equipment, valves, controls, and anything else that operating and maintenance personnel need to access. Designs that provide good access to equipment, while ideally avoiding bending, climbing, and stretching, are inherently safer from an ergonomic standpoint than designs that require those movements. It is also more likely that safety equipment will be properly operated in an emergency, particularly if additional highlighting is provided such as through signage, color-coding, labeling, and/or lighting.

Consider a Task Analysis to determine if the intended operation will improve or hamper human performance. Can the operators actually do what the process design and operating procedures have asked them to do? Tasks and facilities should be designed with knowledge of ergonomic considerations and performance shaping factors (PSF) so that operator reliability and improved performance can be designed into the task. Performing a Job Safety Analysis (JSA) or Job Hazard Analysis (JHA) can help identify any barriers to safe, effective and efficient job task operations.

Designs and systems should minimize potential harmful exposures in both normal and emergency operations. This consideration affects the location of normal and emergency drains and vents.

> **Example 6.1:** One six-floor process unit was designed with the expectation that the operators would catch drainings from sample lines and filters in a bucket and carry the bucket down the stairs to pour it into a tank. Since the streams contained an irritating organic acid, the operators found it easier—and safer, for the one highest in the structure, but not for anyone below—to flush the drained material through the floor grating with a water hose. Raincoats were mandatory for entering the facility.
>
> To reduce employee exposure and environmental incidents, the plant deemed it necessary to add drain pans and piping under the routine samples and filter drains. The final system was safer than the original, but not necessarily inherently safer. If technologically possible, eliminating the need to drain or reducing or substituting the chemical with one with fewer hazards would be inherently safer than the drain pans and piping.

Designers must include consideration of all of the necessary tasks in their designs. Some designs for steady state operation have been reported to lack start-up or shutdown capability. Other plant designs lacked the vents and drains required to empty, flush, and clear the equipment for maintenance. Facilities designed with at least an outline of the operating procedures are inherently safer than those designed without such knowledge.

Designs should be based on knowledge of what the human is willing and able to do. Experienced operators trained in human factors should be included in design reviews. The HAZOP methodology for process hazard analysis offers an excellent opportunity to identify design and procedural opportunities for inherently safer systems (See Chapter 8). This is the intention of the operability considerations, or "OP" in HAZOP—to illustrate the means of achieving this with one common PHA methodology (CCPS, 1992).

Procedures that are easy to understand and follow are inherently safer than those that are not. For example, a safe startup procedure that requires the operator to walk up and down the stairs three times to manipulate valves in the correct sequence—and where a hazard could occur if taken in the incorrect sequence—can be made safer by:

- locating the valves so that operator has to walk up the stairs only once during the startup, reducing the frequency of errors per operation, but not necessarily reducing the frequency of an error to a satisfactory level;
- interlocking the valves so that they cannot be opened in the incorrect sequence, thereby eliminating the human element for valve changes, but still allowing for human error in the maintenance or operation of the valves; or
- eliminating the valves through a piping or process technology change, a First order inherently safer design that considers human factors.

6.3 MAINTAINABILITY

A space station design that requires less "space walk" time is inherently safer than a design that requires more. If the astronauts do not need to go outside their space vehicle, they bear less risk. For chemical plants, designs or operating regimes that reduce or eliminate the need for vessel or other process equipment opening, entries, or maintenance are, in general, inherently safer than those that do not.

Example 6.2: An elevated horizontal storage tank that allows for wall thickness measurements from the outside is safer than a vertical storage tank that requires cleaning and preparation of the vessel for entry in order to perform these measurements. This could be considered a 2nd order inherently safer design. A First order design, from a human factors perspective, would be one that doesn't require entry or measurement. It is also easier to detect a leak or corrosion problem in the elevated tank, making it less likely that it will fail undetected and result in continuing environmental contamination.

During the design phase, the human interactions with the chemical process should be identified and the means provided to make those interactions inherently safer.

Example 6.3: Rail cars, tank trucks, and some reactors and storage tanks were cleaned manually by personnel who entered the vessel; fatalities occurred from unexpected or undetected low oxygen content or toxicity. An inherently safer system is a rotating pressurized water spray head that does the cleaning without vessel entry.

Eliminating filters that must be changed reduces the potential for exposure. This may require a redesigned filter, or a process change that eliminates the need for a filter at all.

Human factors should be considered in the design and location of equipment requiring maintenance and the development of maintenance procedures and frequencies:

- inspection items
- calibration items (on-line, off-line, or shutdown)
- periodic replacement
- repair without shutdown
- failure modes

Equipment that can be reached for inspection, repair, or monitoring from permanent platforms is more likely to be inspected, calibrated, and replaced than equipment that requires climbing with a safety harness or scaffold.

Calibrating equipment usually requires disconnecting it from the process. Equipment that requires less calibration is inherently safer. For example, a furnace oxygen analyzer is not protecting the furnace while it is being calibrated. Equipment that can function in abnormal operating conditions is inherently safer than equipment that fails in those conditions. Another example is an oxygen analyzer that was designed to shut itself down when the oxygen content went below 4%. While the oxygen analyzer shut down tripped the furnace, it left the operators blind during the shut down and delayed the re-start. An analyzer that continued to show the actual concentration during the upset would be inherently safer.

Equipment should be designed so that there is only one right way to reassemble it.

- If it is important for a pipe sleeve to be right side up, then it could be notched or pinned so it will only go in right side up.
- The use of unique fittings on hoses and connections for flushing or purging (i.e., with nitrogen) is inherently safer than using common fittings that could allow the wrong material (i.e., water or air) to be introduced inadvertently into the process.
- One plant found key relief valves installed backwards after testing because the inlet and outlet flanges were identical. They revised the valve and piping flanges so the relief valves could only be installed in the correct orientation.

Maintenance procedures also require the same attention to human factors, as maintenance errors can be just as costly as operating errors. Designing or purchasing equipment that is easily maintained, and that is difficult or impossible to assemble or install incorrectly, improves the inherent safety of the equipment, and thereby the process. Of course, equipment that is designed to require no maintenance, such as seal-less pumps, is inherently safer than equipment that requires periodic maintenance, i.e., seal replacement. It should be recognized that there is frequently a tradeoff between hazards; the pump that eliminates seal leaks may bring its own set of hazards, such as rapid temperature build-up if it runs dry or dead-headed. Additional protection layers may be needed to avoid the new hazardous consequences.

6.4 ERROR PREVENTION

To prevent errors, it is important to make it easier to do the right thing and more difficult to do the wrong thing (Norman, 1988). If the design and layout of procedures do not clearly indicate what should be done, the resulting confusion can increase the potential for error. Likewise, the design of training programs and materials, including verification of knowledge and skills, can increase or decrease the potential for error.

Systems in which it is easy to make an error should be avoided. For example, to reduce the risk of contaminated product and reworked batches, it is generally better to avoid bringing several chemicals together in a manifold. However, manifolding can be done safely, and may be the best design when all factors are considered, particularly when clear labeling and/or color coding is employed. The alternatives to a manifold should be considered systematically and a decision made on the most inherently safe design.

6.4.1 Knowledge and Understanding

Operators and engineers need a correct mental model of how the process is operating to understand the risk and avoid errors. If the operators do not understand the process conditions or means of operation, they may operate the process incorrectly—even with the best of intention (an error of commission). For example, many people adjust their home air conditioning thermostat to a very low temperature setting in the mistaken belief that it will cool the house quicker. They do not realize that the thermostat simply switches the air conditioning unit on and off at a given temperature, and a lower setting will not make it cool faster, but instead will make it run longer to achieve the desired temperature.

Additionally, Norman (1988) discusses **knowledge in the world** versus **knowledge in the head**. With knowledge in the world, one is guided in a task by what one sees. Knowledge in the head requires memorizing the tasks to do it satisfactorily. Knowledge in the world is the sticker on the phone, "For Emergency, Dial 911." Knowledge in the head is one's memorized home phone number.

For a chemical operator, knowledge in the world is the upper and lower limits for a pressure reading shown on the log sheet, along with the response to take. A log sheet with only a blank to fill in the reading requires knowledge in the head. This contrast is shown in Table 6.2.

Training requirements should be based on the task difficulty, frequency, and criticality. The more difficult or critical the task, the more rigorous the verification that the trainee can perform the task correctly should be, particularly under pressure. More infrequent tasks require a higher level of human factors attention, through use of administrative controls such as readily-available checklists, labeling, signage, and/or color coding, in order to minimize the potential for error, because it is more likely that an individual will forget when and how to perform the task correctly over time.

Table 6.2: Knowledge in the Head vs. Knowledge in the World

KNOWLEDGE IN THE HEAD REQUIRED	
Time	**Inlet Temp, °C**
800	
1000	

KNOWLEDGE IN THE WORLD REQUIRED	
Time	**Inlet Temp, °C**
	Never exceed 100° C, Start Emergency Cooling at 95° C, Trip Reactor at 97° C
800	
1000	

Operating, maintenance, safety and emergency procedures must be accessible and up-to-date to be useful. Careful attention to document control is necessary to ensure that only the most up-to-date procedures are available.

This control can be facilitated by the current widespread use of electronic procedure management systems. A Management of Change program must ensure that procedures are updated whenever process or equipment changes are made. These provisions are also driven by requirements for quality management systems (i.e., ISO 9000). Most importantly, procedures must be clear and written in a level of detail appropriate for the operation, and at a reading level consistent with that of their users. Involvement of experienced operating or maintenance personnel in developing and reviewing procedures helps to ensure that these features are addressed. Chapter 6.4.3 offers more detailed guidance from CCPS (1996) for writing inherently safer operating procedures.

Example 6.4: It was common for operators of a unit to keep personal copies of a two-year-old shutdown procedure for a process. The old procedure was significantly different from the updated, controlled procedure. The old procedure, if followed exactly, presents a greater risk, given that it includes less inherently safe practices that used to be allowed. If the controlled procedure and the outdated procedure were intermingled, there could be high temperature and inadvertent catastrophic corrosion may occur. Following discovery of this cultural practice during an audit, the unit management team stopped the use of the bootleg procedure. Operators were limited to the updated shutdown procedure by requiring the automatic use of the current version found on the PSM document management system used whenever a process operation was conducted. This document was maintained as part of their ISO 9002. (ISO 9000, 1994, 2000). Alternate, less automated systems may use an expiration date or require periodic reconfirmation, which may still allow for some human error.

6.4.2 Design of Equipment and Controls

CULTURE

Cultural stereotypes (also termed populational stereotypes) are established in all countries and must be followed when designing equipment and controls. A cultural stereotype is the way most people in a culture expect things to work based on the customary design of equipment in that city, region, country or part of the world. Avoid violation of cultural stereotypes. Designs that include knowledge of the cultural stereotypes are inherently safer than those that do not.

Example 6.5: Common examples of cultural stereotypes include:

- Light switches:
  - in the USA, a common wall light switch is flipped up 🡅 up to turn on.
  - in the UK, it is common to turn the switch down 🡇 to turn on.
- Common convention is to turn valves counterclockwise to open and clockwise to close, but this is only **usually the case**. (Note that "turn counterclockwise to open" is an inherently safer instruction than "turn to the left to open." For a multi-turn valve with a wheel on an upwards-pointing vertical stem, "turn to the left" assumes the

operator is looking at the side of wheel farthest away; the closer side of the wheel is moving to the right.)

- On quarter turn valves, when the handle is perpendicular to the process flow, that indicates that the valve is closed; when the handle is parallel to the flow, it indicates the valve is open.
- The hot water faucet is on the left, cold water on the right – but not always.
- Pump run lights
- US chemical plant: Green = Run, Red = Off
- US power plant: Red = Run (i.e., has power, danger), Green = Off
- Japanese chemical plant pump lights: Red = Run, Green = Off

Example 6.6: A plant startup required heating the circulating gas to remove oxygen from the equipment. After several spurious trips, the operator noticed that all cooling fluid pumps were off, when they should have been on throughout the start-up. The operator turned them all on at the same time. Steam in the equipment condensed and the resulting vacuum imploded six vessels. The run lights and start/stop buttons were color coded differently in the control room from those outside, which contributes to the cooling fluid pumps being inadvertently turned off.

What people expect varies with the culture. While "culture" may be difficult to define, we can make plants inherently safer by designing them with knowledge of the cultural stereotypes of the employees.

Example 6.7: The authors of the first edition of this book, while meeting at a famous San Francisco hotel, discovered the hard way that the water faucets (round handles) in the committee meeting room restroom did not open as expected. Anticipating movement one way and finding the handles actually worked in reverse resulted in splashed clothing (See Figure 6.1 for the layout). We later learned that water valves with lever handles (frequently seen in hospitals or in sinks intended for disabled persons) are designed for the lever to be pushed away from the user to close the valve. Thus, the cold water valve designed for a lever will turn to the left to close. If a round handle is fitted instead of the lever…splash!

Therefore, we need human systems that not only install the correct valves, but can also install the correct handles!

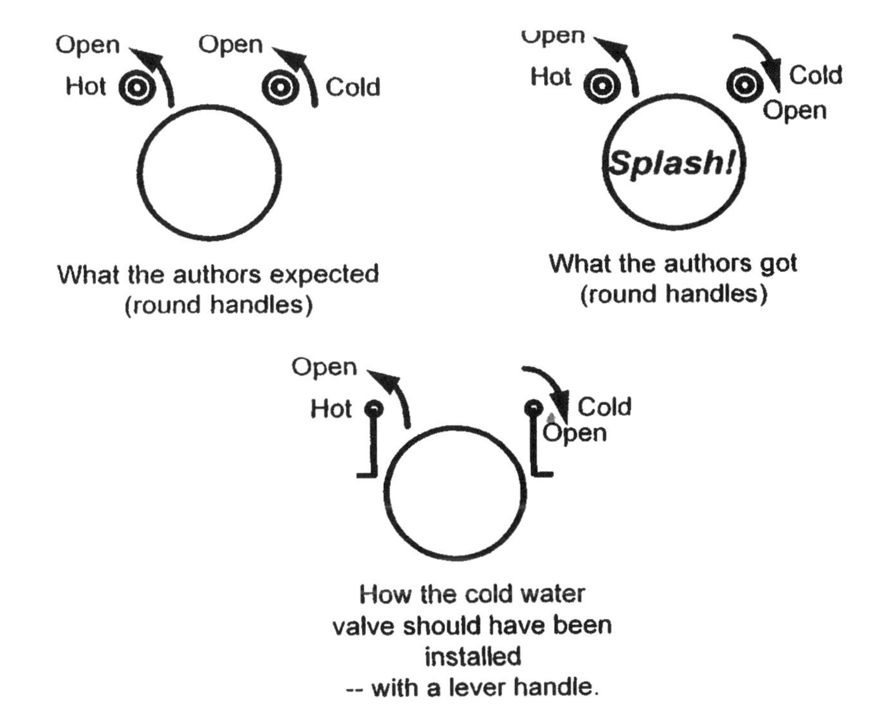

Figure 6.1. Confusing water faucets violate cultural stereotypes

---

Example 6.8: Here is another example that comes courtesy of the authors of the first edition of this book. The authors felt too hot in the aforementioned hotel meeting room. Three of them opened the digital thermostat and studied the complex set of buttons, but they were unable to make the room any cooler. When they asked a hotel employee how to fix the problem, he pointed to the almost invisible switches on the bottom edge of the thermostat and said, "First, you should turn this switch to ON, and second, you should turn this switch to COOL." This incident reminded them of two things about inherent safety:

- Design the key controls so they are easily seen and are clearly labeled.
- Keep related information and controls close together, and leave out information and controls that are not needed (Norman, 1988).

CONSISTENCY

Consistency in controls and response is important. For example, upward movement on a control panel should always causes the valve to open, whether the valve is "air to open" or "air to close." Switches and indicator lights should also be oriented consistently (i.e., the start button should have the same position with respect to the off button).

> **Example 6.9:** A process waste heat boiler was damaged when the control operator inadvertently closed the cooling water valve when he intended to open it. The manual control knob turned in the opposite direction from most of the other valves in the plant — the cooling water valve was air to close, while most control valves were air to open. In an inherently safer design, all of the control knobs would turn the same way to open a valve.

Consistency is also important for inherently safer computer control applications. The widespread use of DCS technology facilitates this consistency within a system. However, it is also important for consistency between systems within a facility where operators may change job positions or operating units. A system learned in one area, particularly over a long period of time, is not easily unlearned in another.

A number of empty or one-page incomplete notes were sent because **F10** was "page down" in one computer e-mail program, and "execute/send" in another application. Other critical information has been accidentally lost because **CTL-X** was "exit" in one program, "cut" in another, and "delete" in yet another.

HUMAN CAPABILITY

Process control systems should be designed with knowledge of the capability of human beings for required tasks. There must be a balance between totally automatic control of the process with operator monitoring versus operator control of key variables. Operators actually need to run the process enough to be able to handle it during abnormal situations or emergencies (CCPS, 1994a). While these considerations may seem obvious, plants that are designed with this knowledge are inherently safer than those that are not.

COLOR BLIND TOLERANCE.

Up to 5% of some male populations are red-green color blind (Freeman, 1996). So, control system and other displays must be designed to transmit information to personnel who may have this trait. If the color red is used for stop (or closed), and green is used for go (or open), an alternate scheme should be used to transmit the same information. One approach for video displays is to design the red symbols with a red outline and black (unfilled) interior and to design the green symbols with a green outline and green (filled) interior (see Figure 6.2). Intensity (brightness) can also be used to convey some status information. There are alternate color schemes that can be seen by most people with red-green color blindness. Some examples are white, black, gray, blue, and yellow. Colored lines on a flowsheet can also be coded with dashes, dots, and crosses (this method also retains the information in a photocopy).

The Abnormal Situation Management (ASM®) Consortium, a research and development consortium of companies and universities concerned about the negative effects of industrial plant incidents, focuses on identifying problems facing plant operations during abnormal conditions, and develops solution concepts. Recommendations developed for the design of distributed control systems include graphics to enable operators to identify, troubleshoot, and manage abnormal situations. ISA-5.5-1985 *Graphic Symbols for Process Displays* presents a uniform scheme for graphic display symbols for process monitoring and control, which provide a more logical and uniformly understandable—and therefore inherently safer—mechanism for modern computer-based process control systems. The recommended scheme is intended to facilitate rapid comprehension by users of the information conveyed through displays, and to establish a uniform practice throughout the process industries.

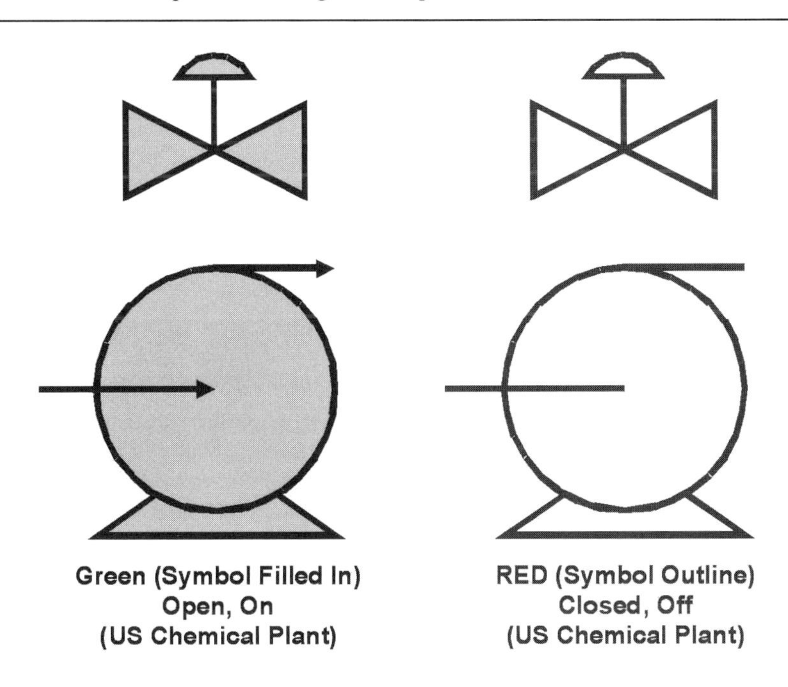

Figure 6.2. Sample Symbols to Overcome Color Blindness

ABNORMAL SITUATION MANAGEMENT

Often, when an abnormal operating situation occurs, alarm "showers" occur due to a deviation causing multiple secondary, and sometimes tertiary, alarms to activate. This situation can make it difficult to identify the initial deviation and its cause, and subsequently troubleshoot the system to take corrective action. Alarm "showers" are prevented by using alarm management, computer logic that can prioritize the alarms in order to highlight the ones that the operator really needs to see in the situation (see Section 4.4). The alarm management system should ensure that valid alarms that the operators should see are not deactivated.

Alarm management is usually necessary in a process manufacturing environment controlled by an operator using a Distributed Control System, or DCS. Such a system may have hundreds of individual alarms that, up until very recently, have probably been designed with only limited consideration of other alarms in the system. Since humans can only do one thing at a time and can pay attention to a limited number of things at a time, an inherently safer control system design ensures that alarms are presented at a rate that can be assimilated by a human operator, particularly when the plant is upset or in an unusual condition. Alarms should also be designed to direct the operator's attention to the most important problem that he or she needs to act upon, using a priority to indicate the degree of importance or rank.

Adhering to this guidance in the design of alarm systems will result in an inherently safer process from an abnormal situation management standpoint.

- An alarm “shower” can occur when high pressure in the reactor trips the feed by closing a valve and turning off the feed pump. In addition to the “high pressure trip” alarm, the operator is frequently “showered” with “low feed flow,” “low-low feed flow” and “feed pump off” alarms.
- With alarm management, the “high pressure trip” would alarm, but the other associated alarms would be suppressed, since they are the expected result of the high-pressure trip. On the other hand, a low flow condition from a different cause would be alarmed. Plants with managed alarms are inherently safer than those without since it is easy to silence and overlook a safety or emergency alarm in the midst of an alarm shower. However, the benefits should be balanced with the increased complexity and maintenance requirements.
- Without alarm management, when the operator turns off the agitator, an alarm will sound for “low agitator speed.” With alarm management, the “low agitator speed” would not alarm when the agitator is switched off by the operator during the proper process condition, but it would alarm for a trip of the agitator motor, a mechanical problem lowering the speed, or if inadvertently turned off by the operator.

After a 1994 fire and explosion at a refinery in the UK that followed an upset where hundreds of alarms went off, the Health and Safety Executive (HSE) issued the following guidance on the use and configuration of alarms:

- Safety alarms must be distinguishable from other operational alarms;
- Alarms should be limited to the number that an operator can effectively monitor;

- Ultimate plant safety should not rely on operator response to a process alarm.

From a broader perspective, the Abnormal Situation Management Consortium is working to apply human factors theory and expert system technology to improve personnel and equipment performance during abnormal conditions. In addition to reduced risk, its goals are economic improvements in equipment reliability and capacity (Rothenberg and Nimmo, 1996). In addition, alarm system performance guidelines have been published in the Engineering Equipment and Materials User Association's (EEMUA's) Publication No. 191 (EEMUA, 1993). EEMUA recommends an average alarm rate during normal operations of less than one alarm per 10 minutes, and peak alarm rates following a major plant upset of not more than 10 alarms in the first 10 minutes. However, a recent study (Reising and Montgomery, 2005) concluded that there is no "silver bullet" for achieving the EEMUA alarm system performance recommendations, and instead suggests a metrics-focused continuous improvement program that addresses key lifecycle management issues.

NUISANCE ALARMS

Another human factors issue associated with control systems is nuisance, or perpetual, alarms caused by poor alarm system design or equipment maintenance. Nuisance alarms often create a "cry wolf" syndrome, where an alarm is ignored and therefore becomes useless as a process control tool for abnormal situation management. The end result is a failure to respond to the alarm in the event of a process emergency. The first step in an alarm management improvement program is generally to measure the alarm rate, and resolve any chronic problems, such as alarms that have no use, often described as one that does not require the operator to take an action.

FEEDBACK

A process control system must be designed to provide enough information to enable the operator to quickly diagnose the cause of the deviation and to respond to it. Feedback can reduce error rates from 2/100 to 2/1000 (Swain and Guttman, 1983).

> **Example 6.10:** For a transfer from Tank A to Tank B, if the operators can see the level decrease in Tank A and increase in Tank B by the same amount, they can be confident the transfer is going to the right place. If the level in Tank A goes down more than it goes up in B, the operator should look for a leak or a line open to the wrong place.

Consider the following in control system design for improving the inherent safety of the system:

- Avoid boredom. If operators don't have anything to do, they go to sleep — mentally, if not physically.

- Display corroborating or verifying information on the same display with, or very near to, the other information. Display the reading from two level sensors for the same tank on the same chart or graphic.
- Put sensibility limits on process control inputs and setpoint changes.
- Limit maximum or minimum setpoint inputs to stay in safe and quality operating regions.
- Limit the maximum step changes to setpoints to prevent upsetting the process.
- Provide bumpless (smooth) transfer and setpoint tracking for switching among automatic, manual, and cascade.
- Catch decimal errors by software or procedure. Have the control system logic trap and prevent setpoint changes, for example, from 6% to 61%, when a change from 6.0 to 6.1% is intended.
- Provide guidance to operators on the magnitude of a specified action to achieve a specified goal. Rather than letting the operator guess at how much to open a valve, suggest opening to 5%, then use minor adjustments to get the desired startup flow. Where needed, give guidance on how to lead or lag in changing setpoints. Advise on how long to blow a line to clear it of liquid.
- "Take out the guess work." Good operators will figure these tips out, document the information, and make it available to all the operators (Collins, 1978).

6.4.3 Standard Operating Procedures

Operating procedures, including emergency procedures, are a key element in the human factors aspect of inherently safer design. Procedures that are not followed due to obsolescence, inaccuracy, unavailability, or difficulty in implementing often present safety risks. Proper design of procedures requires consideration of the following (CCPS 2006):

- *Completeness and accuracy:* Does the procedure have enough information for the user to perform the task safely and correctly?
- *Appropriate level of detail:* Has the level of detail considered the experience and capabilities of the users, their training and their responsibilities?
- *Conciseness:* Conciseness demands eliminating detail and language that does not contribute to work performance, safety, or quality. Conciseness also means including only "need-to-know" and omitting "nice-to-know" information.

- *Consistent presentation:* This element insures that the procedure is readily comprehensible. It demands the use of:
    - A consistent terminology for naming components and operations.
    - A standard, effective format and page layout.
    - A vocabulary and sentence structure suitable for the intended user.
- *Administrative control*: All procedures must be reviewed thoroughly before use and periodically thereafter.

The use of Information Mapping, a technique for organizing and labeling information for easy comprehension, use and recall, can be a useful tool in the development of inherently safer procedures. Information Mapping is a structured approach for the creation of clear, concise and highly usable information focused on the target audience. It does this by analyzing, organizing and presenting the information based on the needs of the user and the purpose of the information, and is both subject matter and media independent. It is a way of thinking and communicating in which information developers approach the content with a set of systematic principles and techniques to ensure that it can be readily used.

CCPS (2006) includes guidelines for when a procedure is required, as well as an example procedure checklist.

6.5 ERROR RECOVERY

Feedback that confirms "I am doing the right thing!" is important for error recovery, as well as for error prevention. It is important to display the actual position of the control device that the operator is manipulating (i.e., remotely operated shutoff valve), as well as the state of the variable he/she is worried about.

> **Example 6.11:** In the Three Mile Island incident, the command signal to close the reactor relief valve was displayed, not the actual position of the valve (Kletz, 1988). Since the valve was actually open, the incident was worse than otherwise.

Systems should be designed with knowledge of the response times for human beings to recognize a problem, diagnose it, and then take the required action. Humans should be assigned to tasks that involve synthesis of diverse information to form a judgment (diagnosis) and then to take action (Freeman, 1996). Given adequate time, humans are very good at these tasks and computers are very poor. Computers are very good at making very rapid decisions and taking actions on events that follow a well-defined set of rules, for example, safety instrumented functions. If the required response time is less than human capability, the correct response should be automated. Unless the situation is clearly shown to the operators, the response has been drilled, and is always expected, anticipate from

10-15 minutes (Swain and Guttmann, 1983) up to one hour (Freeman, 1996) minimum time for diagnosis.

For key operating variables, post, train, and drill the responses for safety operating parameters. If the process variable approaches the Mandatory Action level, the operator should take the Never Deviate action (see Figure 6.3). It is important that operators be trained and drilled, and that supervision supports taking the Never Deviate action to avoid reaching the Never Exceed limit.

The response should be thought out and written down in operating procedures for abnormal conditions before it is required, and personnel should be trained and drilled. In the past, operating instructions would often say, "If high temperature occurs, notify the engineer (foreman)." Written instructions should tell the operator whether to increase the cooling water flow, set off the deluge system, or evacuate! The operator should not have to "wing it" when all the alarms are going off and the relief valves are lifting.

For high consequence events, simulators are useful to practice diagnosis and correction of errors and abnormal conditions in emergency conditions (CCPS, 1994a).

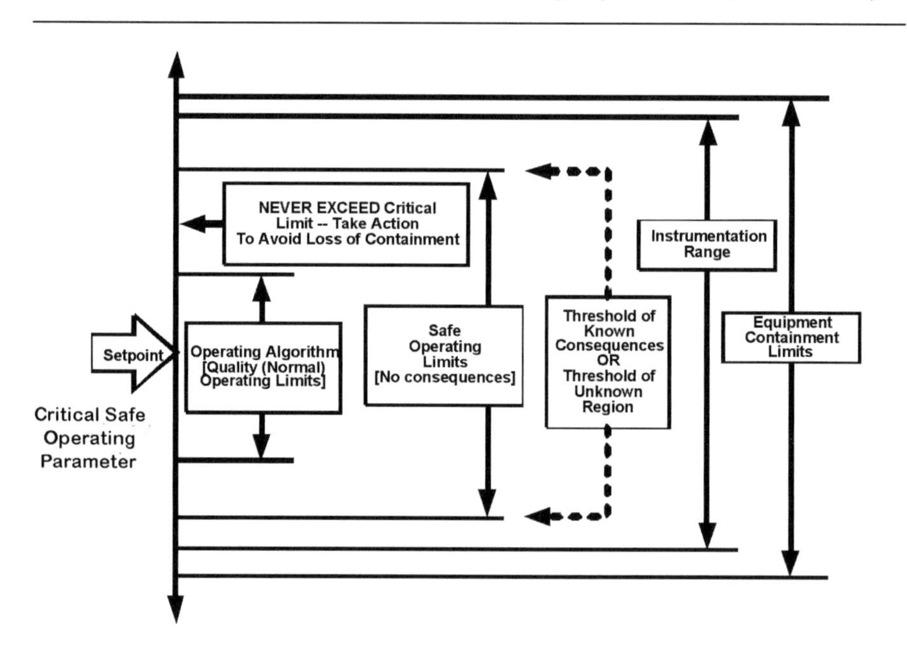

Figure 6.3 Operating Ranges Illustrating "Never Exceed" Limits

The operating philosophy should also address how to effectively use personnel in response to a process upset. Without such a system, the most knowledgeable person(s) in the unit frequently rushes to attend to the perceived cause of the emergency. While this person is thus engaged, other problems are developing in the unit. Personnel may not know whether to evacuate, resources may go unused, and the ultimate outcome may be more serious. The Incident Command System, used by fire fighters and medical personnel for responding to emergencies, should be considered for application to a process incident (CCPS, 1995c). Using this system, the knowledgeable person assumes command of the incident, designates responsibilities to the available personnel, and maintains an overview of all aspects of the incident. Thus, as resources become available, the process corrective actions, emergency notifications, perimeter security, etc., can be attacked on parallel paths under the direction of the incident commander.

Similarly, unit operating staffs can be trained to work together during a process upset using all the skills and resources available. An inherently safer system would have personnel trained to use all of the resources for error recovery. Such training is part of nuclear submarine training ("Submarine!," 1992) and cockpit flight crew training for commercial airlines. This training helps overcome the "right stuff" syndrome. The test pilots in the book *The Right Stuff* (Wolfe, 1979) would rather crash and burn than declare an emergency, since an emergency was an admission that they were not in control, and therefore didn't have the "right stuff."

Error recovery by the operators is only one of several layers of protection to prevent undesired consequences (see Figure 2.1). Process and equipment designs (discussed in previous chapters) that prevent undesired process excursions are inherently safer than designs that require operator intervention. Likewise, designs that enable the operators to intervene before an upset becomes serious are inherently safer than those that do not.

6.6 SYSTEM AUDITS

An inherently safer system should have inspection and reliability testing of safety systems, practices, and equipment (CCPS, 2007). Develop methods to measure the effectiveness of inherent safety efforts and to provide feedback to personnel to improve performance.

> **Example 6.12:** At one site, the project team installing a DCS (Distributed Control System) carefully developed and tested techniques to make the displays clear for red-green color-blind personnel (see discussion in 6.4). The displays were effective and were applauded by the operators. However, in subsequent DCS installation projects at the same site, different project teams made no provision to make displays visible to red-green color-blind personnel. This inherently less safe condition was found during a design review at one unit and at the pre-startup safety review for another.

Audits give a snapshot in time; corrective actions are identified and executed, but the audit must be done again later to see if the system has changed. Using an analogy to process control, a continuous measurement system is desired to give warning of problems in the process management system. CCPS has developed ProSmart, a PSM program evaluation software tool, which uses key, real-time measurable variables of performance (in terms of quality and thoroughness of program implementation) to enable users to monitor and improve process safety management.

6.7 ORGANIZATIONAL CULTURE

The performance of human beings is profoundly influenced by the culture of the organization (see discussion of the "right stuff" above). Culture is generally defined as a set of shared values and beliefs that interact with an organization's structure and management systems to establish norms of behavior, or, "the way we do things around here." Poor safety culture has been identified as a contributing factor in many major accidents, including the Chernobyl nuclear accident in 1986 and the Space Shuttle explosions of Challenger in 1986 and Columbia in 2003.

One area in which unit/plant/company cultures vary is in the degree of decision making permitted by an individual operator. Cultures vary in their approach to the conflict between "shutdown for safety" versus "keep it running at all costs." Personnel in one plant reportedly asked "Is it our plant policy to follow the company safety policy and standards?" In an organization with an inherently safer culture, people would know how to answer that question. A safety culture that promotes and reinforces safety as a fundamental value is inherently safer than one that does not.

An operating philosophy that trains and rewards personnel for shutting down when required by safety considerations is inherently safer than one that rewards personnel for taking intolerable risks. Likewise, a culture that values safety and encourages the raising of safety concerns and suggestions for improvement—and acts on them—is inherently safer than a culture that does not. A. Hopkins provides an excellent discussion of how organizational culture affects safety in his book *Safety, Culture and Risk: The Organizational Causes of Disasters* (2005), including the role of risk reduction (inherently safer) vs. risk management (safer).

There are several models for defining and assessing an organization's safety culture. Table 6.3 provides a set of features that help define a positive (inherently safer) organizational culture.

6.8 SUMMARY

The subject of Human Factors plays an important role in inherently safer design. While not eliminating or reducing hazards, it seeks to improve the reliability of human systems, both in reducing the potential for human error, as well as in providing safeguards for detection and mitigation of abnormal situations. Careful attention to Human Factors in the design of such areas as instrumentation and controls, workplace layout, procedures, and overall safety culture will make the process inherently safer. Due to the role that inattention to human factors has played in major accidents, much effort has gone into understanding this emerging science, designing with these factors in mind, and improving existing operations to minimize risk. Those interested will find a large library of information available, including CCPS (2006), which details the subject in more depth and provides strategies for designing for Human Factors and improving existing operations.

Table 6.3: Factors and Characteristics Associated with a Positive Safety Culture *(CCPS 2006)*

Factor	**Characteristics**
Hardware	• Good plant design, working conditions and housekeeping • Perception of low risk due to confidence in engineered systems
Management Systems	• Confidence in safety rules, procedures and measures • Good job communication • Safety prioritized over profits and production • Good organizational learning • Satisfaction with training
People	• High levels of employee participation in safety • Trust in workforce to manage risk • High levels of management safety concern, involvement and commitment
Behavior	• Acceptance of personal responsibility for safety • Willingness to speak up about safety • Frequent informal safety communication • A cautious approach to risk
Organizational Climate Factors	• Low levels of job stress • High levels of job satisfaction

---

7.

Inherent Safety and Security

7.1. INTRODUCTION

This chapter is concerned with the application of inherent safety (IS) concepts as a component in addressing chemical security risk. With increasing worldwide concern over terrorism and the potential for other intentional attacks on chemical facilities and assets,[1] those responsible for chemical facilities must approach the security of their processes and facilities from a holistic viewpoint. Increasingly, chemical engineers are being called upon to apply their chemical process expertise and techniques to chemical process *security* management because it:

1. minimizes the risk of harm to the public, employees, and environment from intentional acts against a process facility and;
2. protects the assets within a process facility by maintaining ongoing integrity of operations and preserves the value of the investment.

7.2 CHEMICAL SECURITY RISK[2]

As discussed in previous chapters, process safety risk is defined as a function of the *consequence* of an event and the likelihood, or *probability*, of its occurrence.

$$\textbf{Risk (R)}_{\textbf{Safety}} = f[\textbf{Consequence (C), Likelihood (L)}]$$

In both safety and security assessment, the potential consequences of an event can generally be estimated. Likelihood, however, cannot be estimated for security in the same way it can be estimated for safety.

---

[1] In a security context, an "asset" is a target of potential interest for an intentional attack. As such, an asset may be an individual production unit, a specific piece of equipment, a particular area of the facility, or even the facility as a whole.

[2] Definitions of "consequence," "threat," "vulnerability," and "attractiveness" used in this chapter are taken from the Center for Chemical Process Safety (CCPS) *Guidelines for Managing and Analyzing the Security Vulnerabilities of Fixed Chemical Sites,* August, 2002, which offers an in-depth discussion of chemical security.

For accidental scenarios, the likelihood (probability or frequency) of occurrence may be estimated by relying on statistically valid data, such as historic failure data or an assumption that the event is random and can be described by a normal probability distribution.

Such an approach is *not* appropriate for events potentially arising from intentional acts, such as a terrorist attack. An intentional act is not random and therefore cannot normally be estimated by a probability distribution. In addition, intentional acts may be discrete and independent of other acts – one act does not necessarily contribute to or influence another. With an intelligent and adaptive adversary, terrorists may learn from one another's methods, and those methods may evolve in response to site-specific conditions, goals of the terrorist or even target security measures. As a result, a history of intentional acts cannot be combined to create a body of data analogous to equipment failure data or other statistically valid data sets. Therefore, a combination of the following factors must be used as a proxy for likelihood in security risk assessment:

- the ***threat*** posed by an adversary to an asset;
- the ***vulnerability*** of an asset to at attack; and,
- an asset's ***attractiveness*** as a target.

Security risk can therefore be considered a function of consequence (C), threat (T), vulnerability (V) and attractiveness (A) or:

$$\textbf{Risk (R)}_{\textbf{Security}} = f\textbf{[Consequence (C), Threat (T), Vulnerability (V), Attractiveness (A)]}$$

Two approaches exist – to assume the T and A are unity and to simply evaluate the consequences and vulnerability, or to consider the T and A specifically and variably, in order to analyze the likelihood in a more absolute vs. conditional sense.

With regard to security, it is appropriate to define the terms of "consequence," "threat," "vulnerability," and "attractiveness."

Consequence is the severity of loss or damage that can be expected from a successful attack against an asset. Examples of relevant consequences include:

- Injuries to the public or to workers
- Significant environmental damage (such as contamination of drinking water)
- Direct and indirect significant financial losses to the company
- Disruption to the national, regional, or local operations and economy
- Loss of business viability

The US Department of Homeland Security (DHS) has identified five potential consequence categories or security issues associated with the production, use, storage or distribution of chemicals:[3]

1. ***Release***—toxic, flammable, or explosive chemicals or materials that, if released from a facility, could create significant adverse consequences for human life or health.
2. ***Theft or Diversion***—chemicals or materials that, if stolen or diverted, could be used as weapons or easily converted into weapons using simple chemistry, equipment or techniques.
3. ***Sabotage or Contamination***— chemicals or materials that, if mixed with readily available materials, could release poisonous gasses or create other significant adverse consequences for human life or health.
4. ***Critical Relationship to Government Mission***—chemicals, materials, or facilities that, if unavailable, could compromise the government's ability to deliver essential services during an emergency.
5. ***Critical Relationship to National Economy***—chemicals, materials or facilities that, if unavailable, could create significant adverse consequences for the national or regional economy.

The first three issues (i.e., release, theft/diversion and sabotage/contamination) relate to the properties of the chemicals and the potential for adverse human health effects. The last two issues–government mission and the economy–reflect the critical uses of some chemicals that may or may not be hazardous.

Inherent Safety (minimization, moderation, or substitution) can affect consequences by either reducing or eliminating the hazard, or by moderating the hazard. Simplification may reduce the opportunities for an event to escalate to a larger degree of consequence.

Threat can be defined as any indication, circumstance, or event with the potential to cause loss of, or damage to, an asset. It is also the intention and capability of an adversary to undertake actions that would be detrimental to valued assets. Threats are manifestations of an adversary's malevolent intent directed at the chemical asset, or use of the chemical asset as a means to attack a different target, such as stealing a chemical for use in producing an improvised explosive device. The magnitude of a threat is influenced by the intention and capability of an adversary to undertake actions that would be detrimental to valued assets. Sources of threats may be categorized as terrorists (international or domestic); current or former disgruntled employees or contractors; activists, pressure groups, single-issue zealots; or criminals (i.e., white collar, cyber hacker, organized, opportunists).

---

[3] Chemical Facility Anti-Terrorism Standards, Interim Final Rule published April 9, 2007 (72 *Fed. Reg.* 17696).

Adversaries may be categorized as "insiders" (internal threats), "outsiders" (external threats) or a combination of both insiders and outsiders (internal-external collusion). Government law enforcement and security agencies may provide additional information on potential adversaries, with regard to motives and tactics. Companies may also choose to acquire such information through other channels, including commercial intelligence services.

Inherent Safety can affect threat. For some adversaries, the reduction or absence of the hazards may lessen or eliminate their interest in a target, or their opportunity to commit an attack. However, inherent safety strategies do not necessarily remove the threat, as a determined adversary may not be persuaded, especially if the inherently safer elements result in a safer operation, but doesn't completely eliminate the opportunity for them to inflict damage.

Vulnerability is weakness that can be exploited by an adversary to gain access to an asset. Vulnerabilities can include, but are not limited to:

1. structural characteristics
2. equipment properties
3. personnel behavior
4. locations of people, equipment and buildings
5. operational and personnel practices

Vulnerabilities are estimates of an asset's ability to withstand specific attack scenarios, considering any existing security elements. The scenarios are usually derived from a combination of facility brainstorming ("knowing what I know about this asset, this is how I'd bring it down") and intelligence estimates or other information about the motivations and tactics of specific adversaries.[4]

Vulnerability is less affected by inherent safety than other factors unless sources are consolidated or otherwise changed. It is more likely that the vulnerability is the same following the inherent safety application to the hazard itself (1st or 2nd order), but, as mentioned before, the consequences may be different. Inherent safety applied to layers of security may be effective in making the layers robust, but this isn't necessarily reducing the hazard.

Attractiveness is the perceived value of attacking a given asset, considering the deterrence value of the security measures and the robustness of the potential target. The perceived attractiveness of a specific asset depends upon the goals of the attacker, and how well the potential consequences of a successful attack align with those goals. Other factors that may increase an asset's attractiveness include:

- presence or absence of visible security measures and activities, including regular presence of security patrols;

---

[4] The Department of Homeland Security provides specific scenarios to facilities required to conduct SVAs under CFATS.

- potential "newsworthiness" of a successful attack, or the ability to garner significant publicity from an attack;
- potential for using a target to create collateral damage to an adjoining target through the proximity of an asset to a desirable target; and
- proximity of an asset to a publicly accessible vantage point from which an attack could be launched, such as a public transportation route or highway overpass.

The attractiveness of an asset is heavily influenced by threat, consequence and vulnerability. As a result, targets with similar potential consequences or vulnerabilities may not be equally attractive, depending on the goals and motivations of the adversary. An asset owned or operated by a large, well known global company may be more attractive to some adversaries than facilities with similar or even greater potential consequences or vulnerabilities operated by small or relatively unknown companies. A facility operated by a company known to have a relationship with an adversary's enemy might be a more attractive target than a similar facility operated by a company that is not known to have such relationships. Because of the influence of threats, consequences and vulnerabilities, attractiveness should be considered a dependent, rather than an independent, security risk variable.

Inherent safety may affect attractiveness in a similar manner as it can affect threat. The adversary may be less attracted to a modified process that doesn't offer the same potential consequences. Typically, security practices affect attractiveness of a target by an increase in visible physical security measures that produce a deterrent.

7.3 SECURITY STRATEGIES

Security strategies for the process industries are generally based on the application of four key concepts:[5]

Deter: A security strategy to prevent or discourage the occurrence of a breach of security by means of fear, doubt, or reduction in facility attractiveness. Physical security systems, such as warning signs, lights, uniformed guards, cameras, and barriers, are examples of systems that provide deterrence.

Detect: A security strategy to identify an adversary attempting to commit a malicious act or other criminal activity in order to provide real-time observation, interception and post-incident analysis of the activities and identity of the adversary.

---

[5] From the CCPS book *Guidelines for Analyzing and Managing the Security Vulnerabilities of Fixed Chemical Sites* (2002). See reference section for full citation.

Delay: A security strategy to provide various barriers to slow the progress of an adversary in penetrating a site to prevent an attack or theft, or in leaving a restricted area to assist in apprehension and prevention of theft.

***Respond*:** The act of reacting to detected criminal activity either immediately following detection, such as notifying local authorities for assistance, or post-incident via surveillance tapes or logs.

A complete security design includes these four concepts in "Layers of Protection" or a "Defense in Depth" arrangement. Ideally, the most critical assets should be located in the center of conceptual concentric levels of increasingly more stringent security measures. Security scenarios often include direct attacks on or near an asset. For this reason, the spatial relationship or proximity between the location of the target asset and the location of the physical countermeasures is important.

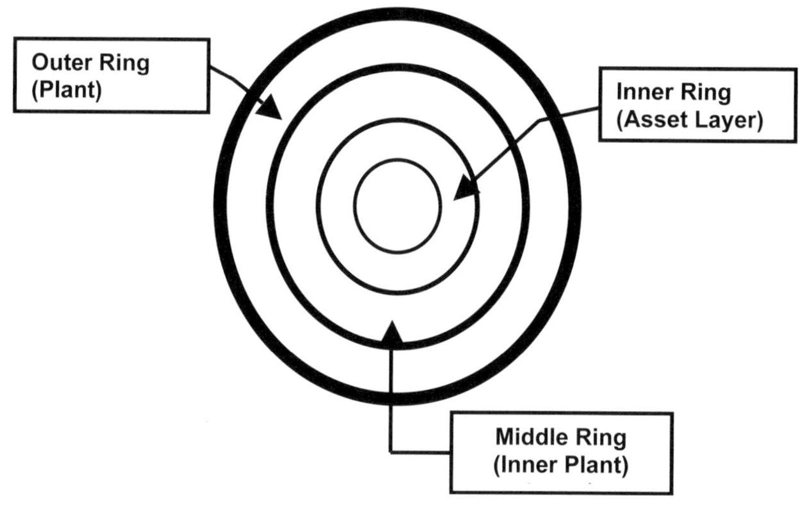

Figure 7.1: Security Layers of Protection or Defense in Depth

The facility has the opportunity to detect, deter, delay, or respond to the event at multiple points in the layers of security. Optimally, security would deter and detect at the outermost layers to provide sufficient time for responders to suppress or neutralize the adversary prior to initiation of the event. The layers of protection for critical assets may need to be quite robust because the adversaries are intentionally attempting to breach the protective features and may use whatever means are available to help ensure a successful attack. This may include the use of explosives or other destructive events that result in widespread common cause failures. Particularly motivated adversaries may go to extreme lengths, including suicide attacks, in an attempt to breach the security layers of protection.

7.4 COUNTERMEASURES

Countermeasures are actions taken to reduce or eliminate one or more vulnerabilities. Countermeasures include security officers, security barriers, technical security systems, cyber security, response, security procedures, and administrative controls to address the following strategies:

- Physical Security
- Cyber Security
- Crisis Management and Emergency Response Plans
- Policies and Procedures
- Information Security
- Intelligence
- Inherent Safety

7.5 ASSESSING SECURITY VULNERABILITIES

Following the September 11, 2001 terrorist attacks in the United States, and more recent use of chemicals in terrorist attacks elsewhere, leaders in the chemical industry recognized the potential for chemical facilities, or chemicals themselves, to be used as weapons by terrorists or other criminals. The industry also understood the need to expand existing security programs to address these new and serious threats. U.S. processing industries built on existing process safety management systems to develop *security* management systems that included requirements to assess and prioritize potential security risks posed by chemical facilities, and to implement measures to address those risks.[6]

The American Institute of Chemical Engineers supported the development of such security management systems by publishing *Guidelines for Analyzing and Managing the Security Vulnerabilities of Fixed Chemical Sites* in August 2002. The *Guidelines* describe a risk-based approach to managing chemical facility security that allows assessment of a wide range of threats, ranging from vandalism to terrorism. With the recognition of threats, consequences, and vulnerabilities, and the evaluation of security event risks, a security management system can be organized to more effectively mitigate the risks identified in the assessment.

---

[6] Examples include the Responsible Care® Security Code, adopted by the Chemical Manufacturers Association in June 2003 (http://www.responsiblecaretoolkit.com/security.asp); and API's *Security Guidelines for the Petroleum Industry* (http://new.api.org/policy/otherissues/upload/SecurityGuideEd3.pdf).

Several trade associations[7] and states[8] require companies to use security vulnerability assessment (SVA) methodologies that meet the design criteria set forth in the CCPS *Guidelines*. The U.S. Department of Homeland Security also adopted elements of the CCPS methodology as part of the Chemical Facility Anti-Terrorism Standards (CFATS).[9] The elements of a typical SVA approach are shown in Figure 7.2, which describes the methodology published by CCPS in *Guidelines for Analyzing and Managing the Security Vulnerabilities of Fixed Chemical Sites* (CCPS, 2002).

Upon completion of an SVA, a report or other documentation is prepared to communicate its results to management for appropriate action. The report should be carefully protected as company-sensitive information, because it includes descriptions of identified vulnerabilities and related countermeasures recommended to address those vulnerabilities.

7.6 INHERENT SAFETY AND CHEMICAL SECURITY

Application of inherent safety (IS) principles may allow a facility to go beyond traditional *physical* security measures. Physical security includes such considerations as security officers, barriers, and surveillance equipment. However, some assets may lend themselves to security evaluation using IS concepts, in addition to physical, and even cyber security options. A combination of strategies, including physical security, cyber security and changes in the process itself could be evaluated to create a robust, effective and individualistic facility security program.

The IS knowledgeable chemical engineer has an opportunity to influence security, as well as safety, through the process life cycle. As is true for safety, security issues can be addressed in the concept and design phases of a project, allowing for more cost-effective considerations that could eliminate or greatly reduce security risks. As described in Chapter 5, inherent safety concepts may be applicable to both new and existing/modified processes by answering questions such as:

1. First, could a potential consequence be eliminated altogether? (First order IS measure)

---

[7] One of these trade associations is the National Petrochemical and Refining Association, whose publication, *Security Vulnerability Assessment Methodology for the Petroleum and Petrochemical Industries* (http://www.npradc.org/publications/general/SVA_2nd_Edition.pdf) presents strategies based on the CCPS SVA design criteria. For more on these strategies, go to (http://www.aiche.org/CCPS/Resources/Simple.aspx?id=3888).

[8] New Jersey allows facilities to use CCPS or CCPS-equivalent SVAs to meet the requirements of Administrative Order 2005-05 (see Chapter 10); Maryland and the City of Baltimore also allow facilities, subject to their chemical security requirements, to use the CCPS or a CCPS-equivalent SVA methodology.

[9] 72 *Fed. Reg.* 17732

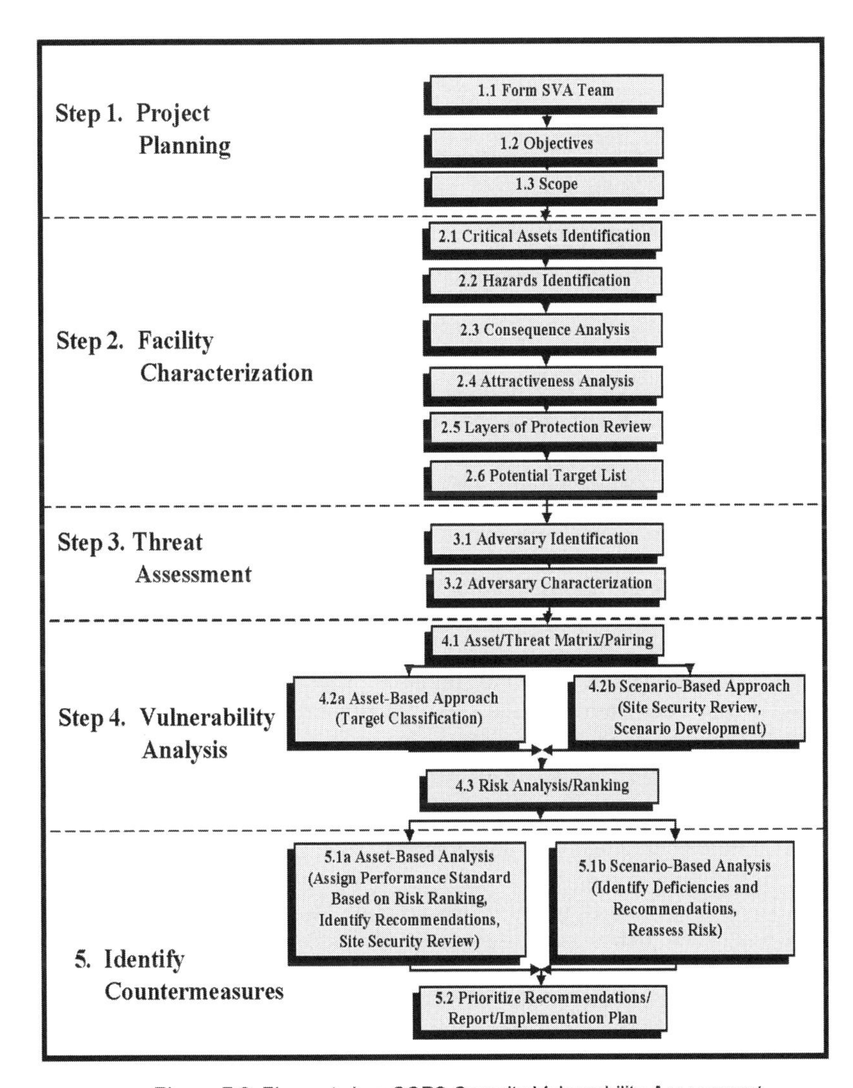

Figure 7.2: Elements in a CCPS Security Vulnerability Assessment (CCPS, 2002)

2. Second, could the magnitude or severity of the potential consequence be reduced? (2nd order IS measure)

3. Finally, could the remaining vulnerabilities be reduced through consideration of the hierarchy of passive, active, and procedural measures? (Layers of Security)

Consequence. As described earlier, chemical security consequences typically result from one of five security issues:

- *Release* of a toxic, flammable or explosive chemical
- *Theft or diversion* of a chemical that could be used as or used to produce a chemical weapon
- *Sabotage or contamination* of a shipment to cause a toxic release during transportation
- Loss of a government *mission critical* chemical
- Loss of an *economically critical* chemical or economically critical facility

Inherent safety (IS) principles are more applicable to potential consequences from some of these security issues than others and may be most readily apparent in the context of security release scenarios. If feasible, substituting a less hazardous chemical, reducing the amount of hazardous chemicals used, or operating under more moderate process conditions may contribute to a facility's security program, as well as provide a viable safety option.

Opportunities to apply First and Second order IS concepts may be limited. There may be no viable alternative for a critical reagent or reducing inventory may make a process difficult or unsafe to operate, such as the need for a surge tank. Moreover, some security issues do not lend themselves to an IS solution. Some chemicals of concern because of potential theft may not be acutely toxic, flammable or otherwise hazardous. In such situations, IS concepts may be more appropriately applied in the consideration of additional layers of protection to reduce the vulnerability of the chemicals to theft, diversion or destruction.

Vulnerability. Similar to the way that layers of protection are used to reduce the probability or frequency of a chemical accident, layers of protection may be implemented to reduce an asset's vulnerability by making the asset less attractive, and by increasing the difficulty of a successful attack. Depending on the security concern associated with an asset, such layers may include:[10]

- Facility or asset perimeter security
- Access controls for persons, vehicles and packages
- Measures to deter, detect and delay an attack
- Measures to secure chemicals and equipment to prevent theft, diversion, contamination or sabotage
- Securing business, process control and other computer systems

---

[10] See 6 C.F.R 27.230 for a list of security performance measures required by the Department of Homeland Security for high risk chemical facilities.

- Monitoring, communications and warning systems
- Emergency response plans

Traditional security measures such as physical security, background checks, administrative controls, access controls, or other preventive or protective security measures may also be considered from an inherent safety perspective. Applying IS concepts to layers of security, particularly minimization and simplification strategies, can make them more robust, reliable, effective, available, and maybe even less expensive to install or maintain. Ideally, layers of protection should be:

- Independent: not sharing common-cause failures (i.e., loss of power, single human operations, or maintenance acts that would cause more than one layer to fail);
- Effective: capable of detecting, delaying, and/or deterring an adversary;
- Auditable: each layer can be checked for its effectiveness;
- Passive: requiring no energy or human input to function properly.

However, passive countermeasures may not be possible for all security elements. Gates will need to open and close, lights and cameras will need power, a human will likely be required to implement a visitor/guest vetting and escort procedure.

In some instances, security countermeasures may also increase safety at the facility. For example, a facility may create a buffer zone between the public areas and the plant fence, and another between the plant fence and critical process equipment. These security measures will create:

- *detection zone(s)* if they are free of obstacles and have sufficient depth to allow for adversaries to be detected while attempting unauthorized entry;
- *standoff zone(s)* provided they have sufficient depth to keep adversaries from using explosives or standoff weapons effectively from the perimeter;
- *delay zone(s)* allowing intervening force time to respond, or time for operators to take evasive action before an adversary reaches a target following detection.

Such buffer zones may also reduce potential off-site impacts from an accidental release by increasing the distance between the potential release and an off-site receptor.

7.7 LIMITATIONS TO IMPLEMENTING IS CONCEPTS IN SECURITY MANAGEMENT

As discussed in Chapter 9, identifying and implementing inherently safer measures may entail evaluating risk tradeoffs to help assure that risk is actually being reduced and not shifted elsewhere or converted to a different one. This is especially true in the context of security in which an inadvertent shift in risk from one population to another could have significantly severe consequences.

Balancing competing security or other concerns: Consider a batch process that includes a large storage tank for a hazardous intermediate product to allow the facility to switch among different uses of the intermediate. While the facility could redesign the process to remove the intermediate storage and, thereby remove or reduce the consequence, the manufacturing flexibility provided by the intermediate storage may be necessary for the facility's continued operations. In order to remain in operation, a more viable security strategy would be to establish or enhance other security countermeasures. Continued operation itself could be a security issue if the facility produced a mission or economically-critical chemical, or if the facility itself was critical to the regional economy.

Creating new security concerns: As described in Chapter 4, phosphorous oxychloride could be used in place of phosgene, and 1,1,1-trichloroethane could be used in place of hydrochloric acid gas (Kletz, 1998). Phosphorus oxychloride and trimethyl phosphite are listed as chemical weapons precursors under the Chemical Weapons Convention,[11] and export controls which create additional recordkeeping and reporting burdens for the facilities. However, phosphorus oxychloride and trimethyl phosphate are listed as Chemicals of Interest under CFATS.[12] From a security standpoint, these chemicals present a concern for theft or diversion because any of them can be converted to chemical weapons using simple chemistry, equipment and techniques. A facility converting to this alternative process could still be required under CFATS to implement security measures to assure that these chemicals are not stolen from the facility, and that shipments are not misdirected or diverted.[13]

Shifting security risk from one concern to another: Chlorine may raise two different security concerns, depending on the quantity and container – *release* of chlorine gas from a vessel at a facility could have the potential to create serious off-site consequences, or *theft or diversion* of chlorine in transportation packaging for use as a chemical weapon at an unknown time and location.

---

[11] *Convention On The Prohibition Of The Development, Production, Stockpiling And Use Of Chemical Weapons And On Their Destruction*, Organization for the Prohibition of Chemical Weapons, (http://www.opcw.org/)

[12] Chemical Facility Anti-Terrorism Standard, 6 CFR Part 27.

[13] Phosgene is also listed in the CWC and as a CFATS chemical of interest for theft, but little phosgene is shipped, meaning that a CFATS facility with phosgene would be less likely to be required to implement anti-theft/anti-diversion measure.

A facility using chlorine from 90-ton rail cars, a potential security release concern, could reduce the potential consequences from an attack on the facility by switching to chlorine from smaller cylinders. However, chlorine in smaller containers could create a security theft/diversion concern not only for the facility itself but, by increasing the number of chlorine cylinders, could create a transit threat as well. Increasing the number of cylinders at the site also provides greater opportunity for a shipment to be diverted, or individual cylinders stolen, for nefarious purposes.

A facility with multiple potential targets for terrorists may adopt a security policy to minimize vehicular traffic into or out of the facility due to concerns that a vehicle coming into a facility could be carrying an explosive to attack an asset within the facility, or could be used to smuggle weapon or bomb precursors out of the facility. This could contrast with a recommendation to reduce the hazardous inventory on-site by increasing the number of smaller shipments into the facility. Either the reduction in vehicular traffic or the reduction in inventory could be more desirable, depending on other conditions at the site or in the surrounding community (see examples below). Regardless, both would need to be identified and evaluated to assure that the increase in the vulnerability created by the increased vehicular traffic did not create a greater overall security risk than the potential consequence created by the larger inventory.

Shifting security risk from one location to another: As described in Chapter 9, a plant might reduce the size of a hazardous material inventory on the site such that the facility would receive shipments by truck (typically about 30,000 pound shipments) rather than rail (typically about 300,000 pound shipments for many materials) thus increasing the number of shipments by a factor of 10. Not only would the increase in the number of shipments potentially increase the risk of shipment diversion (discussed earlier), but a truck route past a terrorist's target (such as a school or government building) might increase the attractiveness of the shipment itself as a target, particularly if the rail line did not run past iconic or other soft targets. This is not to say that the reduction in inventory should not be considered, only that security factors outside the facility must also be considered before a decision is made.

If the security of alternative storage is less robust than at the facility, the overall security risk would not have been reduced, but merely shifted to another location. Knowing whether moving the inventory reduces or increases security risk would depend on an evaluation of the relative security postures of the facility and the alternatives. If the facility security were more robust, then increasing the amount of hazardous material stored within the facility's security perimeter might reduce the overall security risk to the community.[14]

---

[14]As promulgated in 2007, CFATS (6 CFR 27) requirements for security at chemical facilities does not extend to chemicals of interest stored in rail yards.

7.8 CONCLUSION

Inherently safer systems can be beneficial for security as well as for safety risk management. Inherent safety can be applied for security as risk deems it to be necessary to reduce the hazards.

However, inherent safety can be a complex issue. Inherent safety may or may not be the most effective security strategy, particularly if a new security concern is created or the security risk is simply shifted. It may be unrealistic to expect that all hazards are eliminated by inherent safety. Instead, all options must be examined in a complete context to assess risk reduction benefits and costs and to consider risk/risk tradeoffs that may unintentionally occur. They also must be weighed against the benefits of the operation, as well as alternatives, such as more traditional security measures, and whether the overall risk is sufficiently managed.

8.

Implementing Inherently Safer Design

8.1 INTRODUCTION

To be most effective, implementation of inherently safer designs/technologies requires a systematic management approach, sound technical basis, and management and cultural emphasis on inherent safety as an important organizational value and tool for reducing the risk of process-related incidents. At a corporate level, it first requires management commitment and leadership to provide resources and establish policies and procedures to integrate the use of inherent safety into the framework of the company's overall process safety management program. As a general philosophy, its concepts should be woven into the way that facilities are designed, constructed, operated and maintained, so that company practitioners can continuously look for and identify ways to maximize the inherent safety of an operation. This is very similar in concept to the current industry emphasis on "Lean Manufacturing," which involves an overall management approach to eliminate waste in over-production, waiting time, transportation, processing, inventory, motion and scrap, and on "Kaizen," a philosophy of continual improvement in business processes to accomplish this objective.

It is now more common for companies to build inherently safer design principles into their process safety management systems. For example, this can be accomplished by incorporating inherently safer design concepts into existing safety and process hazards reviews. Companies may wish to enhance their existing review systems with inherent safety reviews at key points in the process life cycle. This chapter discusses methodologies to conduct inherent safety reviews at three key stages of the life cycle:

- during product and process development;
- during conceptual facilities' planning early in the process design phase; and,
- during various phases of routine operation, including modifications and incident investigations.

Many companies may elect to include the objectives and features of these reviews into their existing process safety management system. If companies decide to incorporate inherent safety into existing systems and reviews, particular attention needs to be given to the subject in R&D for new product and process development. This is the life cycle stage where application of inherently safer design concepts can have the greatest impact.

Overton and King (2006) note that implementation of inherently safer technology will be an evolutionary process occurring over time. For example, a management system that calls for periodic IS reviews for existing processes, such as PHA revalidations, provides on-going opportunities to identify new technology breakthroughs that can be economically applied to increase the level of inherent safety in a process. Significant risk reduction may be achieved in this manner. An important key to incorporating the tenets of IS into the entire spectrum of process safety management activities is making sure that operations, maintenance, and engineering staffs are aware of its technology and benefits.

8.2 MANAGEMENT SYSTEMS APPROACH FOR IS

An effective inherent safety (IS) program requires an integrated management system approach that embodies the "Plan-Do-Check-Act" management cycle. A high level corporate process safety management policy and program that explicitly includes IS as one of the key principles of risk management is generally a starting point. This document would typically outline the management system required, and would include, at minimum, the following elements within the framework of the process safety management program:

- Development of IS policies, procedures, and design standards.
- IS education for affected functions (i.e., research, process design, engineering, operations, maintenance)
- IS reviews of new and existing facilities, including recommendation follow-up tracking (may be incorporated into project design reviews and facility PHAs)
- Implementation of IS concepts in on-going aspects of an operation (i.e., management of change, maintenance, SOPs)
- A systematic management review process, including performance metrics, performance assessment, and review and implementation of improvements.

Amyotte, et al. (2006) describe how IS concepts can be integrated into each element of a process safety management program, including maintenance and operating procedures, training, management of change, and incident investigation. Dowell (2001) emphasizes the need to integrate the many environment, safety and health elements, including those involving process safety management, into a comprehensive management system so that they become part of a company's way of doing business, and not separate management programs. In this way, *inherent* safety becomes an *inherent* part of the company's system for improving process safety and reducing risk.

Like other management systems, an IS management system moves ISD from the conceptual stage to implementation in relevant process safety-related activities. Without a management systems approach, IS will remain as a concept only, and not a functional element of the organization's safety efforts.

8.3 EDUCATION AND AWARENESS

8.3.1 Making IS Corporate Philosophy

Once management has established its commitment to the principles of inherent safety, including the development of a management system through policies and procedures, training is required for those who have responsibilities that impact the consideration and use of IS in process safety-related activities. Management promotion and adoption of the principles of inherent safety, wherever possible, is necessary to provide a visible demonstration of leadership and commitment to effective process safety management (Amyotte, et al., 2006). However, this is not sufficient by itself to establish the IS program. Education and awareness of IS concepts and application is also necessary.

As a general philosophy, one of the objectives of an IS education and awareness program is that all employees responsible for design or engineering decisions become aware of inherently safer concepts and "automatically" look for opportunities in their everyday work activities. This is more proactive than any review process. As technologists do their design and operations tasks, they understand the value of IS and implement the concepts where appropriate at all levels of detail, from basic process chemistry selection by research chemists, to the writing of operating procedures by plant personnel.

8.3.2 IS in Education

In recent years, the basic concepts of chemical process safety have increasingly been integrated into the undergraduate chemical engineering curriculum at many colleges and universities. As the profile of inherently safer technology increases, particularly during the post-9/11 focus on security in the chemical industry, interest in IS is expected to likewise increase. In order to better prepare students for work in the chemical process industries, this education should be extended to both chemistry and technical management programs. Basic chemistry is a key to ISD, and chemists need to be taught to recognize this and look for IS chemistries when conducting research into new or modified processes. IS has a parallel in current efforts to institute the philosophy and practice of "green chemistry," which is the design of chemical products and processes in such a way as to reduce or eliminate the use and generation of hazardous substances, in order to minimize or prevent pollution and the creation of hazardous waste. Green chemistry, which has inherent safety as a parallel benefit, is an overarching philosophy of chemistry now becoming an important part of the chemistry education curriculum.

Another parallel to IS that exists in the environmental field is the current emphasis on pollution prevention, which calls for designing processes to minimize the amount of waste generated, rather than relying on "end-of-pipe" means to treat wastes generated. This approach also has its basis largely on basic process design and operating conditions, as well as modifying raw materials and, in some cases, products. This philosophy has been adopted by many companies, and has, in fact, been instituted by regulation in many states through requirements for Pollution Prevention plans.

CCPS sponsors SACHE (Safety and Chemical Engineering Education), whose major activities include the development of instructional materials for chemical process safety education in the chemical engineering curriculum, as well as sponsorship of workshops for chemical engineering faculty on process safety topics. SACHE has developed an instructional module on inherently safer processes, as has the Institution of Chemical Engineers in the UK in conjunction with the International Process Safety Group (IChemE/IPSG, 1995) (see Section 8.10).

Teaching inherently safer technology concepts in undergraduate chemistry, chemical engineering, and related disciplines will be of great benefit as students move into industry after graduation, and will help further the practice of IS.

8.4 ORGANIZATIONAL CULTURE

An organization's Safety Culture, which is the collection of its values, beliefs, and behaviors—or "the way the organization conducts its business"—flows from visible management commitment and leadership. As an active strategy for eliminating or reducing—as opposed to just managing—risk, inherently safer operations and the associated innovation that is often required must become ingrained in the comprehensive process safety culture, in the same way that the commitment to Total Quality Management (TQM) or Lean Manufacturing has become ingrained—through management leadership and commitment. In most companies today, there must be a well-understood business benefit in addition to the safety benefit in order for management to be convinced of the need for inherently safer operations. This often comes through effective training and communication throughout the organization on the concepts, applications, particularly in process development and early design stages, and benefits of ISD—such as reduced inventories, smaller equipment, and fewer add-on safeguards to be maintained, etc.—as well as its limitations. The evaluation of both initial and on-going costs associated with inherently safer designs can often help enhance this understanding. Many IS improvements can be implemented cost-effectively, particularly incremental improvement in existing plants. Kletz (1999) discusses cultural and management barriers to the implementation of inherently safer designs, as well as the actions needed to overcome these barriers. Amyotte and Khan

(2003) also provide guidance for making the use of inherently safer design principles more routine.

Turney (2001) lays out five steps, taken from the European Process Safety Centre's Statement of Good Practice, which are necessary for effective adoption of inherent safety within an organization:

- Support by a champion
- Suitable training
- Application from the earliest stage of a project
- Reviews throughout project development
- Recognition and reward of those involved in the project

By incorporating inherently safer design into the organizational culture, inherent safety becomes an on-going way of examining and addressing processes and their hazards, and this philosophy then permeates all aspects of the process safety management systems. For example, once its application is fully understood, the concept of simplification with respect to human factors can become an inherent aspect of the writing of standard operating procedures.

8.4.1 Multiple Demands of IS in the PSM program

Dowell (2001) points out that excellence in operations, which must include personnel and process safety, requires a comprehensive management system approach. The challenge is to integrate the many ESH compliance standards (PSM, RMP, Responsible Care®, RP750, company requirements, etc.) and special program activities (ISO 9000, TQM, etc.) into the fabric of the system of making chemicals, which, in turn, is grounded in company culture. These activities must go beyond doing the minimum required to give the appearance of compliance. Figure 8.1 illustrates the required foundation of management leadership (Auger, 1995). Managing process safety is based on a corporate mission which management implements. By their words and actions, managers at all levels of the corporation must understand the benefits of IS and show their commitment to such programs in tangible ways. The latter should include providing the necessary resources, such as personnel, time, and funding, for IS-related activities like reviews and evaluation/ implementation of follow-up recommendations that arise.

Figure 8.1: Management leadership is the foundation for process safety management

8.4.2 Incorporating IS Into Normal Design Process

As stated previously, IS should be made a part of the normal design process, by incorporating IS thinking into design processes and standards. Only by implementing IS consistently throughout the organization—from Research through Operations—will it become "the way we do business."

Consideration of IS should begin at the start of the process life cycle, which is the initial process research and development. Because the basic level of inherent safety is shaped at this stage, attention to IS at this point can have the greatest impact on process safety—even more so than at design stage reviews where piping and instrumentation drawings (P&IDs) have already been developed. It is imperative that designers incorporate IS into their thinking in the initial process selection, design, and equipment specification, as the ability to modify the design becomes more difficult as the project progresses into the detailed design and construction stages.

When a hazard is identified at any stage of development, the designer should ask the following questions in this order (Gupta and Hendershot in Lees 2005):

1. Can the hazard be eliminated?
2. If the hazard cannot be eliminated, can the magnitude of the hazard, or potential incidents arising from the hazard, be significantly reduced?

3. Do the alternative designs identified from the answers to Questions 1 and 2 increase the magnitude of other existing hazards, or introduce any new hazards to the process?
4. If the alternative designs increase other hazards, or introduce new ones, develop an understanding of these hazards and their relative importance.
5. Identify passive, active, and procedural safeguards appropriate for management of all hazards.
6. Use a logical decision process to select the optimum design, considering inherent safety and other safety characteristics, and all other relevant design parameters.

It is important for a designer, upon identifying a hazard, not to skip past the first four questions and go directly to question 5—identifying "add-on" safety features to manage and control a hazard whose existence and magnitude the designer accepts. While it will not always be possible or feasible to eliminate or significantly reduce all hazards, it is certain that it will never happen if the question is never asked. This is the heart of the inherently safer design philosophy – to first ask if hazards can be eliminated from the process, or if they can be significantly reduced. Designing of safety systems to manage and control hazards should not proceed until it has been firmly established that it is not feasible to eliminate or reduce these hazards from the start.

Basic process selection is at the core of inherent safety at the pre-design stage, for this is the point where the basic chemistry and selection of unit operations is established. Where one or more alternative processes for producing the desired product exist, they should be compared in terms of their inherent hazards, such as raw materials used, intermediates and wastes produced, and operating conditions including temperature and pressure. Selection of the inherently safer process is not always straightforward, and a number of conflicts may exist that need to be reconciled to optimize the level of risk against other factors (see Chapter 9). Ankers (1995) describes a software application for identifying inherently safer process options, with an emphasis on early hazard identification. A number of methods have been developed for measuring and comparing the relative level of inherent safety between two or more processes. These are discussed later in this chapter.

The use of corporate design standards, which incorporate the use of applicable inherently safer design features can also be beneficial in providing guidance for designers of new processes, and helping to establish company expectations and decisions regarding risk minimization. A number of companies have established internal design guides that incorporate the use of IS designs for new facilities, as well as for major modifications to existing facilities.

Many companies rely on engineering contractors for design work at various levels, from detailed engineering of a process developed by the customer, all the way to "turnkey" licensed technology and plants. Companies should clearly define expectations for IS considerations in the design process, *before* contracts are awarded.

Some contractors may not be aware of IS concepts and how to implement them, so providing an engineering standard on how these concepts are to be considered is an important factor. The company should provide guidance as to how IS should be considered in layout and residual hazard management, as well as how it will be documented.

The 1999 Control of Major Accident Hazards (COMAH) regulation in the UK requires the consideration and documentation of inherently safer design alternatives during the initial design stage. In addition to the standard IS strategies, it includes plant layout as part of the overall approach, particularly to prevent the potential for domino effects by fire, explosion (pressure wave and missiles) or toxic gas cloud that could cause loss of operations control in another location. Plant layout, which is established during the design stage, is often a compromise between a number of factors such as:

- The need to keep distances for materials transfer between plant/storage units to a minimum to reduce costs and risks;
- The geographical limitations of the site;
- Interaction with existing or planned facilities on-site, such as existing roadways, drainage and utilities routings;
- Interaction with other plants on-site;
- The need for plant operability and maintainability;
- The need to locate hazardous materials facilities as far as possible from site boundaries and from people living in the local neighborhood;
- The need to prevent confinement where release of flammable substances may occur;
- The need to provide access for emergency services;
- The need to provide emergency escape routes for on-site personnel; and,
- The need to provide acceptable working conditions for operators.

These issues are partially dealt with in the PSM and RMP regulations under Facility Siting, which involves the spatial relationship between process hazards and people, but can also be applied to the spatial relationship between hazards and other buildings, equipment, or processes. In its most basic form, inherently safer siting calls for locating hazardous operations in remote areas, away from inhabited areas. For this reason, the manufacture of explosives is conducted in unpopulated areas of the country.

Other IS considerations dealt with primarily at the design stage include:

- Designing vessels to withstand the maximum temperature and pressure to which they are likely to be subjected in operation.
- Selection of materials of construction that are compatible with the process fluids to be handled and the chosen design conditions, in order to

minimize corrosion/erosion potential rather than relying on inspection programs to identify and correct.

- Minimizing the size of vessels, piping and other equipment to limit the potential release quantity.
- Addressing issues associated with reactive chemicals, such as eliminating the potential for water contamination of the process if a water-reactive material is being used.

In most cases, an assessment of alternatives will involve consideration of cost—both initial and ongoing, such as inspections, testing and maintenance—and other factors, including relative risk reduction using other risk management approaches, such as emergency relief/mitigation system vs. higher vessel design pressure. For complex, technically challenging, or inherently higher risk projects, a formal risk comparison should be carried out using tools such as ranking methodologies or quantitative risk assessment. Ankers (1995) describes a software application for identifying inherently safer process options, with an emphasis on early hazard identification. Other methods are discussed later in this chapter.

8.5 INHERENT SAFETY REVIEWS

The quantity of literature on company inherent safety policies and procedures has continued to increase during the past several years, indicating that awareness of inherent safety principles is improving. Besides presenting an opportunity to reduce risk, IS reviews are now mandated by regulation for high-risk facilities in New Jersey and Contra Costa County, California. The discussion of IS in relation to reduced security risk, and its consideration for inclusion in chemical security legislation, have also raised the visibility and awareness of IS in the U.S. Companies which have chosen to incorporate inherent safety principles into their existing hazard management programs may want to consider highlighting, or otherwise identifying, inherently safer design aspects of those programs. This will promote an increased awareness of the concepts among a broad population of chemists, engineers, and business leaders. In turn, this heightened awareness will encourage the application of inherently safer design principles as a part of the normal, everyday work processes.

A number of companies have published descriptions of their inherent safety review practices:

- Bayer (Pilz, 1995) uses a procedure based on hazard analysis, focusing on the application of inherent safety principles to reduce or eliminate hazards.
- Dow (Sheffler, 1996; Gowland, 1996b) uses the Dow Fire and Explosion Index (Dow, 1994b) and the Dow Chemical Exposure Index (Dow, 1994a) as measures of inherent safety, while applying inherently safer design principles to reduce hazards.

- Exxon Chemical (French, et al, 1996; Wixom, 1995) has developed an inherent safety review process that includes applications at various points in the process life cycle. Its expansion of IS concepts to safety, health, and environmental reviews is documented in *Inherent Safety, Health and Environmental (SHE) Reviews*.
- ICI (Gillette, 1995; Turney, 1990) describes a 6-step hazard review procedure, which occurs at specific points in the life cycle of a process, and focuses on inherently safer designs.
- The Rohm and Haas Major Accident Prevention Program (Renshaw, 1990; Berger and Lantzy, 1996; Hendershot, 1991a) is based on potential accident consequence analysis, and uses checklists based on inherently safer design principles to identify ways to eliminate or reduce hazards.
- BP has published *Inherently Safer Design Guidelines for New Projects and Developments*.
- Nova Chemicals' Loss Prevention Standard 2.2 requires project and design teams to look conscientiously for inherent safety opportunities.
- DuPont (Clark, 2007) describes how ISP (Inherently Safer Processing) is integrated into the overall corporate PSM program, based on a checklist and a corporate training program. A semi-quantitative ISP scoring system is used by corporate R&D to ensure that it is appropriately considered at the earliest stage of a process life cycle.

In addition, the Contra Costa County, CA, Health Services Department has issued a guidance document for the IS review of existing facilities, and new facilities at the chemistry-forming, facilities design scoping and development, and basic project design stages. IS analyses must be performed for all situations where a "major chemical accident or release"—as defined in the standard—could reasonably occur. This document also includes guidance for evaluating the feasibility of recommendations, and for IS review documentation. IS reviews for existing processes can be conducted as part of an initial/five-year PHA or as a separate study, using a checklist[1] or guideword analysis[2] that incorporates IS.

8.5.1 Inherent Safety Review Objectives

The objectives for an inherent safety review are to employ a synergistic team to:

- Understand the hazards.
- Find ways to eliminate or reduce those hazards.

---

[1] See Appendix A for a detailed example of an IS Checklist

[2] Table 8.3 that appears later in this document offers sample Guidewords, while Table 8.5 shows a sample Guideword Matrix

The first major objective for the inherent safety review is the development of a good understanding of the hazards involved in the process. Early understanding of these hazards provides time for the development team to implement recommendations from the inherent safety effort. Hazards associated with flammability, pressure, and temperature are relatively easy to identify. Reactive chemistry hazards are not. They are frequently difficult to identify and understand in the lab and pilot plant. Special calorimetry equipment and expertise are often needed to fully characterize the hazards of runaway reactions and decompositions. Similarly, industrial hygiene and toxicology expertise is desirable to help define and understand health hazards associated with the chemicals employed.

Reducing and eliminating hazards and their associated risks is the second major objective. Applying inherent safety principles early in the product/process development effort provides the greatest opportunity to achieve the objectives of the inherent safety review process for the project at hand. If these principles are applied late in the effort, the results may have to be applied to the "project after next," as the schedule may not permit implementation of the results.

Experience with inherent safety reviews for a new process or project indicates that project investment costs are often reduced as a result of this exercise. Eliminating equipment and reducing the need for safety systems are typically the main contributors to investment reduction. Capturing these potential savings largely depends on the timing of the reviews within the process or project development cycle.

The same two primary objectives apply to an IS review of an existing process. Generally, these reviews are designed to identify either incremental improvements, primarily in the area of simplification, possible major improvements that can be made in the event of a process renovation or major equipment replacement, or IS opportunities for future facilities manufacturing the same product. Often, large capital expenditures to improve IS are difficult to justify if the process has been operating without major incident. However, when major projects are planned, they provide a good opportunity to identify opportunities for incorporating inherently safer design features.

8.5.2 Good Preparation is Required for Effective Inherent Safety Reviews

Any IS review should begin with a brief overview of IS concepts, as well as the process being studied and the approach to conducting the review (HAZOP, What-If?, Checklist, etc.).

The completeness of the information described in steps 1 to 9 (Chapter 8.5.5) prior to the review will influence the quality of the inherent safety review. The chemist needs to define the desired reactions and develop an understanding of potential side reactions. Effects of mischarges or process deviations on reaction chemistry should be developed. These process chemistry information requirements are discussed in Section 5.2.

Reaction kinetics are often difficult to develop in the normal experimental laboratory or pilot plant. Specialized reaction calorimeters and reactive chemistry testing, such as differential scanning calorimeter (DSC), Accelerating Rate Calorimeter (ARC™), and a vent sizing package (VSP) are required to obtain thermodynamic and kinetic information (CCPS, 1995e). Some chemical companies have established reactive chemistry laboratories with reaction engineering specialists to assist in developing this information. Specialized analytical laboratory testing firms also provide this expertise and experimental capability.

Industrial hygiene/toxicology professionals need to develop acute and chronic toxicology information on all the materials used and produced in the process. This information should also include the potential products of abnormal reactions. The industrial hygiene member of the team should be prepared to explain the toxicology information on the material safety data sheet (MSDS) to the review team.

Another necessity is physical property data for the process chemicals. This information includes such items as melting point, boiling point, vapor pressure, water solubility, flammability data, and odor threshold, to name a few. All this information cited above must be organized and coordinated to satisfy the company's process and facilities documentation, and regulatory requirements.

For IS reviews at the design stage, and for existing facilities, accurate P&IDs are also necessary, as well as updated standard operating instructions, and process safety information that includes the size of the equipment and the materials of construction. Operations and maintenance personnel with experience in the process or similar processes should be included in the review, as they often have valuable knowledge and experience to contribute.

8.5.3 Inherent Safety Review Timing

Inherent safety reviews should be considered during appropriate stages of a typical project's life cycle. These may be stand alone reviews or incorporated with other process hazard analysis or appropriate project review. Suggested timing includes:

- During the chemistry forming (synthesis) stage for product/process research and development to focus on the chemistry and process (refer to Section 5.2).
- During the facilities design scoping and development prior to completion of the design basis to focus on equipment and configuration (refer to Sections 5.3 and 5.4).
- During the basic design phase of the project (i.e., design stage PHA).
- During regular operations to identify potential improvements in existing facilities, as well as those that can be incorporated into the design of the next plant.

The timing of IS reviews in the first three stages is critical. For each stage, there is a window of opportunity. If carried out too early, essential information may not be available. If carried out too late, the design will be too far advanced and changes may cause significant expense as well as project delays (Kletz, 1999).

It is recommended that an IS review be revalidated periodically, like the five-year PHA cycle. The revalidation should include and document the following:

- IS and other process safety improvements made since the last review was conducted or a new method was selected to perform the IS analysis;
- An IS review for all changes that have been made since the last analysis, unless specifically reviewed for IS under Management of Change;
- Review of any significant incidents that occurred in the process since the previous review; and,
- Review of any new and existing technologies not previously reviewed that can potentially make the process inherently safer.

8.5.4 Inherent Safety Review Team Composition

The composition of the inherent safety review team will vary depending upon the stages of the development cycle and the nature of the product/process. The team composition is generally four to seven members selected from the typical skill areas checked in Table 8.1. Knowledgeable people, with an appropriate matrix of skills, are required for a successful review effort. The industrial hygienist/toxicologist and chemist play key roles on the team to ensure understanding of the hazards associated with reactions, chemicals, intermediates, and products, and to explain hazards, especially where there may be choices among chemicals or processes.

Often, an organization will strive for the elimination of a specific toxic material from a given process, such as chlorine from a water or wastewater treatment process. Alternatives will also have other hazards and risks that require an informed choice. The industrial hygienist, chemist, and safety engineer play an important role in generating the information needed to select between alternatives.

Table 8.1: Inherent Safety Review Team Composition

	Product Development	Design Development	Design Stage PHA	Operations
Industrial Hygienist/Toxicologist	√	√	√	√
Chemist	√	√	√	√
Process Design	√	√	√	√
Safety Engineer	√	√	√	√
Process Technology Leader	√	√	√	√
Environmental Scientist/Engineer	√	√	√	√
Control Engineer		√	√	√
Operator			√	√
Operations Supervisor		√	√	√
Maintenance			√	√

8.5.5 Inherent Safety Review Process Overview

As previously stated, good preparation is very important for an effective inherent safety review, particularly for a new process. Preparation for the review is summarized in Figure 8.2 and includes the following background information:

1. Define the desired product.
2. Describe optional routes to manufacture the desired product (if available), including raw materials, intermediates, and waste streams.
3. Prepare simplified process flow diagram.
    - Include alternative processes
4. Define chemical reactions.
    - Desired and undesired
    - Determine potential for runaway reactions/decompositions.
5. List all chemicals and materials employed.
    - Develop a chemical compatibility matrix.
    - Include air, water, rust, etc.
6. Define physical, chemical, and toxic properties.
    - Provide NFPA hazard ratings or equivalent
7. Define process conditions (pressure, temperature, etc.).
8. Estimate quantities used in each process system (tanks, reactors, etc.).
    - State plant capacity basis
    - Estimate quantities of wastes/emissions
9. Estimate quantities used in each process system (tanks, reactors, etc.).

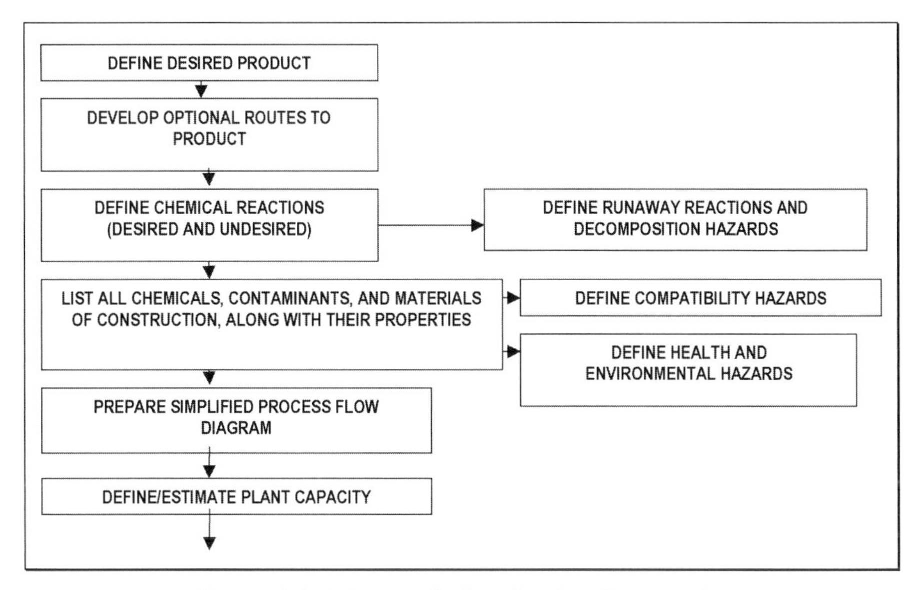

Figure 8.2: Inherent Safety Review Preparation

After the background information is developed, the inherent safety review can be arranged. The review steps are summarized in Figure 8.3. Steps proposed for the review are as follows:

1. Review the background information from steps 1 to 9 above.
2. Define major potential hazards.
3. Systematically review the process flow schematic, looking at each process step and hazardous material to identify creative ways to apply inherent safety principles to eliminate or reduce hazards.
4. During the inherent safety review, at the design development stage, identify potential human factors/ergonomics issues that should be addressed by the design team.
5. Document the review and follow-up items.

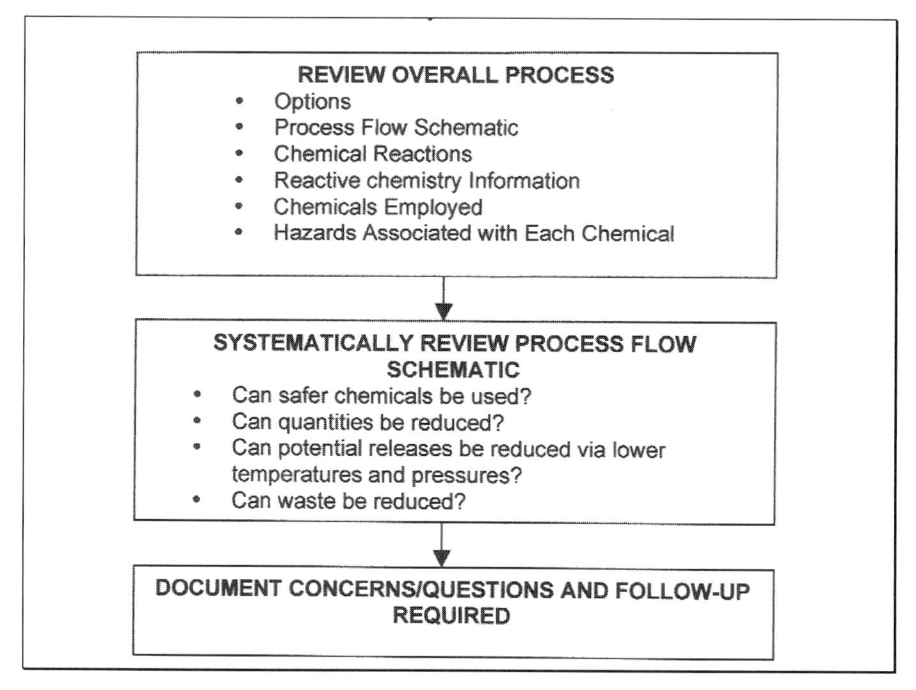

Figure 8.3: Inherent Safety Review Process

During the systematic review of the process flow schematic (step 12) the team will examine some of the following questions:

- Can safer chemicals be used (i.e., non-toxic/non-flammable or non-volatile reactant)?
- Can quantities be reduced (particularly intermediate requiring storage)?
- Can potential releases be reduced via lower temperatures or pressures, elimination of equipment or by using sealless pumps?
- Can waste be reduced (i.e., by using a regenerable catalyst or by recycling)?
- What additional information (toxicology information or reactive chemicals data) is required?
- Can the equipment or process be simplified, be made mistake proof, or at least mistake tolerant, by applying ergonomic/human factors principles?
- Have trace materials or contaminants in the process been identified, along with their effects?

It is not unusual to have several inherent safety reviews during the product/process development effort. Early reviews will often not have all of the information required for steps 1 to 9. The follow-up items will describe what is necessary to obtain the missing information, such as toxicity data on new intermediates or products of undesired reactions.

A more comprehensive IS review process example is provided in Appendix B. This example includes information on pre-study preparation, team training, and IS review conduct and documentation.

8.5.6 Focus of Inherent Safety Reviews at Different Stages

The focus of an inherent safety review will change from the chemistry-forming stage through regular operations. While each review is unique, Table 8.2 shows which areas receive emphasis at a particular stage. In the regular operations review, the team is identifying inherent safety opportunities not only for the existing process, but also for the next major expansion or grass roots facility. This is particularly true if the process was not originally designed with inherent safety in mind. The check marks indicate the relative emphasis each area receives.

In the "chemistry-forming" inherent safety review, the team will cover such things as:

- Understanding the hazards
- Choosing the best route to produce a given chemical or product
- Process improvements
- Reactor types and conditions
- Intermediate storage optimization
- Waste minimization
- Separation technology
- Requirements for additional information

During design scoping, the team will concentrate on minimizing equipment, reducing inventories, simplifying the process, reducing wastes, and optimizing process conditions. Inherent safety concepts should also be considered during process hazards reviews, such as HAZOP, for both new and existing processes. The initial design should be "mistake-proofed," and each safety device and procedure examined to see if there is a way to eliminate the need for it.

When the inherent safety process has been expanded to review regular or routine operation, the team should look at all aspects of inherent safety to provide suggested improvements for both the existing facility and for the next plant. Even if the process was originally designed with inherent safety in mind, these improvements may arise from advances in technology, changes in product specifications or application, or lessons learned from incidents and near-misses, both in the facility being studied or in similar facilities elsewhere.

Table 8.2: Focus of Different Inherent Safety Reviews

Note: The number of check marks indicates the relative importance of the strategy

	Chemistry and Process Selection	Design Scoping	Regular Operation
Minimize • Reduce quantities	✓	✓✓✓	✓✓
Substitute • Use safer materials	✓✓✓	✓	✓✓
Moderate • Use less hazardous conditions • Reduce wastes	✓✓	✓✓	✓✓
Simplify	✓	✓✓✓	✓✓

8.5.7 Stage in the Process Life Cycle

The following stages in a process life cycle have been previously defined:

Process Research and Development
- Chemistry

Conceptual Design Phase
- Process engineering
- Pilot plant operation

Detailed engineering
- Plant design

Construction and Startup

Routine operations
- Operations
- Inspection
- Maintenance

Plant and Process Modification

Decommissioning

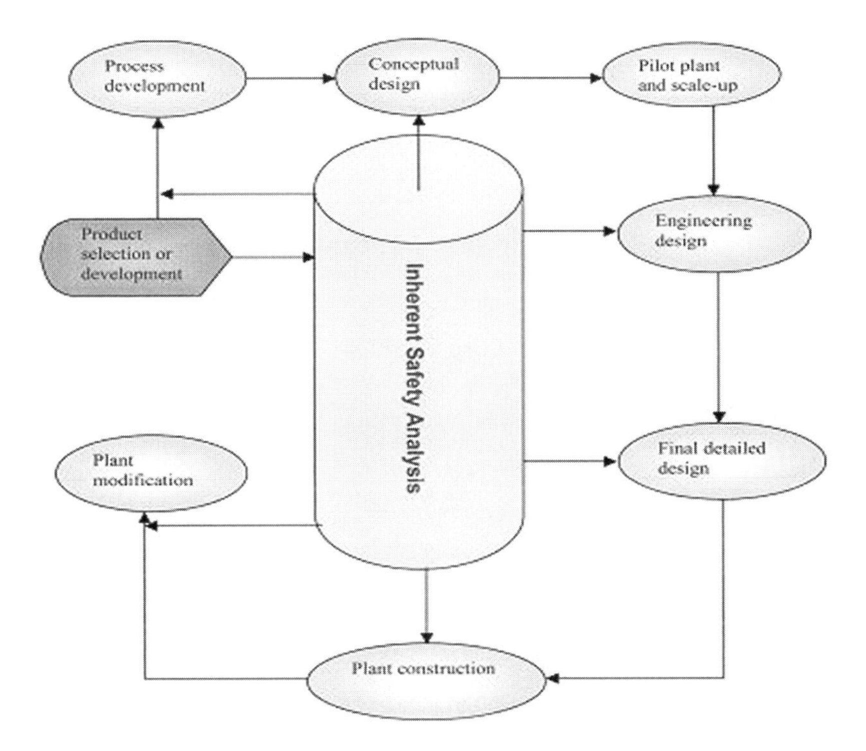

Figure 8.4. Applicability of inherently safer design at various stages of process and plant development.

As depicted in Figure 8.4 above, the types of IST reviews described in this chapter should be conducted at each process life cycle stage, using the most appropriate method.

What follows are examples of the kinds of decisions to which inherently safer design concepts should be applied by the designer or operator of a process at different stages of its life cycle (Gupta and Hendershot in Lees 2005):

> CONCEPTUAL PROCESS RESEARCH AND DEVELOPMENT: selection of basic process technology, raw materials, intermediate products, by-products and waste products, chemical synthesis routes
>
> PROCESS RESEARCH AND DEVELOPMENT: selection of specific unit operations, types of reactors and other processing equipment, operating conditions, recycle streams, product purification, waste treatment processes

PRELIMINARY PLANT DESIGN: location of manufacturing facility, relative location of units on a selected site, size and number of production lines, size of raw material, intermediate, and product storage facilities, selection of specific equipment types for the required unit operations, process control philosophy

DETAILED PLANT DESIGN: size of all equipment, pressure rating and detailed design of all equipment and piping, inventory in processing equipment, location of specific equipment in the plant, length, route, and size of piping, utility design, layout of equipment, detailed control system design, plant-operator interface design

OPERATION: identification of opportunities to modify plant to enhance inherent safety (reduce inventory, upgrade with more modern equipment, identify opportunities for inherently safer operation based on improved process understanding), consideration of inherently safer design when making modifications and changes, "user-friendly" operating instructions and procedures

New chemical processes will eventually replace the processes of today, and when these new facilities are designed, fundamental changes in chemistry and materials can be made more cost effectively. Process design depends on the physical and chemical properties of the materials being handled; changes can have far reaching implications in the equipment design and are best applied early in the engineering process (Overton and King, 2006). IS reviews at the design stage of a new process are most critical for the implementation of inherently safer alternatives.

IS reviews should be conducted early in the development phase of a new process and then reviewed throughout the different project design phases, including the following, where applicable:

- During the chemistry forming (synthesis) phase for product/process research and development to focus on the chemistry and process
- During facilities design scoping and development, prior to completion of the design basis, to focus on equipment and configuration
- During the basic design phase of the project

However, the emphasis on the importance of early consideration of inherently safer design (ISD) alternatives can lead engineers and managers responsible for the operation of existing facilities to conclude that ISD is not relevant to them. They have an existing plant with established technology that is not easily or inexpensively changed. With existing plants, it is often not economically feasible to apply new inherently safer technology. If the proposed change involves different chemicals or significantly different operating conditions, then it is likely that a major capital expenditure will be required to replace existing equipment (Overton and King, 2006).

But ISD is relevant to existing plants and processes. Opportunities often exist to apply ISD principles to an existing plant, making it inherently safer, and many such examples have been reported. Furthermore, tools such as release consequence modeling and quantitative risk analysis (QRA) can be beneficial to understanding the potential benefits of ISD options, helping to gain management support for implementing ISD modifications to existing facilities (Hendershot, et al., 2005).

Additional guidance on conducting IS reviews at the following major process life cycle stages is provided in Section D of the Contra Costa County Industrial Safety Ordinance (ISO) guidance document. [3]

- Chemistry-forming phase
- Facilities Design Scoping and Development
- Basic Project Design
- Existing Process Units

8.6 IS REVIEW METHODS – OVERVIEW

8.6.1 Three Approaches

Inherent safety reviews of new and existing processes form the foundation of an IS management program. IS reviews for new processes (or modifications to existing processes) present the best opportunity to identify possible design features that are inherently safer than those proposed. For existing processes, these opportunities are more limited, due to the cost and feasibility involved in making major modifications. However, whenever a change or even an in-kind replacement is planned for an existing facility, an IS review should be conducted to identify any potential IS alternatives that can be incorporated into the design.

Many of the more qualitative hazard analysis tools can be adapted to incorporate the principles of inherent safety, and three basic approaches to conducting IS reviews have recently emerged in the chemical process industries. These are similar to PHA approaches that have been in use for many years:

- HAZOP
- What-If?
- Checklist

The Checklist approach can be used in combination with both the HAZOP and What-If methodologies. Each of these techniques is described below.

Because of the importance of considering inherent safety early in the design sequence when changes can most readily be made, inherent safety considerations are

---

[3] *Industrial Safety Ordinance Annual Performance Review & Evaluation Report*, Contra Costa Community Health Hazardous Materials Programs, December 4, 2007. See Chapter 10 for a detailed description of this ordinance and its application.

particularly important in conducting hazard reviews (i.e., a preliminary hazard analysis). The above methods can be applied to these reviews.

8.6.2 Formal IS Reviews

MULTIPLE APPROACHES FOR INHERENTLY SAFER REVIEW

Although, in most cases, a separate, formal IS review of new or existing processes is the best way to identify opportunities to improve the level of inherent safety, there are a number of valid ways to consider IS in the design of a process.

The INSIDE Project (INSIDE Project, 1997), which is discussed later in this chapter, has developed the Inherent SHE (Safety, Health, Environment) Evaluation Toolkit (INSET) to identify and evaluate inherently safer design options throughout the life of a process.

An outline for conducting an IS review procedure is provided in Appendix B, including pre-study preparation, training of the team on IS, dividing the process up into sections, and conducting and documenting the review.

FACTORS TO CONSIDER IN SELECTING AN APPROACH

The factors used to select an IS review method are similar to those for selecting a PHA methodology. Each method has its benefits and drawbacks, and selection is based largely on the process life cycle stage that is being reviewed, along with the purpose, scope and objectives of the study. Factors that influence the choice of IS review method include the following:

- Process size
- Process complexity
- Stage in life cycle
- Process operating/incident history
- Phase of process development
- Degree of risk
- Review team's familiarity with techniques
- Opportunities for risk reduction

In general, the larger, more complex, and more hazardous the process (and the more experienced the team with IS), the more detailed the analysis, using perhaps the HAZOP or HAZOP/checklist approach. For smaller, simpler, and less hazardous processes, consider using a checklist, such as the one in Appendix A. Early in the design stage, a high-level HAZOP approach using the matrix shown in Table 8.3 can be used in a Preliminary Hazards Analysis-like approach. If the review is being conducted as part of the PHA for the process, of course that method would be used, with the IS guidewords (Table 8.3) used as deviations for each node or subsystem. In general, combining the use of HAZOP, or What-If? Methodology with a checklist provides for creative brainstorming as well as a detailed means to ensure that most issues have been covered. It should be pointed out that no checklist is perfect and there may be opportunities not identified in the checklist that can only be discovered through a more subjective analysis.

8.6.3 IS Review Methods

The following review methods can be used to ensure that inherent safety is considered and documented for hazardous processes:

- **An independent IS analysis done in addition to a PHA, either in tandem or separately.** This analysis should review the process for ways to eliminate or reduce hazards present in the covered process and may be achieved using an IS checklist (Appendix A) or guideword analysis (Table 8.3).
- **An IS analysis that is incorporated into the existing PHA review process.** In most cases, an initial stand-alone IS analysis should be conducted for the entire process to ensure it receives adequate attention. Again, this may be achieved using a checklist (Attachment A) or guideword (Table 8.3) approach. This type of analysis would review the processes for ways to eliminate or reduce hazards, as well as to reduce risks using the other risk management strategies (passive, active, and procedural).

8.6.4 Research & Development Application

There are significant benefits to applying IS concepts and methodologies within Research and Development (R&D) laboratory and pilot plant operations. IS requirements should be mandated both when a process hazards review is done, and when a project progresses toward full development:

- A process safety review should be required for each new or significantly modified R&D "process" (a semi-works, laboratory experiment, etc.). This review should include hazards identification and evaluation, facility siting, consequence analysis, human factors, and, importantly, inherently safer design. The emphasis should be on the *safety* of the pilot plant operation or experiment being proposed. Thus, the IS review is primarily geared toward the actual performing of the experiment or running of the pilot plant test.
- The maturation of a new or modified "process," from concept to experiment to pilot plant to production facility, can be done using a series of "gate keeping steps." Each gate should be designed to ensure that important issues are resolved *before* progressing to the next step, so as to resolve problems as early as possible. IS issues are no exception and, in fact, are paramount.

The evaluation should address material factors, quantity factors, and process factors. *Material factors* are related to the toxicity, flammability, and reactivity of the materials to be used and created. *Quantity factors* are related to the quantities of chemicals that would be present in the production facility. *Process factors* are related to the process variables, such as pressure, temperature, and ease of operation.

The IS evaluation should include identification of alternative process technologies to reach the same or similar end product. For each alternative, the researcher should be asked to evaluate the material, quantity, and process factors. The end result would be a list of alternatives offering various degrees of IS implementation, and ranked in order of that degree. It is recognized that additional risks of personal injury, environmental harm, lost production, and loss of "right to operate" may be involved. It is also recognized that the most inherently safe processes may not be economically feasible from a business perspective. The evaluation should include the identification of the most significant benefits and barriers to the adoption of the inherently safer process alternatives, and of a less than optimum inherently safe alternative, and the identification of alternative risk management strategies (i.e., passive, active, procedural) that may, in some way, mitigate concerns and reduce risk, albeit perhaps not in strictly inherently safer ways.

8.6.5 PHA - Incorporation into HAZOP or other PHA Techniques

HAZOP

Inherently Safer reviews can be conducted as separate studies, using a HAZOP-like technique, or incorporated into HAZOP studies or revalidations. This involves the application of inherent safety guidewords as "deviations," using, for example, the Guideword Matrix provided in Table 8.3. Recommended inherent safety guidewords and their descriptions are shown in Table 8.4 (Goraya, et al., 2004). As previously mentioned, these guidewords are simply the four most general and widely applicable principles of inherent safety. The description for each is purposely brief and is focused on materials, process routes, equipment and procedures. Use of these guidewords as "mind triggers" for each section of a process, as shown in Table 8.3, or during a particular process safety activity, such as management of change or incident investigation, as subsequently described, will help ensure that the concepts of inherent safety are visible within the process or activity. These guidewords are intended as a supplement to existing tools that may already be in use within a specific process safety protocol.

As with the use of HAZOP methods for conducting regular PHAs, it can be used at any level of subdivision of the process, including node-by-node. While thorough, this can also increase the amount of time required to conduct the study. It also requires a degree of brainstorming by a team that understands IS concepts well and can apply them to a process based on the P&ID and knowledge of the operation. Appendix C provides an example of a combination HAZOP/checklist worksheet that includes review of potential IS opportunities applicable to the particular node and its hazardous materials.

Table 8.3: Guideword Matrix *(Contra Costa County, CA, Industrial Safety Ordinance Guidance Document, Appendix B)*

	Minimization	Substitute/ Eliminate	Moderate	Simplify
Raw Material				
In-process Storage				
Product Inventory				
Process Chemistry				
Process Controls				
Process Piping				
Process Equipment				
Process Conditions				
Maintenance				
Siting				
Transportation				
Misc. Material (i.e., solvents)				

Documentation for incorporating inherent safety review guidewords into a Hazard and Operability Study should be consistent with the documentation used during any HAZOP. As for any other inherently safer system analysis, the documentation should include IS already implemented, IS alternatives considered, the alternatives implemented and not implemented, and the rationale for rejection.

8.6.6 "What-If?" Method

Use of the "What-If?" (W-I) method for IS reviews has the same benefits and pitfalls as its use as a PHA technique. "What-If?" analysis provides an opportunity for a team to brainstorm possible process deviation scenarios in order to help identify safeguards, or in this case, IS improvements that can reduce or eliminate the potential for the scenario to develop in the process, such as through elimination of equipment, rather than by adding additional safeguards.

Table 8.4: Inherent Safety Guidewords
(Amyotte, Goraya, Hendershot and Khan, 2006)

Guideword	Description
Substitute	Replace a substance with a less hazardous material, or a processing route with one that does not involve hazardous material. Replace a procedure with one that eliminates or presents less of a hazard.
Minimize	Use smaller quantities of hazardous materials when the use of such materials cannot be avoided. Perform a hazardous procedure as few times as possible when the procedure is unavoidable.
Moderate	Use hazardous materials in their least hazardous forms or identify processing options that involve less severe processing conditions.
Simplify	Design processes, processing equipment, and procedures to eliminate opportunities for errors, i.e., by removing the possibility that two chemicals that may be incompatible under certain conditions can react unfavorably, such that the chemicals will not be together in the process. Designing equipment that cannot be over pressured by process conditions.

8.6.7 Checklist Method

Checklists can be used either independently or in combination with the W-I (or HAZOP) methodologies. In this manner, the checklist analysis is combined with the W-I (or HAZOP) analysis to yield a method that combines the creative brainstorming features of W-I with the systematic, rigorous features of a checklist. In addition, the inherent safety HAZOP guideword/checklist approach helps to ensure that inherent safety considerations are explicitly considered in identifying both hazards and safety measures (Amyotte, et al., 2006).

Appendix A contains an IST checklist based on CCPS (1996) and additional items from Hendershot (2000). This checklist has been incorporated into guidance for conducting IS reviews under the Contra Costa County, CA Industrial Safety Ordinance (ISO) (2002). The ISO requires IST evaluations for both new and existing covered processes, as well as for certain modifications where a Major Accident risk has been identified. This checklist is similar to one provided in the CCPS book, *Essential Practices for Managing Chemical Reactivity Hazards* (2003).

In addition, a set of excellent checklists for a number of specific types of chemical processing units (vessels, reactors, heat transfer equipment, mass transfer equipment, etc.) can be found in CCPS *Guidelines for Design Solutions to Process Equipment Failures* (CCPS/AIChE, 1998). These checklists include suggestions for inherent, passive, active, and procedural approaches to risk management for a number of incident scenarios. An example for reactors is provided in Figure 8.5.

Operational Deviations	Failure Scenarios	Potential Design Solutions		
		Inherently Safer/Passive	Active	Procedural
Overpressure (Batch, Semi-batch and CSTR Reactors)	Loss of agitation resulting in runaway reaction, or hot bearing/seals causing ignition of flammables in vapor space.	Vessel design accommodating maximum expected pressure Use different type of reactor (plug flow) Alternative agitation methods (e.g., external circulation eliminates shaft seal as a source of ignition in vapor space)	Agitator power consumption or rotation indication interlocked to cutoff feed of reactants or catalyst, or to activate emergency cooling Uninterrupted power supply backup to motor Emergency relief device Pressure or temperature sensors activating bottom discharge valve to drop batch into a dump tank with diluent, poison, or short-stopping agent, or to an emergency containment area Inerting of vapor space Provide nitrogen buffer zone around seal using seal enclosure Automatic agitator trip on low agitation (velocity) sensor, low seal fluid, or low shaft speed	Operators to visually check mechanical seal fluid on regular basis In-vessel agitation (velocity) sensor with alarm Mechanical seal fluid reservoir low level sensor, with alarm Speed or vibration sensor with alarm Manual activation of bottom discharge valve to drop batch into dump tank with diluent, poison or short-stopping agent, or into an emergency containment area Manual activation of inert gas sparging of reactor liquid to effect mixing

Figure 8.5 Example of Potential Design Solutions for Reactor Failure *Scenarios from* CCPS Guidelines for Design Solutions to Process Equipment Failures, *1998.*

Other checklists useful for ISD evaluations can be found in such references, as the CCPS book, *Human Factors Methods for Improving Performance in the Process Industries* (2006).

8.6.8 Consequence-Based Methods

A number of consequence-based methods have been devised to measure the level of inherent safety in a process, based on potential incident consequences. These include the following:

- Rohm and Haas Major Accident Prevention Program
- Dow Fire and Explosion Index
- Dow Chemical Exposure Index
- Mond Index (ICI)
- EPA RMP modeling
- CCPS SVA methodology (screening process)

The *Dow Fire and Explosion Index* (Dow Chemical Company, 1994) and Mond Index (ICI, 1993) are used primarily for determining the hazard level of a process plant. However, their use requires that the process design be fully in place since much of the information required is based on this design. Therefore, they are unsuitable for measuring the level of inherent safety at the conceptual design and preliminary process development stages, where the use of such indices is most useful (Gupta and Hendershot in Lees, 2005).

Shah, Fischer and Hungerbuhler (2003) outline a hierarchical approach in which a chemical process is divided into different layers (substance, reactivity, equipment and safety technology), each representing a different level of analysis. Using an automated tool, these layers are successively, and also possibly repeatedly, assessed for non-idealities of the process with regard to inherent safety. In the case of non-idealities, a worst-case scenario is determined and analyzed with the help of available data, and recommendations are made for possible preventive and protective measures. The assessment of different process aspects—substances, reactivity of substances, operating conditions and worst-case scenarios in equipment units—and the selection of safety technologies for a given chemical process within one single framework provides a clear qualitative and quantitative picture to design inherently safer chemical processes and to prevent accidents.

8.6.9 Other Methods

Traditionally, consequence modeling has been used to estimate the potential impact of potential accidents based on worst-case scenarios (EPA RMP) or other criteria. A number of programs, both proprietary and public, have been developed for this purpose (PHAST, ALOHA, DEGADIS, CAMEO, RMP*Comp, etc.). These methods are the most direct way of measuring the level of inherent safety, as they do not consider likelihood as part of the risk equation. Using these programs, plant personnel can estimate the impact of reduced quantity and changes in some process conditions, such as pressure, on potential consequences.

The concepts of inherent safety can also be applied to health, to eliminate or reduce risk of personnel exposure, and environmental goals, to eliminate or reduce risk associated with routine emissions or generation of waste. Measurement tools which can be quickly and easily used early in the life cycle of the process design are particularly important, because that is the time at which the designer has the greatest opportunity to change the basic process technology. In recent years, a number of tools have been developed to measure combined inherent SHE characteristics of processes, and the applicability of several existing tools, such as the Dow Indices, to the understanding of inherent SHE has been recognized (Hendershot, 1997).

One major chemical company decided that special emphasis should be placed on reducing the risk of low probability, high consequence events, and developed the Major Accident Prevention Program (MAPP). This program systematically examines the handling and processing systems for hazardous chemicals that could present toxic or explosive vapor cloud threats to employees or neighboring communities.

Traditionally, chemical systems with the potential for high consequence events have been studied using hazard analysis techniques that concentrate on reducing the frequency of such events. In contrast, MAPP places the emphasis on reducing the consequences. An advantage of this approach is that damage reduction techniques are passive in nature and lead to inherently safer systems. Risk reductions that reduce the frequency of an event tend to be active systems that require devices that must be operated and maintained to reduce risk. MAPP follows an established protocol to identify hazards, understand potential consequences, evaluate means to reduce the potential consequences and manage risks appropriately (Rohm and Haas, 2002).

Analyzing the potential consequences of an accident is a useful way to understand the relative inherent safety of process alternatives. These consequences might consider, for example, the distance to a benchmark level of damage resulting from a fire, explosion, or toxic material release. The Major Accident Prevention Program (MAPP) (Renshaw, 1990) encourages inherently safer process development by requiring accident consequence analysis of a standard list of potential chemical process accidents for a specified list of high hazard chemicals. Accident consequence analysis is of particular value in understanding the benefits of minimization, moderation, and limitation of effects. Table 8.5 illustrates the use of consequence analysis as a metric for understanding the benefits of a number of inherently safer process options (Hendershot, 1997).

Table 8.5: Examples of Potential Accident Consequence Analysis as a Measure of Inherent Safety

Inherent Safety Strategy	Description	Consequence
Minimize	Chlorine transfer line rupture — distance to 20 ppm atmospheric concentration, D atmospheric stability, 3.4 mph wind speed	2-inch line - 3.4 miles 1-inch line - 1.2 miles
	Explosion overpressure 100 feet from a reactor	3000 gallon batch reactor - 1.1 psig 50 gallon continuous reactor - 0.3 psig
Substitute	Concentration in the atmosphere 500 feet downwind from a large spill from a storage tank, D atmospheric stability, 3.4 mph wind speed	Methanol - 1000 ppm Butanol - 130 ppm
Moderate	Distance to 500 ppm atmospheric concentration for a large spill of monomethylamine, D atmospheric stability, 3.4 mph wind speed	Stored at 10°C — 1.2 miles Stored at 3°C — 0.7 miles Stored at -6°C — 0.4 miles
	Concentration in the atmosphere 500 feet downwind from a large spill from a storage tank containing methyl acrylate, D atmospheric stability, 3.4 mph wind speed	2500 sq. ft. concrete containment dike — 1100 ppm 100 sq. ft. concrete containment pit — 830 ppm

The INSIDE (INherent SHE In DEsign) Project is a European government/industry project sponsored by the Commission of the European Community, to encourage and promote inherently safer chemical processes and plants (INSIDE Project, 1997). The INSIDE Project expands the inherent safety concept to include inherent approaches to Safety, Health, and Environmental (SHE) aspects of chemical processes. The project has developed a set of tools, the INSET Toolkit, to identify and evaluate inherently safer design options throughout the life of a process. Table 8.6 summarizes these tools, and they are described in more detail by the INSIDE Project (1997).

The tools of particular interest with regard to metrics for measuring the inherent safety of chemical processes are the ISHE Performance Indices listed in Table 8.6 (Tools I.1 through I.11). The ISHE Performance Indices are relatively simple, intended for hand or simple calculator computation, so that a large number of process options can be rapidly evaluated. The various inherent safety, health, and environmental aspects of a process are evaluated using separate indices, and no attempt is made to combine the indices into a single overall measure.

Table 8.6: Summary of INSET Tools (INSIDE Project, 1997)

Tool	Description
A	Detailed Constraints and Objectives Analysis
B	Process Option Generation
C	Preliminary Chemistry Route Options Record
D	Preliminary Chemistry Route Rapid ISHE Evaluation Method
E	Preliminary Chemistry Route Detailed ISHE Evaluation Method
F	Chemistry Route Block Diagram Record
G	Chemical Hazards Classification Method
H	Record of Foreseeable Hazards
I	ISHE Performance Indices
I.1	Fire and explosion hazard index
I.2	Acute toxic hazard index
I.3	Inherent health hazard index
I.4	Acute environmental hazard index
I.5	Transport hazard index
I.6	Gaseous/atmospheric emissions environmental index
I.7	Aqueous emissions environmental index
I.8	Solid emissions environmental index
I.9	Energy consumption index
I.10	Reaction hazard index
I.11	Process complexity index
J	Multi-Attribute ISHE Comparative Evaluation
K	Rapid ISHE Screening Method
L	Chemical Reaction Reactivity - Stability Evaluation
M	Process SHE Analysis - Process Hazards Analysis, Ranking Method
N	Equipment Inventory Functional Analysis Method
O	Equipment Simplification Guide
P	Hazards Range Assessment for Gaseous Releases
Q	Siting and Plant Layout Assessment
R	Designing for Operation

The INSET Toolkit instead recommends a multiattribute decision analysis technique to evaluate the overall inherent SHE aspects of the various process options. The INSET Toolkit is particularly interesting as an Inherent SHE measurement tool because it represents the consensus combined expertise of a number of companies and organizations, and because it is intended to consider safety, health, and environmental factors in one set of tools (Hendershot, 1997).

Gupta and Edwards (2003) also propose an ISD measurement procedure for use in differentiating between two or more processes for the same end product. Palaniappan, et al. (Palaniappan, Srinivasan and Tan, 2002; Palaniappan, 2001) have developed an i-Safe index, which includes five supplementary indices: Hazardous Chemical Index (HCI), Hazardous Reaction Index (HRI), Total Chemical Index (TCI), Worst Chemical Index (WCI) and Worst Reaction Index (WRI). Lastly, CCPS (1995) presents a large range of decision aids for risk analysis and decision making in industry, including cost-benefit analysis, voting methods, weighted scoring methods and decision analysis, so that critical risk decisions can be made in a more consistent, logical, and rigorous manner.

8.7 INCORPORATING IS CONSIDERATIONS INTO EXISTING PSM ACTIVITIES

In addition to the PHA program, IS principles should be incorporated into other elements of the process safety management program as appropriate. As a general philosophy, there are many opportunities to implement IS in the way that process safety programs are designed, including simplification of procedures, permits, and checklists (SOPs, MOC, MI, HWP, PSSR, etc.) to ensure that complexity is minimized and implementation is as straightforward as possible. The use of hazardous work permits, for example, are administrative controls that serve to reduce the potential for human error through omission, total dependence on training, and the need to recall the precepts of the written procedure. Requiring that someone independent of the operation confirm and sign off on a statement that the permit's provisions have been satisfied also reduces the potential for error and associated risk. These approaches fall under the definition of Human Factors, which involves the design of systems (both technical and social) that minimize the potential for human error (see Chapter 6). Although incorporating IS concepts into other process safety elements does not generally reduce or eliminate the hazard, it reduces the likelihood of an incident occurring, or improves the reliability of an independent protection level (see discussion in Chapter 2).

Besides PHA, the other process safety activities that most lend themselves to enhancing the level of inherently safer design are incident investigation and management of change. Consideration of IS during incident investigations requires that investigators ask whether future incidents could be avoided through the four basic concepts of minimization, substitution, moderation, and simplification. This requires that investigation teams have a good understanding of the principles of IS design and the charter from management to employ them during incident investigations. For example, questioning the need for an expansion joint that failed causing a release should be raised during the incident investigation, since hard piping is inherently safer than an expansion joint, since it is less likely to fail. This will eliminate the hazard, as well as reduce the likelihood of failure. If the expansion joint is required, for instance, due to a need to isolate equipment vibration, additional considerations could be given to more robust designs or elimination of the vibration issue itself. More frequent inspections or replacement (administrative or procedural controls) would be the least attractive option from a reliability standpoint.

When modifications and expansions are first considered by an organization, many of the same opportunities to make the plant inherently safer are present as at the initial concept and development stages. New knowledge or technology may now be available that will make it possible to operate the facility with fewer hazards, lower inventories, or less severe operating conditions (i.e., lower temperatures or pressures). An appropriate level of inherent safety review should be built into the facility's management of change system, to prompt those responsible for proposing or reviewing changes to consider inherently safer alternatives. An Inherently Safer Process Checklist, such as the one in Appendix A, may be helpful in this regard (CCPS, 2003).

A generic inherent safety-based protocol for management of change (MOC) is shown in Figure 8.6. This sequence of steps is based on the MOC process presented by Kelly (2000). By recommending the use of inherent safety guidewords in the first step, "identify need for change," the protocol, as previously stated, recognizes that inherent safety is both a driving force for MOC and an opportunity during MOC. Use of the guidewords and a checklist during the hazard review and control steps is again a recognition that the techniques used for these purposes are easily adaptable to explicit incorporation of inherent safety (such as the "What-If?"/Checklist approach described in the next chapter).

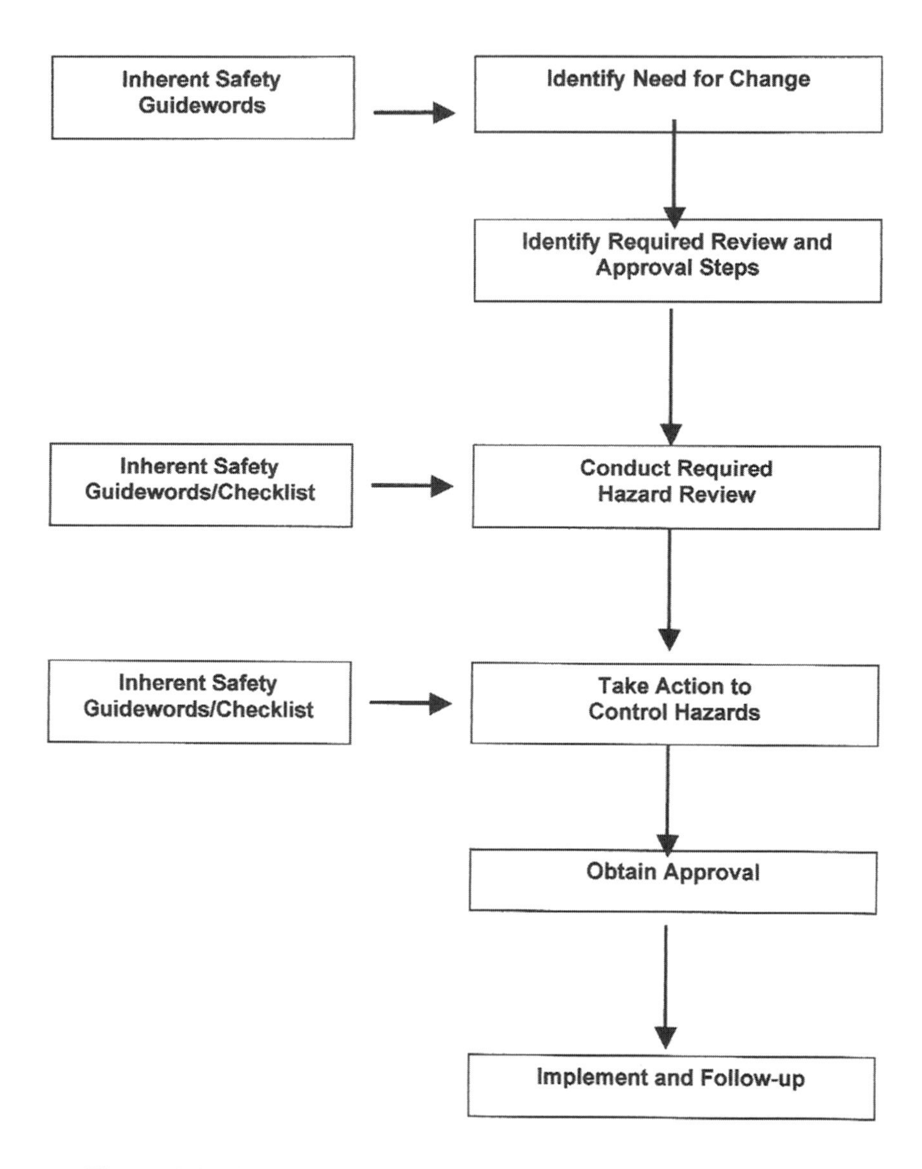

Figure 8.6: An Inherent Safety-Based Management of Change Protocol *(Amyotte, et al., 2006)*

Particular attention should be paid towards consideration of IS alternatives as part of planned "like-for-like" (in-kind) replacements. These replacements require that the project review process include this element, since MOC programs generally do not cover replacements in-kind. The Management of Change program must also protect against compromise of inherently safer features (Bollinger et al., 1996) or the introduction of new hazards. Documentation of IS features in a process is particularly important to help ensure that their purpose is clearly understood and communicated in order to minimize the potential for compromise.

Incorporation of IS into training programs requires that programs follow good training practices, including establishment of valid learning objectives and verification that trainees have met those objectives. Although Human Factors primarily involves the design of facilities, equipment, and the work environment to meet the capabilities of the worker, effective training is also a vital aspect in the objective of minimizing the potential for human error.

Emergency response is made inherently safer by minimizing process hazards and by implementing means to minimize the hazard to which responders may be exposed during emergencies. This is accomplished by providing secondary containment and active mitigation means to address an emergency from a safe distance, without requiring that responders rely on PPE to enter the hot zone. For example, remotely actuated/failsafe isolation valves at the outlet of storage tanks allows isolation of downstream piping and equipment without exposing response personnel to the hazard involved in closing a manual valve. Fixed deluge systems with automatic detection, while considered an active safeguard, also lowers the hazard exposure to responders.

As discussed earlier in this chapter, consideration of facility siting, including but not limited to, spacing and layout, is directly linked with IS design, as it relates to the separation of hazards and people, or implementation of protective barriers between the two. For example, relocating a control building or modifying it to withstand the maximum potential overpressure from a vapor-cloud explosion improves the inherent safety of this control building.

8.8 HUMAN FACTORS AND INHERENT SAFETY REVIEWS

As pointed out in Chapter 6, human factors play an important role in the design of inherently safer facilities. Although attention to human factors does not eliminate or reduce hazards, processes that are designed with these concerns in mind are inherently safer than those that are not, due to the reduced potential for, and impact of, human error that can lead to a hazardous situation. Ergonomics, although generally considered an occupational health issue, such as designing to avoid carpal tunnel syndrome or other musculoskeletal disorders due to vibration or repetitive motion in an awkward position, does relate to human factors in that the easier a task is to do physically, the less likely it is that it will be skipped or done incorrectly. A process in which manual valves are easily accessible and operable reduces risk of a valve not being operated when needed, including in an emergency situation. Human factors are of particular interest during the detailed design stage for both new facilities and existing facilities.

Incorporating Human Factors issues into IS review checklists can be helpful for assessing this element. Alternatively, a separate Human Factors checklist can be used to conduct an independent HF review. As noted above, in step 13 of the inherent safety review at the design development stage, the team reviews the process to develop a list of potential ergonomics/human factors issues. This list provides input to the design team so that these issues can be addressed as the design progresses.

Involving operations and maintenance personnel knowledgeable in human factors and ergonomics is important to help identify optimum solutions to these challenges. Applying inherent safety principles can reduce risks associated with:

- Materials handling of bags, drums, solids, etc.
- Tankcar, tankwagon, and hopper car loading, unloading, and cleaning
- Manual operations such as filter cleaning, sampling, batch operations
- Hose handling at manifolds, docks, etc.
- Safe operating procedures
- Equipment isolation for maintenance or emergencies

Good ergonomic design can reduce the risk of both process safety incidents and personnel injury. For a more complete discussion of human factors issues as they relate to inherent safety, see Chapter 6, or the book, *Human Factors Methods for Improving Performance In the Process Industries* (AIChE/CCPS, 2006). The latter provides a thorough discussion of the topic, including a detailed human factors checklist.

8.9 REACTIVE CHEMICALS SCREENING

Experience has shown that reactive chemistry hazards are sometimes undetected during bench scale and pilot-plant development of new products and processes. These chemistry hazards must be specifically identified so they can be addressed in the inherent safety review process. Chemists should be encouraged and trained to explore reactive chemistry of abnormal operations. Simple reactive chemicals screening tools, such as the interactions matrix described in Chapter 5.2, can be used by R&D chemists.

The CCPS book, *Essential Practices for Managing Chemical Reactivity Hazards* (2003), outlines a process for identifying and managing reactive chemistry hazards throughout the life cycle of a facility. Adequately managing chemical reactivity hazards requires routine knowledge of:

- the potential for uncontrolled reaction(s) to take place within the facility.
- how such reactions might be initiated (i.e., heat, contamination, inadvertent mixing, impact, friction, electrical short, lightning).
- how to recognize when an uncontrolled reaction is taking place.
- what the consequences would be if such a reaction took place (i.e., toxic gas release, fire, explosion).

Once the materials to be used in the process have been identified, they must be screened for potential reactive hazards. CCPS (2003) provides a detailed methodology for reactive chemicals screening, both for intentional and unintentional reactivity issues, as well as guidance on how to manage those hazards. From an IS perspective, opportunities to eliminate or minimize, rather than manage the hazards, are preferred. This includes avoiding inadvertent reagents that can create a hazard through unintentional mixing or contact. For example, not using water in any process involving water-reactive materials is inherently safer than trying to prevent contact between the two. Many of the questions in the IS review checklist provided in Appendix A can be adapted for consideration of reactivity hazards, such as:

- Can reactive raw materials inventory be reduced?
- Is it possible to completely eliminate reactive raw materials, process intermediates, or by-products by using an alternative process or chemistry?
- Is it possible to eliminate reactive components (i.e., water, iron, air, etc.) from the process to prevent inadvertent contact?
- Is it possible to design operating conditions such that materials that become unstable at elevated temperatures or freeze at low temperatures will not operate in those ranges?
- Is the equipment designed to totally contain the materials that might be present inside at the maximum attainable process temperature?
- Can equipment be designed such that it is difficult or impossible to create a potential hazardous situation due to an operating or maintenance error?

CCPS (2003) also offers the following as a typical agenda for an inherent safety review that includes consideration of reactive hazards, either at the concept or development stage of a new facility.

- Review what is known of the chemical reactivity and other hazards that will need to be contained and controlled in the proposed process. This existing level of knowledge might come from past experience, suppliers, literature reviews, incident reports, etc.
- Based on the level of knowledge of chemical reactivity hazards, determine if additional screening will be needed for these hazards. Having reactive functional groups might indicate the need to perform literature searches, access databases or run differential scanning calorimetry.
- Discuss possible process alternatives and their relative hazards, including such topics as alternative solvents and possible incompatibilities to avoid.
- Brainstorm and discuss possible ways to reduce the hazards, such as a checklist that can be used as an aid to the brainstorming process.
- Obtain consensus on significant unknowns that will need to be addressed.

- Document the review, including attendees, scope, approach, and decisions.
- Assign follow-up items with responsibilities, goal completion dates, and a closure mechanism, such as a specific time to reconvene.

An inherently safer process checklist is provided in Appendix A.

8.10 INHERENT SAFETY REVIEW TRAINING

IS training should include, at minimum, the following topics:

- The hierarchy of risk management strategies, from inherent to passive to active to procedural levels
- Various approaches to applying inherently safer systems, including minimization, substitution, moderation, and simplification
- Company design standards, examples, and other information that provides guidance on evaluation and implementation of IS design features
- Potential IS conflicts

A sample IS training program outline is provided below (Hendershot, 2003).

- History of Inherently Safer Design
- Basic Concepts
- Chemical Process Safety Strategies
- Inherently Safer Design Strategies- Minimize, Substitute, Moderate, Simplify
- Inherent safety and reliability
- Robust processes
- Inherent safety in other industries
- Inherent safety conflicts and measuring inherent safety
- When to consider inherently safer design
- Inherently safer design tools

Training in IS should also include company requirements for evaluating and implementing inherently safer technology during the life cycle of a process, as well as how this technology should be applied to the various process safety management program elements. As with most elements of safety, including process safety, employee participation is very important, as employees at all levels will be asked to help identify, evaluate, and implement changes to operations that may improve the inherent safety of the process. All relevant functions must have a good understanding of IS and its possible conflicts, so that good judgments can be made on the issue of risk reduction vs. cost, and the potential impact of changes on other important operating parameters, including quality, yields, operating costs, and waste generation, etc. It is also important to emphasize that the proper application of other risk management strategies, like active and procedural controls, can often satisfactorily manage risk at an acceptable level.

Training requirements for inherent safety should be merged into existing management systems. In general, inherent safety training objectives should be established for all personnel, including engineers, scientists, and business leaders, as well as for operations and maintenance personnel. To be truly effective, training should be provided at two levels: "awareness" level training to provide a basic understanding of IS principles and application, and "technical," or specific, training for those whose job function calls for the evaluation and implementation of IS. Awareness training should include the basic concepts of inherent safety, along with how these concepts are to be implemented in the organization's process safety management, new product development, and project execution programs. More specific training should be considered for those people who play key roles in inherently safer processes such as:

- R&D chemists involved in product and process development.
- Process development and process design engineers involved in process scoping, development, and design activities.
- Functional specialists, such as safety engineers and industrial hygienists.
- Method leaders.

If an organization chooses to include inherent safety reviews in their implementation approach, training should be included on how to conduct such a review.

The Institute of Chemical Engineers (UK) Inherently Safer Process Design (IChemE and IPSG, 2005) is an excellent ISD program available on CD-ROM. Updated by the International Process Safety Group, the program provides the training support needed to ensure that the principles of inherent safety are understood and applied to both new plant designs and proposed changes to existing plants, with an emphasis on past experiences and lessons learned. It contains a CD-ROM with trainer's notes, 150 illustrated slides, video and a variety of case studies on best practices. New materials include a checklist section, more information about inherent safety assessment and design tools, and an extended bibliography of more than 400 inherent safety references. The program includes a comprehensive review, management issues, process intensification, sustainability, waste minimization and inherently safer processing, barriers to inherent safety and its relationship to cost and legislation, inherent safety and plant reliability, human factors, chemical reactivity hazards, and case studies.

Companies may wish to develop their own workshops to train potential team members in the inherent safety review process. The workshop can provide background information on inherent safety concepts, the extensive systems required to manage hazardous materials, and the elements of the inherent safety review process. Videos, problems, examples, and team exercises should be included to maximize student interest and learning.

The training management system should also define requirements for initial and refresher training. Definition of these requirements will assist in maintaining the IS culture and skill level by ensuring appropriate training for new employees, transferred and relocated employees and for experienced personnel.

8.11 DOCUMENTATION OF THE INHERENTLY SAFER DESIGN FEATURES OF A PROCESS

An important factor in implementing inherently safer design is the issue of company memory with respect to the management of process safety information. The intent of this component is to ensure that knowledge and information gained from plant experience that is likely to be important for the future safety of a facility is well-documented so that it is not forgotten or overlooked as personnel and organizational changes occur (CSChE, 2002). Hendershot (2005) argues that this is especially critical when dealing with inherent safety and inherently safer design (ISD) features. He gives several examples where ISD features were essentially put at risk because the reasons they were implemented were not clearly and adequately documented. Potentially, this could compromise facility safety when future modifications are made by people who do not understand the original designer's intent, or what is involved in a particular safety feature. Examples include why a certain size feed line for limiting maximum reagent flow is part of the safety design basis, or how the routing of a pipe is intended to minimize consequences of a spill.

ISD features are particularly susceptible to lapses in corporate memory given that, unlike an add-on device such as a high-pressure alarm, they are such a fundamental part of the design that their purpose may not be obvious (Hendershot, 2005; Amyotte et al., 2006). All these inherently safer design features must be documented in original design manuals, and appropriate process safety information, including P&IDs and SOPs, must be readily available. This issue is also pertinent to Management of Change in that a proposed modification should be reviewed carefully to ensure no ISD features are being compromised.

8.12 IS REVIEW DOCUMENTATION

Whichever type of IS analysis is conducted, a report of the review should be generated to document the study. This report should include, at minimum, the following information:

- A summary of the approach used for the IS review (i.e., methodology, checklist used, etc.).
- Names and qualifications of the team facilitator/leader and team makeup, including positions, names, and any relevant experience or training.
- IS alternatives considered, as well as those already implemented or included in the design.
- If an independent inherently safer systems analysis was conducted, documentation should include the method used for the analysis, what inherently safer systems were considered, and the results of each consideration. If an IS checklist was used, document the reasons why items were not considered, for example, if they were not applicable or had been considered previously.
- Documentation of rationale for rejecting potential IS opportunities (cost, creation of other safety or operability problem, etc.).
- Recommendations/action plans for further evaluation or implementation of IS alternatives identified during the study.

If the IS review was conducted as part of a larger study (i.e., PHA or hazard review), this information should be incorporated into the report of this activity. As with PHA reports completed under the OSHA PSM/EPA RMP standards, this information should become part of the permanent process safety file, and be maintained for the life of the process. Electronic versions in an editable format (i.e., MS Word) should be maintained to facilitate future updates and revalidations.

The rationale for why recommendations from IS reviews were rejected should follow the following guidance, which includes OSHA criteria for declining recommendations from incident investigations and process hazards analyses (CCC ISO):

- The analysis upon which the recommendations are based contains factual errors.
- The recommendation is not necessary. For example, the safeguards may be inadequate, but the consequences are operational, or the consequence or severity of the scenario would not result in a significant release.
- Another IS alternative would provide a sufficient level of hazard reduction. (NOTE: Implementing only one option to address identified hazards may not be adequate to address the greatest hazard reduction or elimination. However, it is not necessary to implement more than one IS alternative if the implementation of a second IS alternative does not add any significant hazard reduction, or has been documented as not feasible.)
- The recommendation is not feasible due to one or more of the reasons listed below:

    - The recommendation is in conflict with existing federal, state, or local laws.
    - The recommendation is in conflict with Recognized and Generally Accepted Good Engineering Practices (RAGAGEP).

- The recommendation is economically impractical, such that the process unit would stop being fiscally feasible. This can include the following factors:
    - Capital investment
    - Product quality
    - Total direct manufacturing costs
    - Operability of the plant
    - Demolition and future clean-up and disposal cost
- The recommendation would have a strong enough negative social impact that the project should not be implemented. Some examples could include an unacceptable visual or noise impact on the community, or increased traffic congestion.
- The recommendation may violate a license agreement that cannot be modified, and so must remain in effect.
- The recommendation may decrease the hazard, but would increase the overall risk.
- An alternative measure would provide more risk reduction than the IS approach rejected.

If the IS alternative recommended is rejected because it will create more risk, or if other non-IS modifications are made that would reduce the overall risk more than if the IS option were implemented, the rationale of how this determination was made should be documented. A qualitative or quantitative risk assessment could be used as a basis for the risk analysis.

8.13 TIME REQUIRED FOR AN INHERENT SAFETY REVIEW

The time required for IS process reviews depends on the type of review (i.e., new facility, existing facility, modification) as well as the level of complexity. Relatively simple or low-hazard process reviews can be completed in several hours. Reviews of complex projects may require one to two days. One key factor is the extent of evaluation of potentially inherently safer options to be done during the review session(s) vs. that done offline. As with PHAs and other types of hazard reviews, the team should be used to help identify the opportunity or alternative, and then assign it to an individual or subteam. This group will evaluate the alternative off-line, and then report back on the results of their evaluation, which can often involve detailed cost estimates and risk comparisons in order to determine feasibility. IS review teams should not engineer solutions as part of the review. A recommendation to implement a particular IS design feature often involves a judgment on cost vs. degree of risk reduction, as well as resolution of other potential conflicts that may arise during any evaluation of potentially inherently safer options. (See Chapter 9 for a fuller discussion of these potential conflicts.)

The key to an effective and efficient review is good preparation, including updated process drawings and the development of IS review protocols that help the review leader to ask the team pertinent questions, and help ensure that the fullest possible range of IS alternatives is considered. Depending on the outcome of the team review, recommendations for further consideration of alternative designs may need to be resolved, with perhaps additional reviews by all or part of the team.

8.14 SUMMARY

Inherent safety is an exciting field that has caught the attention of researchers, plant designers, management and regulators worldwide. A number of methods for measuring inherently safety have been developed. However, a straightforward, generally accepted approach is needed to compare different process designs for a particular end product.

The best way to invent and build inherently safer processes is to make it a part of the thinking of all engineers and chemists as they go about their work, primarily through education. Once they recognize the importance and benefits of hazard elimination, their creativity in inventing new ways to eliminate or reduce hazards as they go about their work activities will go further toward enhancing inherent safety than any particular review methodology could. Once company personnel start applying the principles of ISD to new and existing processes, the obvious benefits would become apparent, prompting further use. Once inherently safer design is successfully applied to process industries, it can be adapted by other accident-prone industries such as mining, construction, transportation, and others. (Gupta and Hendershot, in Lees, 2005).

9.

Inherently Safer Design Conflicts

9.1 INTRODUCTION

In many cases, the inherent safety advantages of one process are clear when compared with the alternatives. One or more hazards may be significantly reduced, while others are unaffected or only marginally increased. Aqueous latex paints, for example, are inherently safer than solvent-based paints, although there are applications where the increased performance of solvent-based paints justifies their use with the appropriate layers of protection.

Unfortunately, many times it is not clear which of several process alternatives is inherently safer. Because nearly all chemical processes have a number of hazards associated with them, an alternative which reduces one hazard may increase a different hazard. Consider the following example that requires a choice between several possible processes. Process A uses flammable materials of low toxicity; process B uses non-combustible, but volatile and moderately toxic materials; while process C uses non-combustible and non-toxic materials, but operates at high pressure. Which process is inherently safer? The answer to this question will depend on the specific details of the process options. In addressing these questions, it is essential that all significant hazards be identified and understood, including:

- Acute toxicity
- Chronic toxicity
- Flammability
- Reactivity
- Instability
- Extreme conditions (temperature or pressure)

- Environmental hazards, including:
    - Air pollution
    - Water pollution
    - Groundwater contamination
    - Waste disposal

The hazards associated with normal plant operations, such as stack and fugitive emissions, as well as those resulting from specific incidents, such as spills, leaks, fires and explosions, must also be considered. Lastly, a designer needs to also consider business and economic factors in making a process selection. These include:

- Capital investment
- Product quality
- Total manufacturing costs
- Operability of the plant
- Demolition and future clean-up and disposal cost

Design strategies that result in an inherently safer design may also tend to improve process economics. For example, minimizing the size of equipment or simplifying a process by eliminating equipment may reduce capital investment and reduce operating costs. However, overall process economics are very complex and impacted by many factors, so it may not always be true that an inherently safer process is also economically more attractive.

An inherently safer process offers greater safety potential, often at a lower cost. However, selection of an inherently safer technology does not guarantee that the actual *implementation* of that technology will result in a safer operation than an alternate process in actual operation. A classic example is alternative modes of transportation. Table 9.1 compares some inherent safety characteristics of airplane and automobile travel. It is apparent that automobile travel, *per se*, is inherently safer, since a car carries fewer people, travels on the ground at a lower speed, contains a smaller quantity of flammable fuel, and operates in an external environment usually capable of supporting life. However, for all trips for which air travel is a practical alternative, it is, in fact safer (Evans, et al., 1990; Sivak and Flannagan, 2003). This is because the hazards inherent in air travel are very effectively managed by passive, active, and procedural risk management systems, including, extensive redundancy in the construction of the aircraft, rigorous training and qualification of all personnel, sophisticated management of airplane movement through the air traffic control system, rigorous maintenance of all equipment, and many other systems. These systems all cost a lot of money to organize, maintain, and operate, but it is worth doing this because of the benefits of air transport—the ability to move people and goods long distances in a short time. Society may make similar choices in other technologies, including chemical manufacturing (Hendershot, 2006b).

Another example is a significant risk in the home: falling down stairs. Recognizing this, Kletz has suggested that, *with respect to the hazard of falling down the stairs,* living in a one-floor ranch house with no stairs is inherently safer. However, McQuaid (1991) points out that only "falling down the stairs" accidents are prevented by living in a one-floor house. He gives an example of a situation in which a two-story house might be considered inherently safer. In the 1970s, in a town near Antwerp, people woke up one morning to find all of their domestic animals were dead. A dense, ground hugging cloud of chlorine had been released from a nearby chemical plant during the night. The animals on the ground floor were killed, while the people sleeping in second story bedrooms were unharmed.

Table 9.1: Comparison of ISD attributes of air travel and automobile travel *(Hendershot, 2006b)*

ISD Characteristic	Air Travel	Automobile Travel
"Inventory" of passengers	Up to 800 for Airbus A380	A few people
Elevation above ground	Typically several kilometers	On the ground
Control requirement	Three dimensions (forward-backward; right-left; up-down)	Two dimensions (forward-backward; right-left)
Speed	~ 900 km/h	~ 100 km/h
Flammable fuel inventory	Thousands of liters	Tens of liters
Passenger compartment	Pressure vessel	May be a convertible (i.e., the outside environment is usually capable of supporting life)

One can envision other hazards against which a two-story house might be considered inherently safer. For example, in an area prone to flooding, a second floor provides an escape from rising waters. A drive through many beach resort communities shows that people have adopted this design philosophy, building houses on stilts to avoid damage from high water. An even more inherently safe solution would be to not build in flood prone areas at all, but to build houses on high ground (Hendershot, 1995).

The traditional strategy of providing layers of protection for a less inherently safe process can be quite effective, although the expenditure of resources to install and maintain the layers may be very large. In some cases, the benefits of the more hazardous technology will be sufficient to justify the costs required to reduce risk to a tolerable level. However, even when we determine that the benefits of an inherently less safe technology justifies its use, we should always continue to look for inherently safer alternatives. Technology continues to evolve and advance, and inherently safer alternatives which are not economically attractive today may be very attractive in the future. The development of new inherently safer technology offers the promise of meeting process safety goals more reliably and more economically.

Inherent safety is assessed relative to a particular hazard, or perhaps a group of hazards, but essentially never relative to all hazards. A chemical process is a complex, interconnected organism in which a change in one area of the system can impact the rest of the system, with effects cascading throughout the process. These interactions must be understood and evaluated. Similarly, the chemical industry can be viewed as an ecosystem with complex interactions, interconnections, and dependencies. Understanding these relationships is necessary in order to reach a well-balanced resolution when technological options conflict.

Inherent safety is assessed relative to a particular hazard, or perhaps a group of hazards, but essentially never relative to all hazards. A chemical process is a complex, interconnected organism in which a change in one area of the system can impact the rest of the system, with effects cascading throughout the process. These interactions must be understood and evaluated. Similarly, the chemical industry can be viewed as an ecosystem with complex interactions, interconnections, and dependencies. Understanding these relationships is

9.2 EXAMPLES OF INHERENT SAFETY CONFLICTS

9.2.1 Continuous vs. batch reactor

The use of continuous, rather than batch, reactors is a strategy that is often proposed for improving the inherent safety of a chemical process (CCPS/AIChE, 1993). This modification generally succeeds because a continuous reactor is usually much smaller, reducing the material and energy inventory of the process, increasing heat transfer per unit of reaction mass, and improving mixing. However, batch reactors may also have safety advantages, and, under the right circumstances, may be judged to be inherently safer.

Consider a simple reaction:

$$A + B \xrightarrow[\textit{Solvent S}]{\textit{Catalyst C}} D + E + HEAT$$

The reaction is exothermic and proceeds virtually instantaneously to complete conversion in the presence of Catalyst C. The process hazard of concern is that the reaction mass becomes extremely unstable if Reactant B is overcharged, or Catalyst C is left out. The resulting buildup of unreacted Reactant B may result in a potentially explosive reaction if Reactant B exceeds a known critical concentration. Two processes are proposed for this reaction, a batch process (Figure 9.1) and a continuous process (Figure 9.2).

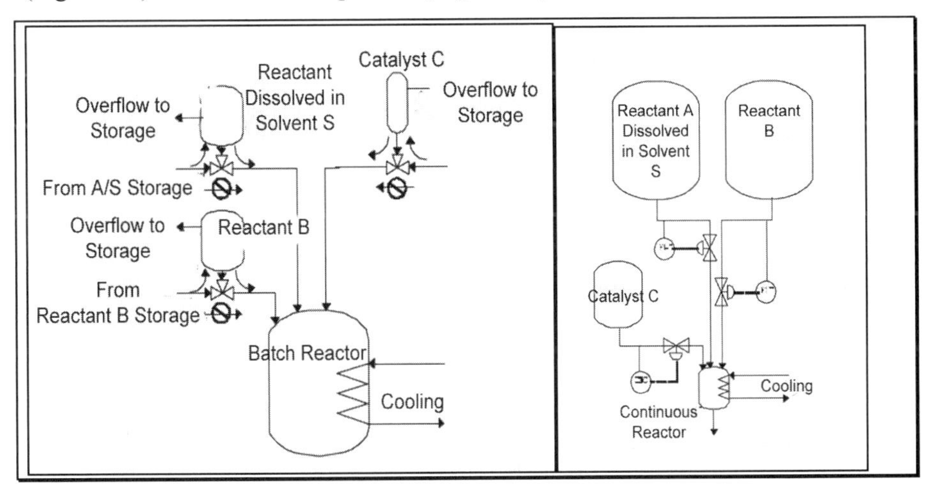

Figures 9.1 and 9.2: Examples of batch (left) and continuous (right) process approaches to the reaction described above.

In the batch process, the correct amount of Reactant A dissolved in Solvent S is fed to the reactor from a weigh tank sized to hold exactly the correct charge for one batch. The apparatus features three-way valves which do not allow flow directly from the storage tank to the reactor, and will return any excess material fed to each charge tank to a tank overflow to ensure the correct charge. Catalyst C is then added, and an exotherm created by the heat of mixing confirms that the catalyst charge has been made. Reactant B is then fed at a controlled rate to maintain the desired reaction temperature. The feed tank is sized for the correct charge for one batch. Even if the entire charge of Reactant B is somehow fed without reacting, the size is such that it is not possible to reach the critical concentration of Reactant B unless the operators filled and emptied the charge tank several times. The inherent safety advantage of the batch process is that it is inherently very difficult to reach a hazardous condition. But, the disadvantage is that the reactor is very large, so the consequence of the potentially explosive reaction would also be large in the event that the critical concentration was somehow reached.

The continuous reactor, on the other hand, can be sized much smaller–perhaps 1/10 the size of the batch reactor for the same production volume. Therefore, the consequence of the potentially explosive reaction is much smaller in case the critical concentration of Reactant B is reached. However, to keep the process running, the continuous process must be connected to a large feed inventory of Reactant A/Solvent S mixture, Catalyst C, and Reactant B. It is possible, with various feed flow rate ratios, to produce almost any concentration of Reactant B in the reactor. The continuous process relies on instrumentation, such as flow meters, ratio controllers, and control valves, to ensure that the materials are fed in the proper ratio. The control instrumentation, logic, and other hardware could fail in a way that results in a concentration of Reactant B in the reactor in excess of the critical concentration. While the control instrumentation could be made highly redundant and reliable, the continuous process still relies on instrumentation to prevent the Reactant B concentration from exceeding the critical value, and should therefore be considered inherently less safe. In addition, with the continuous process operating at steady state and the reactor temperature controlled by cooling the reactor jacket, there is no positive feedback signal to verify that Catalyst C is being fed. As a result, the continuous process has both an inherent safety advantage—the reactor is significantly smaller—and a disadvantage—it relies on instrumentation and control to ensure that the critical Reactant B concentration is not exceeded.

There is no general answer to the question of which of these two proposed processes is inherently safer. The conclusion will depend on the specifics of real process and reaction kinetics, including the specific consequence of exceeding the critical concentration of Reactant B, the size of the various material charges, and other process-specific details. The example illustrates the point that process alternatives can have both inherent safety advantages and disadvantages, and they must be carefully considered before deciding between the two options. Beyond that, many potential variations that can further enhance inherent safety also need to be explored. For example, for the continuous process, one might use metering pumps with a limited range of flow rates—perhaps driven by a single motor and drive shaft—to control the feed ratios, rather than flow sensors, controllers and valves (Hendershot, 1994).

9.2.2 Reduced toxicity vs. reactive hazard

For many processes, a designer can choose from several reaction solvents. These solvents will sometimes differ significantly in their hazardous characteristics, such as toxicity, flammability, and volatility. The CCPS *Guidelines for Engineering Design for Process Safety* (1993) provides an example. For an exothermic batch reaction, there may be a choice between a toxic, but non-volatile solvent, and a non-toxic, but volatile solvent (perhaps water). Each solvent has inherent safety advantages and disadvantages, as summarized in Table 9.2 (Hendershot, 1994).

Table 9.2: Some inherent safety advantages and disadvantages of alternative process solvents

Solvent	Inherent Safety Advantages	Inherent Safety Disadvantages
Non-toxic, volatile	Solvent is non-toxic, reducing hazards resulting from normal handling, and hazards resulting from a discharge due to a runaway reaction. The volatile solvent limits temperature rise in case of a runaway reaction due to the "tempering" effect when the solvent boils.	High vapor pressure of solvent results in potential for high pressure in the reactor in case of a runaway exothermic reaction.
Toxic, non-volatile	Runaway reaction exotherm may not be sufficient to raise the reaction mixture temperature to its boiling point, so there is no risk of overpressurizing the reactor.	Potential exposure of personnel to toxic solvent; environmental damage in case of a spill.

The additional factors that must be considered in order to reach a decision on these solvent options include such issues as: environmental impacts; economic considerations, such as capital costs versus operating costs; solvent availability (for example, for materials that may become difficult to procure due to a changing regulatory landscape). Social and ethical considerations, such as materials that may affect the ozone layer, also enter the decision making.

9.2.3 Reduced inventory vs. dynamic stability

If the strategy of minimization is employed in a chemical process, there could be a tendency to reduce the slack inventory to a minimum in order to reduce the total potential release volume in the event of a containment breach. One inherent safety conflict with that concept is that a process may then actually be *less* tolerant of operational upsets. For example, assume there is a process that ordinarily has a surge drum with an inventory that allows for compensation of a loss of an upstream process by providing continued feed to the downstream process until the system is restored. In the interest of IS, the surge drum is minimized or eliminated, making the process less stable. If the results of a process upset could lead to a major catastrophic outcome, the apparent IS move may have caused an increased overall risk of catastrophic loss.

9.2.4 Risk transfer vs. risk reduction

When discussing the issue of risk transfer versus risk reduction, one's point of view plays a large part in deciding how a particular risk decision is resolved. If no one is tasked with taking into consideration the overall risks to society, or taking a "30,000-foot view" of an issue, then decisions may be made that, while reducing the risk for a portion or subset of parties, results in an increased level of risk for society as a whole.

Different populations of potentially impacted people may perceive the inherent safety of technology options differently. For example, for a process that requires relatively small quantities of chlorine gas, a plant may have a choice between supply in 1-ton cylinders or 90-ton railroad tank cars. Neighbors located several miles away from the plant would consider the 1 ton cylinder supply to be inherently safer because it is unlikely that a leak would affect them at that distance. On the other hand, the plant operators would have to connect and disconnect cylinders 90 times, as opposed to once for a railroad car. They would consider the railroad car to be inherently safer because they are more likely to be exposed to potentially negative impacts by any release, even a small one, and face a much higher frequency of relatively high risk operations by connecting and disconnecting so many hoses that could potentially contain chlorine. Of course, these hazards can be managed with procedures, personal protective equipment, and other safety management systems, but these are not inherent. Both the neighbor and the operator are correct in their perception of the inherent safety characteristics of the chlorine supply options, but they are concerned about different kinds of incidents. The challenge for the designer of the system is to understand these conflicting requirements and make an intelligent choice, including consideration of the entire risk management system (inherent, passive, active, and procedural) (Hendershot, 2006b).

It is important to consider whether an inherently safer option actually *reduces* risk or merely *transfers* it somewhere else. A plant might reduce the size of a hazardous material storage tank, thereby reducing inventory and site risk. Use of the smaller tank, however, may require a change in how material is shipped to the plant from railroad tank cars (typically about 300,000 pound shipments for many materials) to trucks (typically about 30,000 pound shipments) because the smaller tank cannot contain more than a truck load of material. Now, the plant will receive 10 times as many shipments, and they will come by road rather than by rail. Depending on the particular location, road shipments may be inherently more hazardous. Even though the site risk is reduced, the overall risk to society may actually be increased (Hendershot, 2006b).

Minimizing the size of a raw material storage tank will reduce the potential impact of a catastrophic release in the process area, but if the tank is too small to unload an entire tank car of material at once, the potential risk for less than catastrophic spill incidents may increase. If the receiving tank is smaller than the shipping container, the risk of overflowing the tank may increase (Overton and King, 2006).

The alkylation process in oil refineries can be accomplished using hydrofluoric acid (HF) or sulfuric acid. Depending on the concentration, hydrofluoric acid is more toxic and, therefore, inherently more hazardous than sulfuric acid. If released from the alkylation unit, HF would form a toxic cloud, while a release of sulfuric acid would more likely create a pool of liquid. If refineries switched from hydrofluoric acid to sulfuric acid alkylation, it would reduce the process hazard level by use of 2nd order IS (use of less toxic material) at the refinery. But, it may or may not reduce the risk to the broader society. The increased use of sulfuric acid could have a number of negative consequences for safety and the environment:

- HF is regenerated within the alkylation unit, while sulfuric acid must be replaced with waste sulfuric acid regenerated in a separate acid-burning process unit that may be located on or off-site. On average, 0.1 pounds of HF are required per barrel of alkylate produced compared to 42 pounds of sulfuric acid per barrel of alkylate (Myers, et al., 1992).
- Because of the difference in acid usage, the HF alkylation process requires about 12 tank truck deliveries per year, while the sulfuric acid process may require 14 to 15 tank trucks per day (Myers, et al., 1992).
- Regenerating the sulfuric acid in a separate acid burning process unit produces sulfur dioxide and sulfur trioxide, both regulated air pollutants.
- Spent sulfuric acid stored prior to regeneration (if the regeneration unit is located off-site) exists as a mixture with flammable hydrocarbons, creating a flammable waste material (U.S. CSB, 2002).
- The total volume of chemical waste produced increases.

9.2.5 Inherent safety and security conflicts

Inherent safety conflicts may arise as sites assess their security posture (see Chapter 7) if the facility does not consider the potential for shifting risk when making security decisions. For example, a chemical plant may choose to minimize the number of chlorine rail cars it maintains on site; however, this would transfer the risk from the chemical facility to a local rail yard or other storage area, unless security for the railcars at these storage sites was more rigorous than that at the facility. The overall risk to the community could increase if the chemical site had a strong security stance, while the rail yard had open gates and no means of controlling unwanted traffic. In this instance, the chemical company and rail yard would need to work together, and with local law enforcement and emergency responders, to ensure risk is adequately managed and not transferred or increased.

9.3 INHERENT SAFETY – ENVIRONMENTAL HAZARDS

9.3.1 PCBs

Polychlorinated biphenyls (PCBs) were originally introduced in the 1930s as non-flammable cooling and insulating oils for electrical transformers. PCB manufacture was banned in May 1979 due to environmental concerns (Boykin, et al., 1986). This is an example of how new data and information led to a change in the use of a material due to an improved understanding of its hazards and a reevaluation of the relative importance of different types of hazard (Hendershot, 1995).

9.3.2 CFCs

With current concerns about the adverse environmental effects of chlorofluorocarbons (CFCs), it is easy to forget that these materials were originally introduced as inherently safer replacements for more hazardous refrigerants then in use. These included ammonia, light hydrocarbons, such as isobutene, ethyl chloride, methyl chloride, methylene chloride and sulfur dioxide (Jarabek, et al., 1994). These materials are flammable, acutely toxic, or both. A release of one of these substances in the home potentially causes an immediate fire or toxic exposure hazard.

Thomas Midgley, Jr. dramatically introduced CFCs in a lecture to the American Chemical Society in 1937. Midgley filled his lungs with CFC vapors, and then exhaled, extinguishing a candle. This graphically illustrated that CFCs are not flammable or acutely toxic (Kauffman, 1994). Now, after many decades of use, we have discovered that CFCs cause environmental damage by depleting the stratospheric ozone layer, and their use is being phased out. Since it is unlikely that our society will give up refrigeration or air conditioning, substitute refrigerants are needed. Perhaps hydrofluorocarbons (HFCs) and hydrochlorofluorocarbons (HCFCs) will emerge as safe and environmentally acceptable replacements (Wallington, et al., 1994). However, in some cases, we are going back to the refrigerants which CFCs originally replaced. For example, home refrigerators using isobutene refrigerant are now available in Europe. The manufacturers are not producing "frost-free" versions of these refrigerators because of concerns associated with ignition hazards because of the small heater needed for the defrost cycle of a frost free refrigerator (*Chemistry and Industry*, 1994). People must recognize that they are making a tradeoff when they replace CFCs with other materials. While the alternative materials are inherently safer with respect to long-term environmental damage, they are often more hazardous with respect to flammability and acute toxicity (Hendershot, 1995).

9.4 INHERENT SAFETY AND HEALTH CONFLICTS

9.4.1 Water Disinfection

Substituting bleach for chlorine in drinking water and wastewater treatment facilities can reduce risk at the water treatment plant, but may increase the amount of chlorine required at the bleach manufacturing site, depending on the population surrounding the sites. The amount of chlorine needed at the treatment plant—whether from chlorine gas or bleach—will depend on the amount of water to be treated, so the total amount of chlorine required will remain the same. The difference is the way in which the facility receives the chlorine and whether a change from elemental chlorine to bleach will reduce the overall risk, or just shift the risk from one place to another. This well-publicized inherently safer modification can be used to highlight the site-specific challenges to identifying IS opportunities.

Converting from elemental chlorine to bleach will reduce the hazard associated with a release of the material to the population around the water treatment plant, whether from the chlorine/bleach storage vessel or the process of connecting and disconnecting the transported chlorine or bleach to the water treatment process. Due to the economics of sewer and water distribution, such plants are generally located in close proximity to the populations they serve. Because the same amount of chlorine will be needed to treat a given quantity of water, and there is less chlorine in a container of bleach than in the same size container of chlorine gas, more containers of bleach will be required. The increased probability of release due to the increased number of connections/disconnections necessitated by an increased number of shipments must be balanced against the reduced potential consequence from the release of a less hazardous material. The reduction in hazard (in this case, 2nd order IS) could require an increase in the layers of protection surrounding the hazard to reduce the increased probability of release. The goal is for the facility to make sure that the overall risk is reduced—that the increased probability of a release of less hazardous materials presents a lower risk than a lower probability of release of a higher hazard material.

Another example of potential shifts, rather than reductions in risk is the question of whether converting from chlorine gas to bleach shifts risk from the population around the water treatment facility to the facility producing the bleach. If the bleach supplier also supplies chlorine gas by taking large quantities of chlorine and repackaging them into smaller containers, then the facility may be able to readjust the amount of chlorine from repackaging to bleach production. However, a bleach supplier that does not repackage chlorine may be required to increase the amount of elemental chlorine used at that facility in order to meet the increased demand for bleach. If the bleach supplier is also in a more densely populated area, the increased chlorine needed at the facility could increase the risk to that new population. Again, the question of whether overall risk is reduced or shifted will depend on the specifics of an individual water treatment plant, as will as the specifics of the treatment plant's supplier (Overton and King, 2006).

9.5 INHERENT SAFETY AND ECONOMIC CONFLICTS

9.5.1 Existing plants – operational vs. re-investment economics in a capital intensive industry

The following example illustrates the selection of an inherently safer design solution for an existing process.

The design problem was to avoid a significant leak in several water-cooled heat exchangers. These exchangers used material on the process side that reacted violently with water, producing corrosive and toxic by-products. Alternative solutions considered included combinations of passive (double tube sheet or falling film exchangers), active (multiple sensor leak detection with automated isolation), and procedural (a variety of nondestructive testing/ inspection techniques, periodic leak testing with inert gas, improved cleaning procedures) strategies. All of these design alternatives resulted in a lower risk level than the original design. However, none was totally acceptable (see Table 9.3). When management studied the effort and commitment of resources necessary to maintain a less than satisfactory risk level, they chose a design that used a compatible heat transfer fluid, an inherently safer design. Alternatively, a design engineer might choose one of the design solutions offered or choose to combine solutions from more than one category. For example, a combination of reduced inventory (inherently safer) and double tubesheet construction (passive) might produce the optimal risk reduction alternative. Ultimately, design engineers should make decisions based on the prevailing risk tolerability and cost criteria, and their understanding of the operations and maintenance requirements for the design.

Table 9.3: Process safety system design solutions for a heat exchanger failure scenario

Design Basis Failure Scenario	
Tube to tube-sheet failure results in mixing of incompatible fluids, resulting in a system over-pressure and/or the formation and release of a toxic material.	
Design Solution Type	*Description*
1. Inherently safer	A heat transfer fluid compatible with the process fluid
2. Passive	Double tube-sheet construction
3. Active	Pressure relief system with discharge to safe location Multiple sensor leak detection with automated isolation
4. Procedural	Periodic manual sampling of the lower pressure fluid Periodic leak detection with inert gas Improved cleaning procedures

Historically, an overemphasis on minimizing initial capital investment, and on time constraints, which often favor active or procedural systems, has resulted in underutilization of inherently safer solutions. Instead, there is an increased dependency on alarms and SISs to reach acceptable risk levels. Economic analyses in the initial design stages often fail to take into consideration the cost of maintaining and proof-testing these systems, which can be significant for large process facilities. When comparing inherently safer design solutions to other solutions, designers should include the total life cycle cost of each alternative before reaching a decision. For example, Noronha, et al. (1982) describe the use of deflagration pressure containment design in preference to using deflagration suppression or other means of explosion prevention based on life cycle cost and reliability considerations (CCPS, 1998).

9.5.2 Often more economical, but not necessarily

Figure 9.3 presents a comparison of the four categories of design solutions with respect to several cost and functional parameters. While procedural solutions can be less complex, they are often also the least reliable. For active solutions, as compared to inherently safer/passive solutions, reliability is typically lower, and complexity is greater. Inherently safer/passive solutions tend to have higher associated initial capital outlays; however, operating costs are usually lower than those for the other design solutions. Operating costs are also likely to be the greatest for active solutions (CCPS, 1998).

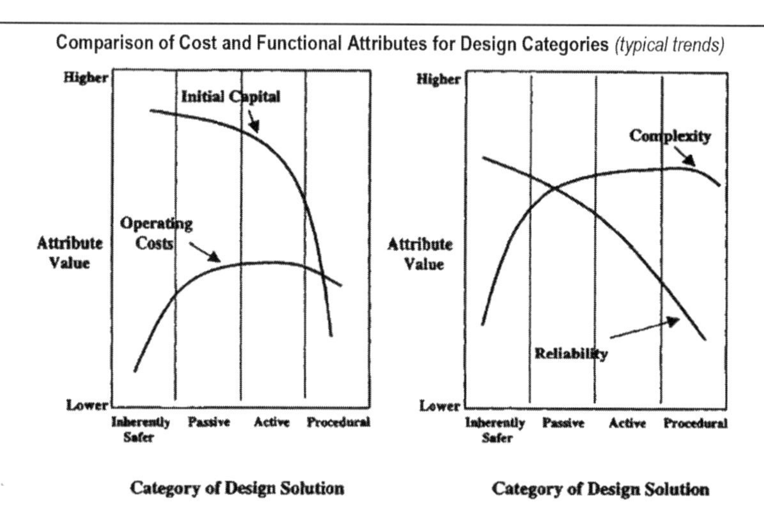

Figure 9.3 Process safety system design solutions for a heat exchanger failure scenario

9.6 TOOLS FOR UNDERSTANDING AND RESOLVING CONFLICTS

Deciding among a number of process options having inherent safety advantages and disadvantages with respect to different hazards can be quite difficult. The first step is to thoroughly understand all hazards associated with the process options. Process hazard analysis and evaluation techniques are appropriate tools (CCPS, 1992). These include:

- Past history and experience
- Interaction matrices
- "What-if"
- Checklists
- "What-if" Checklists
- Hazard and operability (HAZOP) studies

The hazard identification step is perhaps the most important, because any hazard not identified will not be considered in the decision process. For example, the impact of chlorofluorocarbons on atmospheric ozone was unknown for much of the period of their use, and this potential hazard was not considered until recent years. Process hazards analysis (PHA) tools are addressed in Chapter 8. Any change requires a re-evaluation of the entire system to ensure that any new hazards introduced are properly assessed. Changes should be addressed by a Management of Change (MOC) system that assesses the technical basis for, and the impact of, the change on safety and health (29 CFR 1910.119(1) Management of change element).

Sometimes the consequences of all hazardous incidents can be expressed by a single common measure, such as dollar value of property damage, total economic loss, risk of immediate fatality due to fire, explosion or toxic material exposure. If all consequences can be measured on a common scale, the techniques of quantitative risk analysis (CCPS, 1989a, 1995b) may be useful in assessing the relative magnitude of various hazards, and in understanding and ranking total risk of process options.

In many cases, it is not readily apparent how the potential impacts from different hazards can be translated into some common scale or measure. For example, how do you compare long-term environmental damage and health risks from use of CFC refrigerants to the immediate risk of fatality from the fire, explosion, and toxicity hazards associated with many alternative refrigerants? This question does not have a “right” answer. It is not really a scientific question, but instead it is a question of values. Individuals, companies, and society must determine how to value different kinds of risks relative to each other, and base decisions on this evaluation.

Once the hazards have been identified, the process options can be ranked in terms of inherent safety with respect to all identified hazards (see Chapter 9.7). This ranking can be qualitative, placing hazards into consequence and likelihood categories, based on experience and engineering judgment (CCPS, 1992). More quantitative systems can also be used to rank certain specific types of hazard, for example, the Dow Fire and Explosion Index (Dow, 1994b; Gowland, 1996) and the Dow Chemical Exposure Index (Dow, 1994a). Unfortunately, none of these indices consider the full range of hazards.

To get an overall assessment of the process options, it is necessary to use a variety of indices and qualitative techniques and then combine the results. A number of quantitative tools for evaluating inherent safety have been developed, or are in the process of being developed. These include Khan and Amyotte (2005), Heikkilä (1999), and Edwards, et al. (1996).

9.6.1 Tools for understanding and resolving conflicts

The CCPS book, *Tools for Making Acute Risk Decisions with Chemical Process Safety Applications* (CCPS, 1995a), presents a list of factors that, in addition to cost and risk, should be taken into consideration:

- Alternatives available for reducing or eliminating the risks
- Availability of capital
- Codes, standards and regulations and good industry practices
- Company and/or personal liabilities
- Company image
- Costs of implementing available alternatives
- Economic impact of the activity on the local community
- Employment opportunities provided by the activity
- Frequency level(s) of the risk
- Inequities in how the risks and benefits are distributed among members of society
- Number of people at risk
- Perceived benefit of the activity and its impact on the public and/or stockholder image
- Profitability of the activity
- Societal component of the risk, such as the maximum number of people impacted by a single event
- Strategic importance of the activity to the company's growth and survival
- Type(s) of risk, such as human fatality, injury, and acute environmental damage

It is impractical to develop an explicit algorithm for selecting a decision aid. But, decision aids can be flexibly applied, and most can be adapted to an organization's needs and to a wide range of problems. However, an understanding of the key characteristics of the problem, and of the decision aids, is basic to making an appropriate selection.

A key step in selecting a decision aid is to understand the aspects of the problem to be addressed. These aspects include:

- *Resource Availability:* The time and analytic resources to address the problem

- *Problem Complexity:* The number of alternatives, the complexity of the system, and the amount of uncertainty in the problem
- *Importance/Scrutiny:* How sensitive decision makers and stakeholders are to the decision, and how extensively the decision will be reviewed
- *Group Involvement:* The desire of the organization to involve multiple decision makers or incorporate input from multiple stakeholders in the decision
- *Need for Quantification:* The desire to be able to point to a quantitative basis for choosing one alternative over another.

In addition to these five problem aspects, there may be constraints that affect the selection of a decision aid. Constraints can include organizational guidelines on the type of analysis to be performed and the types of issues to be addressed.

Specific criteria can be used to evaluate the appropriateness of various decision aids to a given problem. Some key characteristics that differentiate decision aids include:

- *Resource Requirements:* The time, budget, and effort required to use the decision aid
- *Depth of Analysis/Complexity:* The detail and explicitness with which important aspects of the problem are addressed, and the complexity of doing a complete and thorough analysis
- *Logical Rigor:* The mathematical soundness and logical rigor of the analysis
- *Group Focus:* The ability to incorporate group opinions, handle problems with more than one decision maker, and address competing objectives
- *Quantitativeness:* The ability to provide a quantitative basis for the decision, accommodate sensitivity analyses, and address inherently quantitative decisions such as resource allocation problems
- *Track Record:* How long the decision aid has been available and, preferably, demonstrated on actual risk decisions.

Each decision aid can be evaluated against these criteria to determine how well it addresses these concerns. Depending on the distinguishing aspects of a problem, some of these criteria may be more important than others. The final step in selecting a decision aid is to consider the relative importance of these characteristics, and thereby identify decision aids that are best suited to a problem.

There may not be a "right" answer for the decision-making process. Different people will put different values on the relative importance of different kinds of desired and undesired outcomes. Negotiation and consensus building are required in order reach a decision about what the "right" answer is for a particular situation. In order to reach a decision, agreements must also be reached on how the system boundaries will be drawn. Will you consider transportation? Impact on upstream or downstream technology? Markets? Impact of changes in the product on other parts of the entire system?

In many cases, formal tools for decision making can be useful, particularly if the hazards vary greatly in type of consequence or impact. Many of these tools introduce additional rigor, consistency, and logic into the decision process. Some available methods include:

- Weighted scoring methods, such as Kepner-Tregoe Decision Analysis and the Analytical Hierarchy Process (AHP)
- Cost-benefit analysis
- Payoff matrix analysis
- Decision analysis
- Multi-attribute utility analysis

The chemical process industry is beginning to use these techniques in making safety, health, and environmental decisions. In CCPS (1995a), the reader will find a review of these decision aids and others, with special emphasis on how they are employed in making chemical process safety decisions. Hendershot uses a weighted scoring technique based on Kepner-Tregoe Decision Analysis (Kepner and Tregoe, 1981; Hendershot, 1995a, 1996), as illustrated for a generic process in Table 9.4. Reid and Christensen (1994) describe the use of the Analytical Hierarchy Process to evaluate three alternative technologies considered for a metal fabrication application, with the overall goal of minimizing waste from the process.

PROCEDURE:

- Assign a weighting factor (1 to 10) to the various parameters based on your judgment of the relative importance of this Safety, Health and Environmental, or other issue.
- For each option, assign a performance factor from 1 to 10, based on the relative performance of that option with respect to the particular parameter. This can be based on judgment, or scaled based on some kind of quantitative analysis.
- Multiply the weighting factor by the performance factor for each parameter and process option combination.
- Sum the products for each process option.

The highest total is most desirable.

Parameter	Weighting Factors	Process Options			
		#1	**#2**	**#3**	**#4**
	Performance Factor>	x 2 =	x 9 =	x 10 =	x 1 =
COST	9	18	81	90	9
	Performance Factor>	x 10 =	x 5 =	x 3 =	x 1 =
SAFETY	10	100	50	30	10
	Performance Factor>	x 3 =	x 5 =	x 1 =	x 10 =
ENVIRONMENT	7	21	35	7	70
	Performance Factor>	x 3 =	x 10 =	x 2 =	x 1 =
OPERABILITY	5	15	50	10	5
	Performance Factor>	x 1 =	x 9 =	x 10 =	x 3 =
DESIGN	3	3	27	30	9
	Performance Factor>	x 7 =	x 5 =	x 10 =	x 1 =
OTHER	3	21	15	30	3
	SUM	**178**	**258**	**197**	**106**

Table 9.4: An example of a Weighted Scoring Decision Matrix[1,2]

9.7 MEASURING INHERENT SAFETY CHARACTERISTICS

Current efforts to measure the inherent safety of processes and operations are still in development stage, and are primarily focused on basic process technology and route selection. These measures have not been used routinely in industry and there is, as yet, little data that relates the application of these indices to chemical processing.

[1] See Kepner and Tregoe (1981) or CCPS (1995a) for additional discussion, particularly on how potential negative consequences may impact the scoring matrix.

[2] The weighting factors in this table are for purposes of illustrating the methodology only, and do not represent recommendations on the relative importance of the factors listed.

9.7.1 Dow Fire and Explosion Index [3]

The Dow Fire and Explosion Hazard Index was developed initially for use within company operations, and was later disseminated in a book published by the American Institute of Chemical Engineers (Dow/AIChE 1994). The index quantifies the damage from potential fire and explosion incidents and identifies equipment that would be likely to contribute to the creation or escalation of an incident.

The Dow Index is the product of a *Unit Hazard Factor* and a *Material Factor*. The material factor for a process unit is based on the most hazardous substances or mixtures present that would result in a worst-case scenario, and it quantifies the amount of energy released. A material factor is determined for each process unit. The unit hazard factor for a process unit is the product of both general and specific process hazards. General process hazards address such issues as exothermic chemical reactions, endothermic processes, material handling and transfer, enclosed or indoor process units, access, and drainage and spill containment. Special process hazards include factors for toxic materials, operation in or near flammable range, dust explosions, relief pressure, low temperature operation, corrosion and erosion, joint and packing leaks, fired heaters and hot oil systems, and rotating equipment.

9.7.2 Dow Chemical Exposure Index

Like the Fire and Explosion Index, the Chemical Exposure Index was developed by Dow to help its employees design and operate safer facilities. In 1998, an AIChE volume brought the index—which by then was considered a seminal tool for rating the relative acute health hazard potential of a chemical release to workers and the neighboring community—to the entire process industry. The newest edition uses a new methodology for estimating airborne quantity released, which allows for more sophisticated process analyses (Dow/AIChE, 1998).

9.7.3 Mond Index

The Mond Index was developed by Imperial Chemical Industries (now ICI, an Akzo Nobel Company) after the Flixborough incident, and is based on the Dow Fire and Explosion Index. The Dow Index was modified to address: 1) a wider range of processes and storage installations; 2) processing of chemicals having explosive properties; 3) improved hazard consideration for hydrogen; 4) additional special process hazards; and, 5) inclusion of toxicity in assessment.

The Mond Index divides the plant into individual units and takes into consideration plant layout and the creation of separating barriers between units. The hazard potential is initially expressed in terms of a set of indices for fire,

---

[3] This discussion of the Dow and Mond indices is based on "Inherent Safety in Process Plant Design, An Index-Based Approach" by Anna-Mari Heikkilä, 1999. See Reference section for full citation.

explosion, and toxicity. The hazard indices are then reviewed to determine if design changes reduce the hazard, and the revised values. Factors for preventative and protective features are applied, and then final values of the indices are calculated.

9.7.4 Proposed Inherent Safety indices

The ***Integrated Inherent Safety Index*** (I2SI), developed by Khan and Amyotte (2005), addresses the economic evaluation and hazard potential identification for each option within the process life cycle. I2SI is comprised of sub-indices accounting for hazard potential, inherent safety potential, add-on control requirements, and the economic aspects of the options. The two main sub-indices are a hazard index and an inherent safety potential index. The hazard index measures the damage potential of the process, taking into account the process and hazard control measures. The inherent safety potential index addresses the applicability of inherent safety principles to the process. The two sub-indices are combined to produce the I2SI value.

The ***Prototype Index of Inherent Safety*** (PIIS) for process design was developed by Edwards and Lawrence (1993). The PIIS is based on a chemical score and a process score. The chemical score takes into consideration inventory, flammability, explosiveness and toxicity. The process score addresses parameters, such as temperature and pressure.

When using an inherent safety index, the user should take the necessary steps to ensure that he/she understands the basis of the index. The developers of the indices use their own judgment and experience in deciding what factors are analyzed and in determining the weighting—sometimes transparently, sometimes hidden—of those factors and how they are combined. The user must be sure that these subjective decisions are in line with their organization's philosophy and goals.

Heikkilä (1996) developed an ***Inherent Safety Index*** to classify process alternatives during the process synthesis phase. This method assigns scores to chemical factors (heat of main and side reactions, flammability, explosiveness, toxicity, corrosiveness, chemical interaction) and process factors (inventory, process temperature and pressure, equipment safety, safety of process structure). The final index is a summation of the chemical and process factors.

9.8 SUMMARY

It is not always clear which of several process alternatives is inherently safer. Because nearly all chemical processes have a number of hazards associated with them, an alternative which reduces one hazard may increase a different hazard. It is also necessary to consider business and economic factors in making a process selection. Design strategies that result in an inherently safer design may also tend to improve process economics. For example, minimizing the size of equipment or simplifying a process by eliminating equipment will usually reduce capital investment and reduce operating costs. However, overall process economics are very complex and are impacted by many factors, and it may not always be true that an inherently safer process is also economically more attractive.

An inherently safer process offers greater safety potential, often at a lower cost. However, selection of an inherently safer technology does not guarantee that the actual implementation of that technology will result in a safer operation than an alternate inherently safe process. An inherently safer design also does not necessarily ensure that the technology will be safer in actual operation.

Inherent safety is assessed relative to a particular hazard, or perhaps a group of hazards, but essentially never relative to all hazards. A chemical process is a complex, interconnected organism in which a change in one area of the system can impact the rest, with effects cascading throughout the process. These interactions must be understood and evaluated. Similarly, the chemical industry can be viewed as an ecosystem with complex interactions, interconnections, and dependencies. Understanding these relationships is necessary in order to reach a well-balanced resolution when technological options conflict.

Various means for measuring inherent safety characteristics are available. These include the Dow Fire and Explosion and Chemical Exposure Indices and the Mond Index. Several inherent safety indices have also been developed by Khan and Amyotte, Heikkilä, and Edwards and Lawrence.

10.

Inherently Safer Design Regulatory Initiatives

10.1 INHERENT SAFETY REGULATORY DEVELOPMENTS AND ISSUES

As this book has already demonstrated, Inherent Safety (IS) is a valuable process risk management tool and process safety professionals are voluntarily embracing its concepts and methods. At the same time, regulators and legislators at the national, state, and local levels have recognized the risk reduction potential in IS and have begun a debate as to whether encouraging or mandating these approaches through regulation could improve overall safety or security results. Debated options have ranged from requiring that facilities "consider" IS as one of several choices, to actual mandates that IS be implemented, the latter option backed up by giving agencies the authority to override facility determinations and require installation of specific IS elements. Since the September 2001 terrorist attacks on the United States, IS has been promoted as a possible—and, by some, even the preferred—strategy for reducing the consequences and attractiveness of a chemical facility to a terrorist attack.

To date, mandatory requirements to implement IS have not been adopted as part of US safety or security regulations due primarily to the recognition that IS design and implementation is a function of site and process conditions, and that such site/process specific decisions are less conducive to regulation. In addition, there are recognized concerns that IS may be theoretically possible, but not feasible—or possibly could cause unintended, but potentially threatening, consequences in a specific application.

Several key obstacles or misperceptions have resulted in IS being underutilized by industry:

- a *perception* that IS is technically and economically impractical for all but new processes, though it has been demonstrated to be potentially useful for existing facilities;

- the *lack of an inherent safety infrastructure*, or a framework for evaluating IS systems, including technologies supporting IS approaches, and methodologies that permit IS to be integrated into technical, economic and safety and security design considerations; and,
- the *lack of specific guidance* on how to conduct an inherent safety study, particularly for existing facilities and processes.

Therein lies the dilemma. Policy makers may see implementing IS as a relatively simple and effective way of minimizing or eliminating the hazards or consequences from process incidents, but in practice companies encounter obstacles since these requirements must be integrated with other design and operational considerations. Despite all the attention devoted to it in the political arena, IS remains more of a philosophy than a codified procedure with a well-established and understood framework for evaluation and implementation. Both industry and regulators lack tools and measures to compare the inherent safety of multiple options or to determine what is "feasible." So policy debates over how best to encourage IS continue to be frustrating for all concerned.

10.2 EXPERIENCE WITH IS REGULATIONS

Unlike other safety operations, IS is not easily regulated. When EPA promulgated the RMP rule, some commenters recommended the Agency require facilities to conduct "technology options analyses" to identify inherently safer approaches. EPA declined to do so, stating that:

> "PHA teams regularly suggest viable, effective (and inherently safer) alternatives for risk reduction, which may include features such as inventory reduction, material substitution, and process control changes. These changes are made as opportunities arise, without regulation or adoption of completely new and unproven process technologies. . . . EPA does not believe that a requirement that sources conduct searches or analyses of alternative processing technologies for new or existing processes will produce additional benefits beyond those accruing to the rule already." [1]

The U.S. Congress continues a protracted debate over the merits of including requirements for inherently safer technology as part of a federal chemical security law. In fact, whether to require chemical facilities to implement, or at least consider, inherently safer technology became the major stumbling block to passing comprehensive chemicals security legislation for more than five years. The Congress reached a temporary compromise in 2006, directing the US Department

---

[1] 61 Fed. Reg. 31699 (June 20, 1996).

of Homeland Security (DHS) to regulate security at high-risk chemical facilities for up to three years[2] (see Chapter 7).

The statutory language precludes DHS from requiring specific security measures such as IST, yet the Department recognizes that "facilities are certainly free to consider IST options, and their use may reduce risk and regulatory burdens." [3] This approach is consistent with expert testimony given before the Senate Environment and Public Works Committee in 2006,[4] as well as other references on the topic (Mannan, 2002). Unless new legislation is passed in the interim, DHS's regulatory authority will expire in 2009.

Despite the absence of federal requirements, two jurisdictions in the United States—Contra Costa County, CA, and the State of New Jersey—have already implemented IS regulations. This section explores the lessons learned from these sites.

10.2.1. Inherently Safer Systems Requirements – Contra Costa County, California

Contra Costa County in California forms the eastern boundary of the San Francisco and San Pablo Bays northeast of Oakland and San Francisco. The county is the ninth most populous in the state, with a population of approximately 930,000 as of January 1, 2000. Due to its extensive waterfronts, the county has long been home to processing industry facilities, including oil refineries and chemical plants.

Following a series of serious industrial accidents, the County enacted the Industrial Safety Ordinance (ISO) [5] in 1999. Designed to be "the most stringent in the United States, if not the world,"[6] the ordinance expands on the California Accidental Release Prevention (CalARP) Program for petroleum refineries or chemical plants that are[7]

---

[2] P.L. 109-295, Department of Homeland Security Appropriations Act, 2007, Section 550.

[3] Interim final rule implementing the Chemical Facility Anti-Terrorism Standard, April 9, 2007, 72 Fed. Reg. 17718.

[4] Taken from the testimonies of David A. Moore and Dennis C. Hendershot before the Senate Committee on Environment and Public Works, Full Committee Hearing on Inherently Safer Technology in the Context of Chemical Site Security, Wednesday, June 21, 2006.

[5] County Ordinance Chapter 450-8.

[6] *Industrial Safety Ordinance Annual Performance Review & Evaluation Report*, Contra Costa Community Health Hazardous Materials Programs, December 4, 2007.

[7] The ISO goes beyond ISS. Among other measures, the ordinance also removes the threshold quantities for regulated chemicals so that the CalARP standards apply to every unit on a site, and requires the facility to include human factors in PHAs and PHA revalidations and in the site safety plan.

- Running at least one CalArp Program 3 process;
- Located within an unincorporated area of the County; and
- Required to submit a Risk Management Plan (RMP) to the U.S. EPA and Contra Costa County Health Service (CCHS)

Among the stipulations of the ordinance is that facilities must, within one year, submit safety plans to the CCHS Hazardous Materials Programs, the agency responsibility for its implementation. In turn, CCHS must audit and inspect the covered facilities within the year. The agency could also inspect any facility within 30 days of a major chemical accident or release. Once a safety plan is accepted, it is valid for three years. The facility's safety plan must be audited at least once every three years and CCHS may also perform unannounced audits. Each covered facility must update its accident history information in the Safety Plan annually, including information on how the facility used inherently safer processes within the last year.

The City of Richmond, located within Contra Costa County, adopted the ordinance[8] in 2001, expanding the coverage of the ISO to eight facilities—six in the county and two in Richmond.[9]

INHERENT SAFETY REQUIREMENT OF THE ISO

The ISO explicitly requires covered facilities to "*consider* the use of inherently safer systems (ISS) in the development and analysis of mitigation items resulting from a process hazard analysis, and in the design and review of new processes and facilities." The ordinance goes on to require the stationary sources to:

- select and implement inherently safer systems (ISS) to the greatest extent feasible; and
- submit the basis and documentation for concluding that an ISS is infeasible. [10]

The ISO defines an ISS by referencing the most current edition of this book[11] and explicitly listing and describing the four categories of risk

---

[8] Municipal Code Chapter 6.43, RISO.

[9] See Footnote 6

[10] County Ordinance Chapter 450-8.016(d)(3): For all covered processes, the stationary source shall consider the use of inherently safer systems in the development and analysis of mitigation items resulting from a process hazard analysis, and in the design and review of new processes and facilities. The stationary source shall select and implement inherently safer systems to the greatest extent feasible. If a stationary source concludes that an inherently safer system is not feasible, the basis for this conclusion shall be documented in meaningful detail. This documentation shall include (i) sufficient evidence to demonstrate to the county's satisfaction that implementing this inherently safer system is impractical, and (ii) the reasons for this conclusion.

reduction—inherent, passive, active and procedural applications. "Feasible" is defined as "capable of being accomplished in a successful manner within a reasonable period of time, taking into account economic, environmental, legal, social, and technological factors."[12] However, a claim of financial infeasibility "shall not be based solely on evidence of reduced profits or increased costs, but rather shall include evidence that the financial impacts would be sufficiently severe to render the inherently safer system as impractical."[13]

To help resolve any confusion regarding ISS implementation, CCHS worked with the covered facilities to expand the original ISO program guidance on these requirements. The resulting ISS guidance, published by the CCHS in 2002, clarifies that "the inherent and passive categories should be implemented when feasible for new processes and facilities and used during the review of Inherently Safer Systems for existing processes," and that such an approach "is good risk reduction."[14] The guidance goes on to describe the inherently safer approaches of minimization, substitution, moderation and simplification. Above all, it emphasizes that hazard must be considered for newly covered processes and when conducting a PHA on existing process units. "If the potential hazard (severity) of consequence of a deviation meets the definition of a Major Chemical Accident or Release,[15] an ISS Analysis should be done for those that could reasonably occur."[16] This could include a process or parts of a process. It might also include

---

[11] County Ordinance Chapter 450-8.014(g): "Inherently safer systems" means inherently safer design strategies as discussed in the latest edition of the Center for Chemical Process Safety publication *Inherently Safer Chemical Processes* and means feasible alternative equipment, processes, materials, lay-outs, and procedures meant to eliminate, minimize, or reduce the risk of a major chemical accident or release by modifying a process rather than adding external layers of protection. Examples include, but are not limited to, substitution of materials with lower vapor pressure, lower flammability, or lower toxicity; isolation of hazardous processes; and use of processes which operate at lower temperatures and/or pressures.

[12] County Ordinance Chapter 450-8.014(c)

[13] County Ordinance Chapter 450-8.016(e)

[14] Page D-2

[15] 450-8.014(h): "Major chemical accident or release" means an incident that meets the definition of a Level 3 or Level 2 Incident in the community warning system incident level classification system defined in the September 27, 1997 Contra Costa County guideline for the community warning system as determined by the department; or results in the release including, but not limited to, air, water, or soil of a regulated substance and meets one or more of the following criteria: (1) Results in one or more fatalities; (2) Results in greater than twenty-four hours of hospital treatment of three or more persons;(3) Causes on and/or off-site property damage (including clean-up and restoration activities) initially estimated at five hundred thousand dollars or more. On-site estimates shall be performed by the stationary source. Off-site estimates shall be performed by appropriate agencies and compiled by the department; (4) Results in a flammable vapor cloud of more than five thousand pounds.

[16] Footnote 5 page D-3.

examining each consequence of a deviation, including the severity of each consequence, to ensure inherently safer systems (inherent and passive categories) are considered and documented for the covered process.[17]

Once a PHA has been conducted, the covered facility should document, as part of its recommendations and mitigations, how it used ISS strategies in sufficient detail to satisfy inspectors. This should include:

- Evidence that at least one risk reduction action was taken using ISS strategy for each PHA mitigation item for scenarios that have the potential for a Major Chemical Accident or Release.
- A description of the risk reduction method selected and the inherently safer system strategy used.
- Details of risk reduction mitigation considered using the inherently safer system strategy that was not implemented.
- Reasons the rejected risk reduction mitigation was determined to be infeasible using the inherently safer system strategies.

The facility should also be able to demonstrate that:

- it has a program in place to ensure that the risk reduction measure implemented will incorporate ISS.
- the program incorporates the process safety hierarchy.
- the facility is committed to moving up the hierarchy from procedural to inherent strategies.

The guidance provides additional clarity on what CCHS will and will not accept as a justification of infeasibility. Many of the criteria are similar to those developed by EPA and OSHA. CCHS will provide an affirmative determination that the documentation, calculations and justifications for not implementing an ISS are acceptable.[18]

However, not every PHA action item is an opportunity for Inherently Safer Systems. In some cases, various layers of protection may provide equivalent levels of risk reduction and may offer more cost-effective solutions. Limiting the consideration to "feasibility" without being able to compare the ISS alternative with other approaches creates a potential to force very expensive and significant changes in the name of inherent safety without a concomitant reduction in risk compared with other less expensive alternatives.

---

[17] D-8

[18] D-12

RESULTS

CCHS reported in 2007 that implementation of the ISO has improved and, in most cases, is being done as required by the ordinance. Since ISO covered facilities must annually update the information on their accident history in their Safety Plans and include how they have used inherently safer processes within the last year, the county has significant information on the types of ISS implemented as a result of the ISO. A summary of the inherently safer systems and risk reduction measures implemented since 2003 appears in Table 10.1 at the end of this chapter.

In the 2007 ISO Annual Report, CCHS reported that the number and severity of Major Chemical Accidents or Releases (MCARs) have been decreasing since the implementation of the ISO. However, the small number of MCARs (fewer than a dozen total incidents in any given year since 1999) makes it difficult to demonstrate a linear trend, or to establish a direct causal relationship between the ISO and/or implementation of ISS and the number of incidences. Figures 10.1-10.3, taken from the 2007 ISO Annual Report, display, respectively, the number and severity of MCARs that have occurred in the county since 1999–incidents at all ISO and CalARP facilities,[19] at county facilities subject to the ISO, and facilities in the county and Richmond subject to the ordinance.

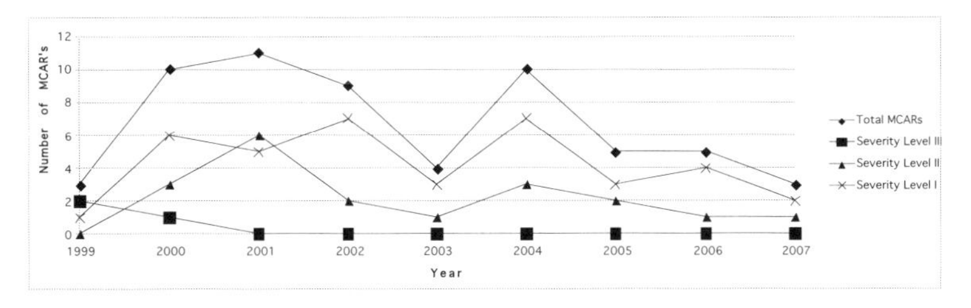

Figure 10.1: Major Chemical Accidents and Releases (MCAR)

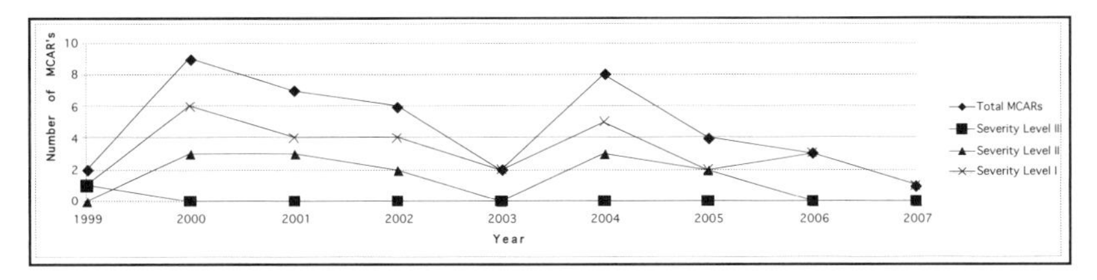

Figure 10.2: ISO Stationary Sources MCARs

[19] Eight facilities are covered by the County and the City of Richmond's Industrial Safety Ordinances. Approximately 50 facilities are covered under the California Accidental Release Prevention (CalARP) Program. (Source: Contra Costa Health Services Hazardous Material Programs)

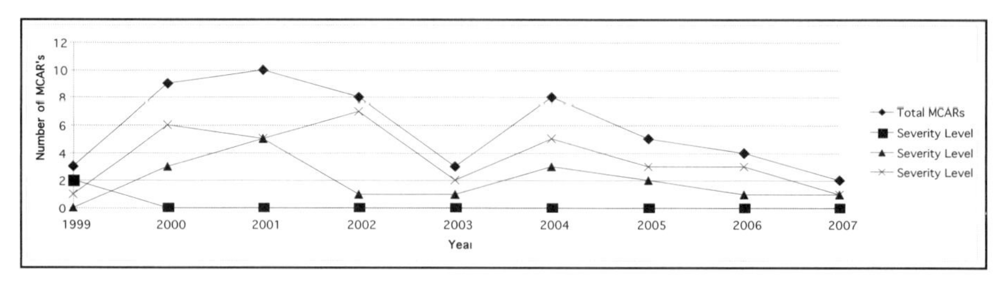

Figure 10.3: County and Richmond ISO MCARs

The ISO program within CCHS, including the costs of the Ombudsman and other staff, public hearings and administrative overhead, is paid for through review, inspection, and audit fees. The fee for the ISO is equal to one-third the total fees charged for the CalARP Program, plus the salary for the Hazardous Materials Ombudsman position. Total ISO fees for the 2006-2007 fiscal year were $325,281 or more than $40,000 per covered facility.[20]

LESSONS LEARNED FROM THE ISO

Applicability: Experience under the ISO confirmed that first- and second-order inherent safety is most readily applicable to new and modified facilities and processes. Third- and fourth-order ISS has been most commonly implemented at existing processes, such as those identified through PHA's. However, broad application of inherent safety across all four strategies of process risk management is more practical and may result in novel risk reduction ideas. The mandate to consider the full spectrum of ISS for these items does seem to have prompted new ways of looking at reducing hazards and risks and appears to have contributed to overall risk reduction at these facilities.

Guidance: Industry needs clear guidance on how and when to apply ISS. The eight facilities subject to the ISO initially found it difficult to apply ISS to existing processes. The 2002 guidance, which provided greater and needed detail on how and when to apply ISS, significantly improved the regulators' and facilities' understanding of what was expected under the ordinance. Before the 2002 guidance, the majority of ISO submissions involving PHA action items did not actually involve application of ISS, and the distinction between ISS and risk reduction in general was often lost. But, the 2002 guidance did not resolve all implementation questions. For example, CCHS and facilities may still disagree about what is reasonable and what is feasible. Nevertheless,

[20] Source: Contra Costa Health Services Hazardous Materials Programs

after the 2002 guidance was developed, many more ISS opportunities were found and applied. Inherent Safety is best applied by those knowledgeable of, and properly trained in, the process. Guidance and training are needed for a team to know how to do this effectively.

Expectations: Both the public and regulators may have higher risk reduction expectations of IS approaches than what is ultimately found to be feasible in a given application. The public and regulators expect to see a realistic effort to apply first- and second-order inherent safety to achieve major risk reduction. However, only considering first- or second-order IS may be at odds with other practical and cost-effective risk reduction options, especially for existing processes. A facility may conclude that IS is infeasible if it does not look at all ISS options. This difference between public and industry perceptions of risk—and a misunderstanding or lack of appreciation of the costs of some ISS modifications—may create a barrier to common expectations, and to an accepted definition of "feasibility." Any move in an inherently safer direction is likely to be a good risk reduction move, and should be encouraged.

10.2.2 New Jersey Toxic Catastrophe Prevention Act (TCPA) and Prescriptive Order for Chemical Plant Security

New Jersey is the nation's most densely populated state. It has a large number of chemical and petroleum facilities, and is also a center of pharmaceutical, biotechnology and other life-science industries. These and other industrial plants that produce or use hazardous chemicals are clustered around the well-developed transportation infrastructure linking the New York and Philadelphia metropolitan regions.

The combination of critical chemical infrastructure in close proximity to high population centers has provided an incentive for the state to enact particularly stringent programs to reduce the risk of release—both accidental and intentional—from such facilities. The state's Toxic Catastrophe Prevention Act (TCPA) was enacted in the wake of the Bhopal incident to protect the public from catastrophic accidental releases of extraordinarily hazardous chemicals. The success of the TCPA program influenced the enactment of Section 112(r) of the Clean Air Act and EPA's subsequent Risk Management Program (RMP). When the September 11 terrorist attacks raised a new potential threat, New Jersey moved twice to add IS requirements to its TCPA program. It has also imposed additional chemical-oriented requirements through an administrative order. As before, the state's actions have been cited by others as a potential model for a Federal chemical security program.

TOXIC CATASTROPHE PREVENTION ACT (TCPA)

New Jersey adopted the Toxic Catastrophe Prevention Act (TCPA) in 1985 to reduce the risk of "potentially catastrophic discharges of hazardous substances in the environment." The legislature concluded that the ability to respond and mitigate a release was critical, but insufficient, and that "*the single most effective effort to be made is toward prevention of those environmental accidents by anticipating the circumstances that could result in their occurrence and taking those precautionary and preemptive actions required.*" (emphasis added)[21] While not using the term "inherent safety," the statute did require that extraordinarily hazardous substance risk reduction work plans document:

> "any *alternative processes, procedures, or equipment which might reduce the risk of a release* of an extraordinarily hazardous substance while yielding the same or commensurate results, and the specific reasons they are not employed."[22]

The state legislature envisioned that facility owners/operators would seriously consider changes in process equipment, and even chemical substitutions, to reduce the risk of accidental releases. Implementing regulations required owners and operators to conduct "state-of-the-art" studies of existing equipment.

The TCPA rules were significantly revised in 2003 to include an explicit definition of, and a requirement to consider, inherently safer technology.[23] Owners/operators of newly designed and constructed covered processes were required to evaluate inherently safer technologies, and to document recommendations from these evaluations, regardless of whether they were incorporated in the new process. The regulations adopted the definition of "inherently safer technology" from the first edition of this book.

> "Inherently safer technology" means the principles or techniques incorporated in a newly designed and constructed covered process to minimize or eliminate the potential for an Extraordinarily Hazardous Substance (EHS)[24] accident that includes, but is not limited to, the following:

---

[21] Toxic Catastrophe Prevention Act (TCPA) NJSA 13:1K-19 et. seq.

[22] TCPA 13:1K-24. Extraordinarily hazardous substance risk reduction work plan; accident risk assessments

[23] 35 N.J.R. 3618(b), August 4, 2003.

[24] TCPA defines an Extraordinarily Hazardous Substance as, "any substance or chemical compound used, manufactured, stored, or capable of being produced from on-site components in this State in sufficient quantities at a single site such that its release into the environment would produce a significant likelihood that persons exposed will suffer acute health effects resulting in death or permanent disability." The statute itself listed 11 compounds as EHS—*hydrogen chloride* (HCl) and *allyl chloride* in quantities of 2,000 pounds or more; *hydrogen cyanide* (HCN), *hydrogen fluoride* (HF), *chlorine* (Cl2), *phosphorus trichloride*, and *hydrogen sulfide* (H2S) in quantities of 500 pounds or more; and *phosgene*, *bromine*, *methyl isocyanate* (MIC), and *toluene-2, 4-diisocyanate* (TDS) – and directed DEP to develop a list of additional chemicals and threshold quantities. (13:1K-22. Extraordinarily hazardous substance list; registration of facilities generating, storing or handling substances).

1. reducing the amount of EHS material that potentially may be released;
2. substituting less hazardous materials;
3. using EHSs in the least hazardous process conditions or form; and
4. designing equipment and processes to minimize the potential for equipment failure and human error.[25]

In explaining the 2003 regulations, the New Jersey Department of Environmental Protection (NJDEP) called for consideration of the full range of inherent safety options, recognizing "that every (inherently safer technology) technique will not be suitable for every process and will rely on the owner or operator's evaluation of which techniques, if any, will be appropriate for a particular newly designed and constructed process."[26] DEP also limited application of inherent safety to "newly designed and constructed processes," concluding that "the most reasonable and effective time for an owner or operator to evaluate inherently safer technology is during the design phase of a process, before it has been built." The Department went on to point out that "in technical literature, the term 'inherently safer technology' is often used interchangeably with the term 'inherently safer design'" as support for the decision to limit the inherently safer technology evaluation to processes that are being newly designed.[27]

These policy positions – describing "inherently safer technology" in terms of principles or techniques, retaining the full range of options, and applying only during new design or construction – are critical. They demonstrate that, as late as 2003, New Jersey recognized inherent safety as a philosophy to be incorporated in the design process, rather than an endpoint to be reached.

In May 2008, NJDEP proposed amendments to the TCPA rules[28] that expands the program in two ways: (1) *all* TCPA facilities—not just those with the potential for off-site consequences—must conduct IST reviews of any process with EHS that meets or exceeds the regulatory threshold; and (2) IST reviews must include both *existing* and new processes, rather than just newly designed and constructed covered processes. The new rule also revises the TCPA application to "releases" not just "accidents." The new rule continues the trend towards a more comprehensive application of inherent safety in New Jersey.

PRESCRIPTIVE ORDER FOR CHEMICAL PLANT SECURITY

In November 2005, the State of New Jersey issued a Prescriptive Order for chemical plants, requiring 157 facilities to comply with "Chemical Sector Best

---

[25] TCPA 13:1K-21. Definitions

[26] 35 N.J.R. 3618(b), August 4, 2003, Comment 13 and DEP response.

[27] 35 N.J.R. 3618(b), August 4, 2003, Comment 11 and DEP response.

[28] 40 N.J.R. 2254(a), May 5, 2008.

Practices" previously adopted by the state's Domestic Security Preparedness Task Force.[29] Specifically, facilities were required to conduct security vulnerability assessments and to develop a security prevention, preparedness and response plan. Chemical facilities subject to TCPA were also explicitly required to "conduct a review of the practicability and the potential for adopting inherently safer technology (IST) as part of the assessment and plan," and to submit a report on the IST review to the state. The order adopted the TCPA definition of IST, but went further than TCPA by requiring that "such review shall include analysis of whether adoption of IST alternatives is practicable, and the basis for any determination that implementation of IST is impractical."[30]

Implementation guidance explained that, "an analysis of IST can be done in all phases of a plant's life, such as during the conceptual phase, design phase, or operational phase. However, an operating plant may have a limited option of IST application, considering the practicability and cost-effectiveness. For an existing facility, an application of multiple levels of engineered safeguards, coupled with practicable and cost-effective technological or design changes, may accomplish the same degree of inherent safety in a process plant." The guidance went on to state that "IST may be applied regardless of release likelihood to reduce the hazard of an EHS."[31]

The guidance calls for the following questions to be answered for each of the four criteria in the IST definition (minimize, substitute, moderate, simplify):

- Are there any available IST alternatives over the current process?
- Is adoption of IST alternatives practicable?
- What is the basis for any determination that implementation of IST is impractical?
- What past IST and risk reduction measures have already been incorporated into the current process? (A complete list of safeguards should be included in PHA reports.)

The prescriptive order specifies that IST reviews were to be conducted by a qualified expert in chemical process safety. While a definition of "qualified

---

[29] The NJ Domestic Security Preparedness Task Force was established by the 2001 New Jersey Domestic Security Preparedness Act as a Cabinet-level body responsible for setting state homeland security and domestic preparedness policy. The Task Force was part of the newly-formed Office of Counter-Terrorism (now the NJ Office of Homeland Security and Preparedness). The Task Force subsequently adopted security standards, guidelines and protocols for 16 critical infrastructure sectors, including the chemical sector.

[30] State of New Jersey Domestic Security Preparedness Task Force, Domestic Security Preparedness Best Practices Standards at TCPA/DPCC Chemical Sector Facilities, November 21, 2005, requirement 5, pg. 6-7.

[31] "Guidance on Inherently Safer Technology," Bureau of Release Prevention, NJDEP, January 12, 2006, pg. 2.

expert" was not provided, the guidance recommended that the review should be performed by a team, comprising representatives from various disciplines including chemistry, engineering, process controls and instrumentation, maintenance, production and operations, and chemical process safety.

A separate IST review was generally conducted for each TCPA-covered process, and each alternative was designated either as "Not Applicable," "IST already implemented," "IST alternative recommended," or "IST rejected due to feasibility." The guidance clarified that the technology should be available at reasonable cost commensurate with the anticipated reduction or elimination of the hazard; it was generally not expected that research into new technologies would be conducted or that a major process modification or replacement be instituted to implement an inherently safer alternative.[32]

Following the IST review, the facility was to prepare a report that documented the findings and evaluation, as well as past IST and risk reduction measures that had already been incorporated into the current process. The facility was also to provide a schedule for implementing any IST alternatives determined to be feasible.[33]

It is important to point out that, while the Prescriptive Order and its IST review requirement were driven by security concerns and the potential for intentional releases, efforts to comply with the Prescriptive Order also served to address the potential for accidental releases. In other words, the required review would address the full range of IST strategies, and not only substitution, minimization, and moderation—the most effective strategies to reduce security-related risk.

RESULTS OF IMPLEMENTING IST UNDER THE PRESCRIPTIVE ORDER

All documents submitted or obtained pursuant to the Prescriptive Order are protected as privileged and confidential information under New Jersey's Domestic Security Preparedness Act.[34] As a result, there is little hard data on the implementation and effectiveness of the IST programs in New Jersey. Nevertheless, the state has provided some qualitative information through public presentations and statements.

According to NJDEP, of the 157 facilities subject to the Prescriptive Order, more than 98% complied within the 120-day deadline established in the Order. All

---

[32] "Guidance on Inherently Safer Technology," Bureau of Release Prevention, NJDEP, January 12, 2006, pg. 2.

[33] "Guidance on Inherently Safer Technology," Bureau of Release Prevention, NJDEP, January 12, 2006, pg. 3.

[34] State of New Jersey Domestic Security Preparedness Task Force, Domestic Security Preparedness Best Practices Standards at TCPA/DPCC Chemical Sector Facilities, November 21, 2005, requirement 12, pg 8.

facilities required to perform an IST analysis documented that they had previously implemented IST or similar risk reduction measures. Of those, 32% provided a schedule to implement additional IST or other risk reduction measures, and 19% identified additional IST or risk reduction measures, but without completion schedules. The remaining 49% of the facilities had no additional recommendations and 80% of the facilities concluded that at least some of the IST or risk reduction measures identified during the IST evaluation were infeasible for their operations.[35]

Based on the results of the IST review program required under the Prescriptive Order, New Jersey believes that evaluating inherently safer technology is not overly burdensome on industry and is an effective tool for critically evaluating the risk reduction opportunities available at a specific facility. Furthermore, Director of Operations Gary Sondermeyer suggests that IST analysis is simply good business practice for any facility storing or utilizing extraordinarily hazardous materials from an economic, worker safety and regulatory compliance standpoint.[36] Experience with IST under the Prescriptive Order served as justification for expanding the IST program for all TCPA facilities.

At this point, the program created by the Prescriptive Order has emphasized the act of conducting an IST review and documenting past and immediately feasible IST options. All 45 facilities reported they had implemented IST or similar risk reduction measures, yet only a third identified additional feasible IST measures for which an implementation schedule could be provided. This is understandable, considering that facilities had only 120 days to conduct the review, and all the facilities that fell under the Prescriptive Order had been subject to the TCPA IST requirements for new or newly constructed processes for several years.

Even with the ability to "take credit" for IST measures previously implemented, industry in New Jersey has expressed concern that a focus on IST as an element of facility security not draw attention and resources from more traditional (i.e., physical) security measures. The experience of at least some companies was that the IST discussions arising from the Prescriptive Order took up the majority of the total review time of the inspectors who largely ignored other facility security measures. "Isolating a single type of security measure for such intense focus at the expense of all the others is not good security practice. Rather, facilities and the state should be considering the full range of security measures."[37]

---

[35] Testimony of NJDEP Director of Operations Gary Sondermeyer, before the US House of Representatives Committee on Homeland Security, December 12, 2007.

[36] See above.

[37] Statement of Clyde D. Miller, Director, Corporate Security, BASF Corporation, before the US House of Representatives Committee on Homeland Security, Subcommittee on Transportation Security and Infrastructure Protection, December 12, 2007.

LEARNINGS FROM NEW JERSEY

In the relatively short period of time from 2003 to 2007, the State of New Jersey expanded its earlier view of IST to one where it is applicable to existing, as well as newly constructed processes, and to security as well as safety. The state was profoundly affected by the September 2001 attacks and, as such, saw IST as an option in their security regulatory arsenal, believing that a facility with less hazardous chemicals will be less attractive to terrorists and, if attacked, have a reduced probability of a serious accidental release and/or reduced consequences.

One way in which the TCPA has driven the concept of inherent safety has been by inspiring companies to seek ways to eliminate or reduce the amount of covered chemicals handled to a level below the threshold quantity to which the regulations apply. In the almost 20 years since the TCPA was enacted, the number of TCPA-regulated facilities, as well as the quantity of EHSs registered in the state, have dropped significantly. This has had the benefit of reducing risk as well as avoiding regulatory requirements.

While it is generally accepted that inherently safer technology has the potential to reduce both accidental and intentional process safety hazards, implementation, and especially regulation, of IST is not generally straightforward. As the New Jersey experience has demonstrated, there is limited feasibility for implementing IST in existing facilities due to costs and other tradeoffs. Industry recognizes that IST is not a "silver bullet" to eliminate security risks, including the recognition that tradeoffs are often involved which can simply change (or increase) the risk and not reduce it. One such example is the use of smaller shipping containers, which may reduce the consequences of a container leak, but will generally increase the risk associated with transportation and connecting/disconnecting the container from the process. Cost is also usually a factor, and the application and management of process safeguards (active, passive, and procedural) often provides an adequate and more cost-effective level of protection compared with alternative process designs.

IST concepts are worthwhile as engineering design guides and as a general philosophy for designing, constructing, maintaining, and operating chemical process facilities. The ongoing concern of industry is that IST alternatives, once identified, will be mandated regardless of feasibility, potentially causing economic or even safety consequences to the facility or to the transportation system. As with other process safety/risk management regulations, the way in which an IST requirement is enforced is the key to the success of such an initiative. Based on the experience in New Jersey, however, it is unlikely that requirements for conducting IST reviews and implementing such technology "where practicable" will necessarily result in large-scale risk reduction against security-related risks.

10.3 ISSUES IN REGULATING INHERENT SAFETY

The New Jersey and Contra Costa requirements are very significant for three reasons: (1) they set precedent for state or local IS mandates within the processing industries; (2) they justify the need to evaluate IS broadly; and (3) the New Jersey Prescriptive Order requires IS as a means to achieve another requirement—reduction of security risks. These approaches may serve as models for future safety and security regulations. As such, learnings from these programs—especially barriers and constraints to implementation—should be fully evaluated in the development of any future IS requirements.

Industry and policymakers have a major opportunity and a major challenge ahead to develop effective programs—whether voluntary industry initiatives or government regulations—to encourage the broad adoption of IS. First, both industry and policymakers need a fuller and more consistent understanding of IS – what it is and how it can be applied. Secondly, new analytical tools for conducting IS reviews and measuring progress, as well as decision-making criteria, will be needed.

10.3.1 Consistent Understanding of Inherent Safety

Misunderstandings or misperceptions about Inherent Safety tend to localize around four concepts – *goals*, *applicability*, *scope* and *economic feasibility*.

- *The goal of both a safety and security program should be to reduce risk.* Inherent safety is an approach to reducing and managing risks; it is not an end in itself. IS policies and regulations will be most successful when they clearly state a risk reduction and management goal with a recognition that some risks are inherent to the production of some critical goods and services, and that such risks can be managed within acceptable ranges.

- *IS may be applicable to existing as well as new facilities and processes*. There may be a perception that IS is only for new facilities and that there are no feasible opportunities once the process is operational. While it is true that the potential for major improvements may be greatest during process development, this book has demonstrated that facilities have reduced or even eliminated hazards, or have managed change to avoid new hazards by applying IS methods throughout the facility lifecycle. The majority of the applications for IS are with the installed industrial base, whereas the feasibility of applying IS to the fullest diminishes as the facility is built. This leaves many companies where new processes (and particularly new technologies) are rarely implemented with fewer occasions to practice the methods.

- *The scope of IS is not limited to hazard reduction*, and its concepts are applicable to the layers of protection surrounding the remaining hazard. A narrow view argues that IS only applies to major changes in the degree of hazard, while a broader viewpoint finds any changes that increase safety through the application of IS principles to be an advantage.

- *Changes to facilities and processes must be economically feasible.* Costs are a primary concern when considering modifications to existing facilities. Inherent and passive approaches are strategic, must be implemented early in the facility development, and have broad and wide ranging impacts on the process design. Tactical approaches, including the active and procedural categories, can be implemented late in the design process and are characterized by repetition and high costs associated with maintenance (Hendershot, 1997). Costs have to be evaluated on a holistic basis, and it has been proven that the life cycle costs and risks may be reduced after implementing IS. Both the Contra Costa and New Jersey programs explicitly include a consideration of economic feasibility as part of an IS evaluation.

10.3.2 Needed Tools

Chapter 9 discusses a number of tools and approaches for evaluating different and potentially conflicting IS. Additional tools and approaches will be needed to support IS regulatory programs including those to support *systematic IS reviews,* and *economic evaluation* and *performance evaluations*.

Systematic review methods: Methods for systematic IS reviews are needed to provide consistent and comparable IS evaluations within and across facilities. Such an approach provides greater assurance that dissimilar facilities will conduct reasonably equivalent IS evaluations, and that the results of those evaluations will be understandable and acceptable to regulators and the public. A number of approaches to develop such a tool have been proposed. *Making EHS an Integral Part of Process Design* focuses on the MERITT (Maximizing EHS Returns by Integrating Tools and Talents) approach for improved process development through better integration of environmental, health, and safety evaluations at the earliest stages of a product's life (CCPS, 2001). The European Community's INSIDE Toolkit provides a comprehensive set of safety, health and environmental indices that are used individually to evaluate processes, but do not combine into a single overall measure (Hendershot, 1997). Khan and Amyotte (2004) have postulated a conceptual framework of an integrated inherent safety index (I2SI) which, if fully developed, would consider the life cycle of the process with economic evaluation and hazard potential identification for each option. The Mary Kay O'Connor Process Safety Center has ongoing research on the development of an inherent safety index based on fuzzy logic theory (Gentile, et al., 2001). While these are excellent developments in the right direction, they are not fully validated or comprehensive enough to assure that the aforementioned issues are satisfied.

Economic valuation methods: Since economic feasibility is a critical factor in identifying viable IS opportunities, better methods for estimating this variable would benefit both regulatory and voluntary IS programs. Methods to estimate the value of IS and to quantitatively assess whether a given process is "as inherently safe as is practicable" are generally unavailable or unproven. Though examples are increasingly available in the literature, case studies documenting the economic benefit of IS projects do not cover a wide array of industrial situations.

How to evaluate performance: A constraint to regulation is the lack of consensus on appropriate IS metrics. Assuming that the regulation is performance-, rather than specification-based, there must be metrics for consistency, and these are very hard to define for such a topic as IS for chemical processes. It is difficult to measure the effectiveness of IS regulations simply by inferences from the rate or severity of incidents.

> "Regulation to improve inherent safety faces several difficulties. There is not presently a way to measure inherent safety. Process plant complexity essentially prevents any prescriptive rules that would be widely applicable. It would seem that legislation could explicitly require facilities to evaluate inherently safer design options as part of their process hazard analysis. But inherent safety would be almost impossible to enforce beyond evaluation because there are unavoidable technical and economic issues" (Mannan, et.al, 2003).

This is particularly problematic when an IS opportunity appears to conflict with another safety, health or environmental policy objective. IS performance metrics could help evaluate such risk/risk tradeoffs.

10.4. SUMMARY

Seeking greater use of inherently safer process and approaches has the potential to improve the safety, security and even economic performance of a process. However, inherent safety should not be seen as an end in itself, or even the preferred strategy to reduce risk. Rather, it must be seen as one strategy to be employed to reach a risk reduction target.

Contra Costa County, California and the State of New Jersey require consideration of IST as part of their chemical security (New Jersey) and process safety (New Jersey and Contra Costa) programs. Facilities have complied with the

IST requirements in both jurisdictions, but in the absence of rigorous metrics, it is difficult, if not impossible, to evaluate the contribution these requirements have made to safety and security.

Policymakers have, and continue to, consider regulation as a means to accelerate the application of inherent safety in some of the higher risk processing industries. Both industry and policymakers will benefit from a deeper understanding of inherent safety – goals, applicability, scope and economic feasibility. Most important is the development of new and improved tools and metrics for evaluating the contribution of IS to risk reduction and economic assessments.

Table 10.1: Summary of ISS Strategies Implemented under CCHS Ordinance

RISK REDUCTION CATEGORY	ISS Strategy	2006	2005	2004	2003	2002
INHERENT:	**Minimize**	Reduction of inventory (4)	Reduction of inventory (11)	• Reduction of inventory (11) • Eliminated a unit and regulated material (1) • Incorporated existing equipment into a new unit and reduced piping (1)	• Eliminated hazard (2) • Reduction of inventory (9) • Process re-design to eliminate need for a second tower (1)	• Reduced the potential inventory • Reduced the inventory • Eliminated tankage (thus reducing inventory) • Eliminated stored aqueous ammonia • Removed excess piping • Installation of a tubular reactor that reduced the inventory of ammonia • Reduced inventory by eliminating tanks or reducing the amount of piping • Eliminated a process furnace
	Substitute		Process change using less hazardous chemical (1)			

RISK REDUCTION CATEGORY	ISS Strategy	2006	2005	2004	2003	2002
	Moderate		Reduced the hazard by changing hazard conditions or form (4)		• Decreased concentration (1) • Less hazard process conditions or form (3)	• Substituted a less hazardous chemical for a more hazardous chemical • Switched from anhydrous dimethyl amine to 60% aqueous dimethyl amine • Substituted a chemical with a less hazardous form of the chemical • Eliminated a process furnace
	Simplify	• Removed out of service equipment or dead leg (8) • Simplified process (1)	• Simplified the process or eliminate equipment (5)	• Revised equipment design features (5) • Simplified the process (5) • Eliminated cross contamination connections (3)	• Eliminated equipment that could cause hazard (2)	• Eliminated tankage that could cause a hazard • Changed the process conditions to less hazardous process conditions • Removed excess piping • Eliminating a process furnace

RISK REDUCTION CATEGORY	ISS Strategy	2006	2005	2004	2003	2002
	Minimize	Revised equipment design features (1)				
	Substitute					
	Moderate	Reduced the potential of a hazard by eliminating the hazard or moving to an alternate location (2)				

RISK REDUCTION CATEGORY	ISS Strategy	2006	2005	2004	2003	2002
	Simplify	• Improved capacity, temperature, and/ or overpressure design rating of equipment (3) • Reduced potential of a hazard or the frequency by changing design features, and/or installation of blinds (8)	• Reduced potential overpressure hazard by equipment elimination (8) • Reduced potential of a hazard by design features, re-routing piping, and increased support (7) • Improved temperature and pressure design rating of equipment (11) • Reduced the potential of a hazard by eliminating liquid accumulation points or vibration stress (5)	• Revised equipment design features (2) • Improved temperature design rating of equipment (2) • Reduced the frequency of a hazard (5)	• Simplified the process or piping (4) • Reduced potential of a hazard by design features, heating media and equipment location (8) • Improved overpressure design rating of equipment (3) • Eliminated hazard to downstream unit without active device by removing contaminants from system (1)	• Hard piping for hydrofluoric acid transfer instead of using a mobile trailer • Installed vessels capable of withstanding the highest pressure scenario • Piping system was modified to naturally restrict flow without use of active controls

Risk Reduction Category	ISS Strategy	2006	2005	2004	2003	2002
Active:						
Procedural:						

11.

Worked Examples & Case Studies

11.1 INTRODUCTION

This chapter illustrates the application of IS principles and concepts in both idealized and actual situations. It also includes a post hoc consideration of IS opportunities as applied to the Bhopal tragedy, dramatically illustrating the potential benefits to both the facility and the surrounding community from identifying and implementing IS opportunities.

11.2. APPLICATION OF AN INHERENT SAFETY STRATEGIC APPROACH TO A PROCESS

As discussed in Chapter 5, inherent safety (IS) concepts can be considered throughout the life cycle of a process. The following example illustrates the concepts described in Chapter 2 (see Figure 2.3), as applied over the life cycle of a process.

Reactive Chemicals, Inc., a fictional coatings industry supplier, is planning to install a new polymerization unit to produce Intermediate C and Final Product Z. The final product goes into various coatings industry applications. Industry expectations are for lower solvent formulations of this type polymer. The following illustrates the processes involved:

$$\text{Intermediate production: } A + B = C$$

In the intermediate reaction, raw material A is reacted with raw material B to produce intermediate C. Current production is in a batch reactor with all materials, including the catalyst, in the initial charge.

Raw material A is flammable (flash point <100°F), toxic, and supplied and stored in bulk. Raw material B is a reactive monomer that is corrosive (to human tissue) and combustible (flash point >100°F), and is typically inhibited with hydroquinone (HQ) or methoxyhydroquinone (MEHQ). Like Raw material A, it is supplied and stored in bulk. The catalyst used for the intermediate production is boron trifluoride (BF_3), a toxic gas supplied in cylinders. The reaction is carried out at 200 °F and at slight positive pressure.

After production of C, the intermediate is batch distilled under full vacuum and the purified C is collected, re-inhibited with MEHQ, and stored at ambient conditions under air. C is a reactive monomer with a flash point above 200 °F. Distillation bottoms are drummed for disposal as a reactive waste.

Final Product Production: C + D = Z

Intermediate C is polymerized with raw material D in a batch reaction producing final product Z, which is diluted in solvent E. Current production is in a batch reactor with all materials including initiator and solvent in the initial charge. The reaction is conducted at atmospheric pressure, and cooling is achieved by solvent reflux and supplemented by a reactor jacket. Available plant cooling water is used.

Raw material D is a reactive, corrosive (to human tissue) monomer, and Solvent E is flammable and considered toxic. Both materials are supplied and stored in bulk. The initiator is a peroxide type which requires refrigerated storage.

The final product Z is a polymer which, by itself, is non-toxic and nonreactive. However, in the current solvent, the product is flammable and toxic (see Figure 11.1).

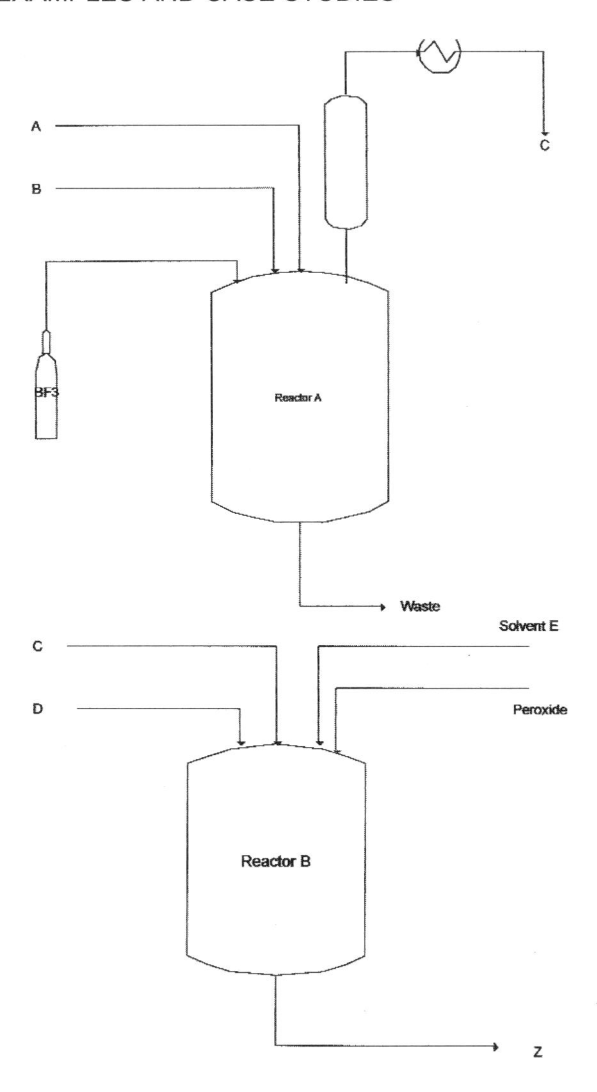

Figure 11.1. Initial Process

Inherent Safety Assessment:

The worksheet that follows demonstrates how different orders of inherent safety could be incorporated, but also how inherent safety *concepts* could be incorporated through the life cycle of the process, as illustrated in Figure 11.2.

Reactor Life Cycle Stages Examples							
	Life Cycle Phase						
Examples of Inherent Safety Order	**Research and Development**	**Conceptual Design Phase**	**Detailed Engineering**	**Construction and Startup**	**Routine Operations**	**Plant and Process Modification**	**Decommissioning**
First order Inherent Safety Examples	Flammable polymerization solvent replaced with non-toxic low VOC solvent and water	Wiped Film Evaporator chosen to replace batch distillation of intermediate C	Elimination of flammable solvents allowed expansion of electrically unclassified areas during design	Use water/ surfactant cleaning methods to eliminate flammable solvent cleanings during startup	Use water/\surfactant cleaning methods to eliminate flammable solvent cleanings during product changeovers, Waste can now go through site waste treatment process	Redesign polymerization catalyst/inhibitor package to eliminate hazardous peroxide use	Complete elimination of hazardous materials (including residuals) from the process using water and surfactants for cleaning.
Second order Inherent Safety Examples	• Initial reaction inhibitor package replaced with phenothiazine (PTZ) which allows reaction to be run under nitrogen (reducing flammability	Decision made to add PTZ inhibitor to monomer B in storage tank (upon receipt) reducing likelihood of runaway in storage	• Batch size calculated to utilize full container of BF3 reducing likelihood of mischarge and inadvertent release of BF3	Minimum inventory in process during startup to minimize any effects	Improve inventory management system of BF3 catalyst to minimize plant inventory (i.e., just-in-time raw material deliveries)	• Replace centrifugal pumps with mag drive pumps to reduce leak potentials and polymerization initiation potential	Identification of technology to immobilize potentially hazardous materials generated during demolition and decommissioning to minimize long term safety, health

Reactor Life Cycle Stages Examples							
	Life Cycle Phase						
Examples of Inherent Safety Order	**Research and Development**	**Conceptual Design Phase**	**Detailed Engineering**	**Construction and Startup**	**Routine Operations**	**Plant and Process Modification**	**Decommissioning**
Second order Inherent Safety Examples	concerns) and reduces likelihood of runaway reaction • Second reaction changed to semi-batch which limits runaway reaction potential		• Monomer storage designed with indirect heating (tank inside a heated room) to prevent overheating as a runaway scenario			• Replace filter press with bag filtration to reduce waste disposal.	and environmental hazards from the demolition waste.
Layers of Protection Examples			• Polymer-ization vessel designed to contain runaway (P & T consistent with adiabatic heat rise) for worst case scenario	Incorporation of human factors features (i.e., accessibility, maintainability, readability)	• Sampling procedures and methods changed to reduce exposures during in-process and product sampling	• Leak detection system installed in BF3 area with mitigation system • Incorporation of human	Mechanical isolation of process to minimize potential for hazardous material release

Reactor Life Cycle Stages Examples							
	Life Cycle Phase						
Examples of Inherent Safety Order	Research and Development	Conceptual Design Phase	Detailed Engineering	Construction and Startup	Routine Operations	Plant and Process Modification	Decommissioning
Layers of Protection Examples			• DCS and SIS systems , secondary containment for all vessels, pressure relief and mitigation systems • Human factors analysis completed during detailed design to facilitate accessibility, maintainability and readability.		• BF3 charging system optimized to reduce potential for exposures during cylinder connection/disconnection activities.	factors features (i.e., accessibility, maintain-ability, readability) • Provision of additional interlocks, alarms, control systems, secondary containment, pressure relief and mitigation systems subsequent to PHA revalidations.	

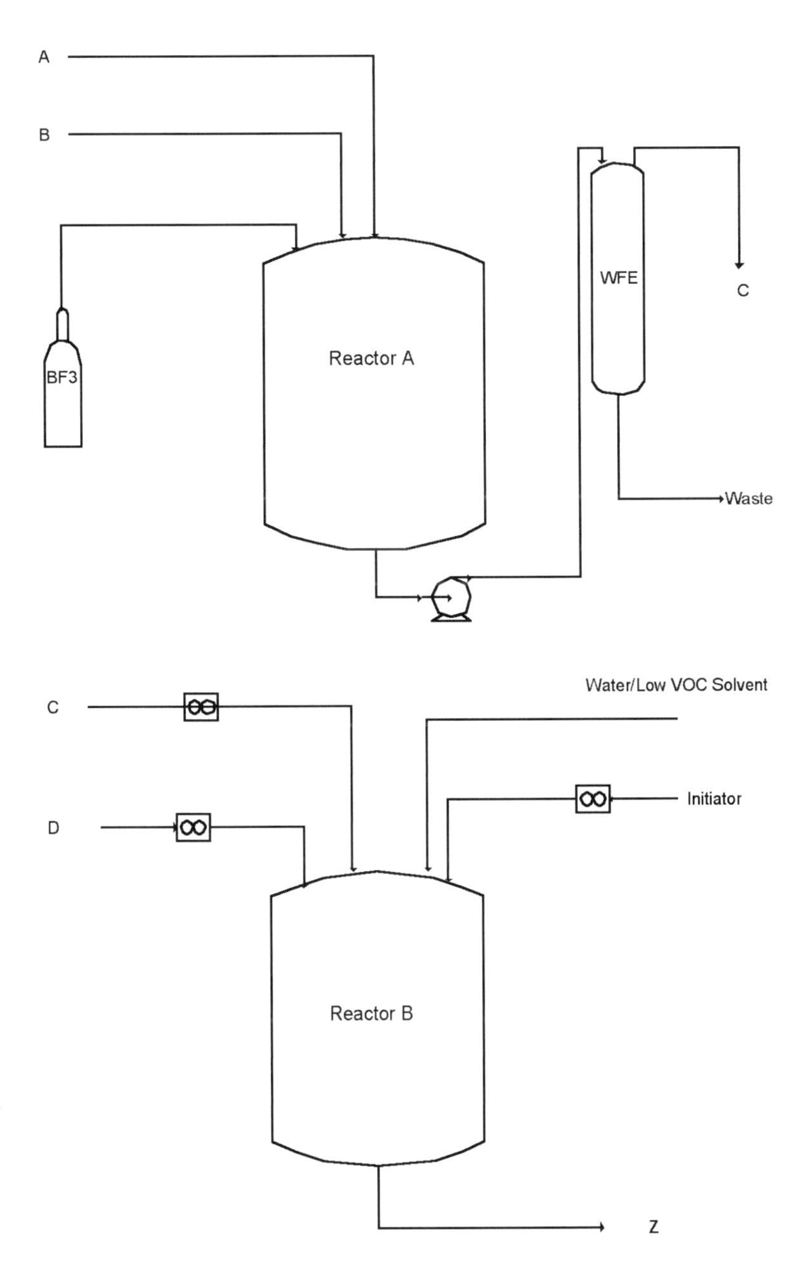

Figure 11.2 Modified process

11.3. CASE STUDIES FROM CARRITHERS, ET AL.

The case studies and examples in the following section are taken from a presentation entitled, "It's Never Too Late for Inherent Safety," by G.W. Carrithers, A.M. Dowell, and D.C. Hendershot. It was prepared for and presented at the International Conference and Workshop on Process Safety Management and Inherently Safer Processes, sponsored by the American Institute of Chemical Engineers, and held October 8-11, 1996, in Orlando, FL.

11.3.1 An Exothermic Batch Reaction

An existing semi-batch process is used to carry out an exothermic reaction:

$$A + B \xrightarrow[\text{Solvent } S]{\text{Catalyst } C} D + \text{Heat}$$

In addition to the highly exothermic nature of this reaction, there is an additional hazard. If Reactant B is significantly overcharged (double charge or more), a side reaction can occur that generates thermally unstable by-products, and a runaway reaction can result. Similarly, if the batch temperature gets too high, there is a potential for a thermal runaway due to undesired side reactions.

A simplified version of the process equipment is shown in Figure 11.3. Reactant A is dissolved in Solvent S in Weigh Tank A, and the solution is then fed to the reactor. Catalyst C is added to the reactor, and cooling is started to the reactor cooling coils. Reactant B, the limiting reagent, is then gradually fed to the reactor by gravity addition from Weigh Tank B, which has been pre-charged with the proper amount of reactant. The feed rate of Reactant B is controlled by the batch temperature. The process has a safety instrumented system (SIS) that takes it to a safe state when abnormal operation is detected. The SIS takes action if the reactor agitator fails, or cooling fails, or high reactor pressure or temperature is detected. All of the SISs stop the feed of the limiting reagent, Reactant B, by closing the feed flow control valve and a dedicated block valve.

As a result of the findings from a PHA, the system was modified as shown in Figure 11.4. The modified system includes several inherent safety features, which were implemented in this existing plant with a relatively small investment.

1. The Reactant B feed tank has been moved to the floor below the reactor level, and the feed is now controlled by a metering pump. In case of high temperature, the metering pump is turned off and a block valve is closed by the SIS on reactor high temperature. It is less likely that there will be a significant leak through the metering pump, as compared to the original

gravity feed where the driving force for the Reactant B flow is always present.

2. The maximum flow rate of the metering pump is not capable of generating more heat from reaction than the reactor cooling capacity can manage. Therefore, it is not possible to overheat the reactor by feeding Reactant B at a rate that exceeds the reactor's capability to remove heat.

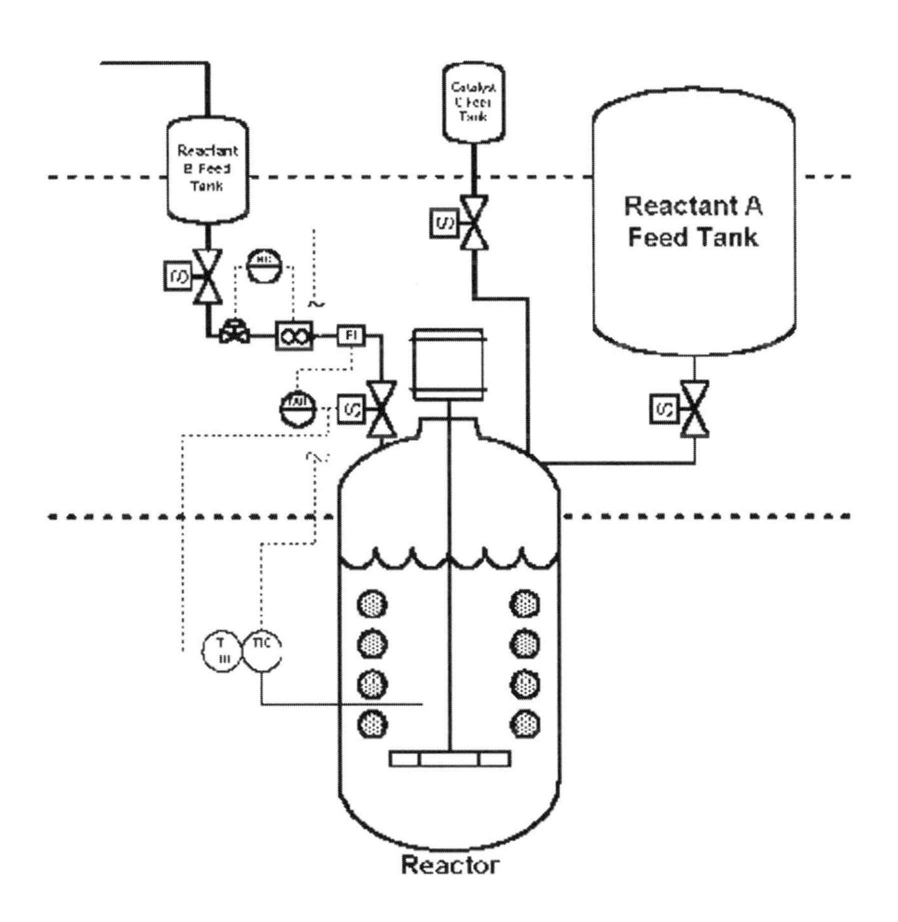

Figure 11.3 Original batch reaction system

3. The maximum capacity of the Reactant B feed tank has been reduced to exactly one batch charge. In this case, the same Reactant B feed tank was re-used, but it was relocated to the lower floor. To reduce its maximum capacity, an overflow was added to the side of the tank at the desired level, with the overflow piped back to the Reactant B storage tank.

4. In the revised system, Reactant B is charged to the feed tank through a three-way valve to the bottom of the feed tank. The three-way valve allows flow either from the storage tank to the Reactant B feed tank, or from the feed tank to the reactor. It is not possible to pump Reactant B directly from the storage tank to the reactor. This system makes it much more difficult to overcharge Reactant B.

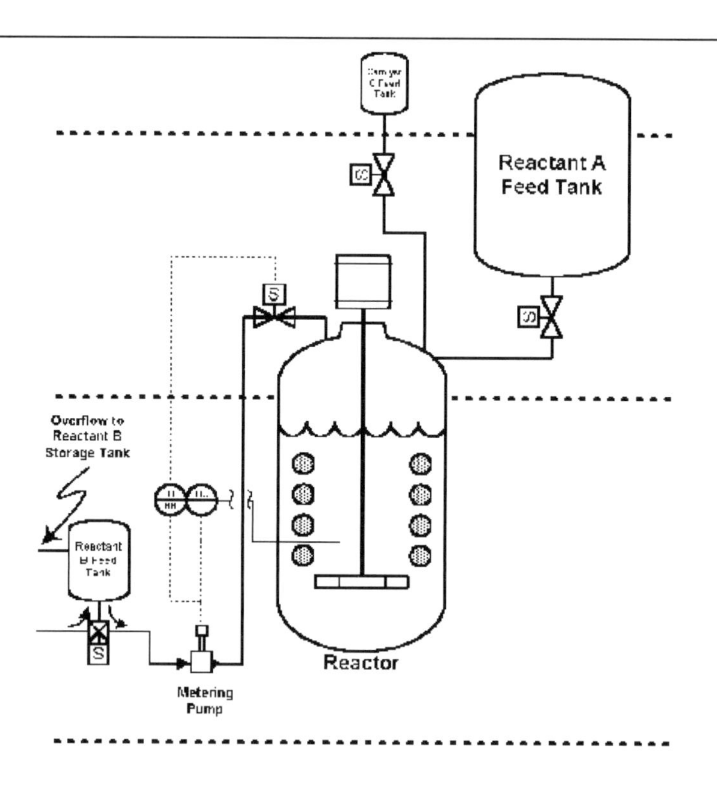

Figure 11.4: Modified, inherently safer batch reactor system

These modifications represent significant improvements to the inherent safety of the existing plant. Following the PHA, a quantitative risk analysis (QRA) of this reactor system was completed. The QRA showed a total risk reduction of about a factor of 1000 when compared to the original system.

11.3.2 Refrigeration of Monomethylamine

Refrigeration can be an effective way of minimizing the effects of a release of a material with a low boiling point (CCPS, 1993). Refrigeration reduces the hazard potential by:

- Reducing the storage pressure, preferably to near ambient, thus reducing the release rate in case of a leak in the system.
- Eliminating or greatly reducing the initial flash of material in case of a leak, because the material is not stored as a superheated liquid above its ambient pressure boiling point.
- Eliminating or reducing the atomization of liquid which does not flash. The very small droplets will not "rain out," but instead will travel downwind in the vapor cloud, where they will eventually evaporate due to heat absorbed from the surrounding atmosphere.

Monomethylamine (H_2NCH_3), a flammable gas with a strong, ammonia-like odor, was stored in a pressure vessel at ambient temperature. The vapor pressure of monomethylamine is about 3.9 atm absolute (42 psig, 57 psia) at 20 °C. Refrigeration of the monomethylamine was identified during a hazard analysis as an option for reducing the hazard zone in case of a leak from the system. The potential impact of this change was evaluated using PHAST®.[1] Table 11.1 shows the results of the consequence modeling study for one example release scenario—the rupture of a 5.1 cm. (2 inch) line between the liquid monomethylamine storage tank and the transfer pump to the manufacturing process. The ERPG-3[2] concentration of 500 ppm was selected as a representative concentration to illustrate the reduction in the hazard distance which could be attained by refrigerating the monomethylamine.

Refrigeration of monomethylamine has a significant impact on the hazard zones. The modified system is also inherently safer because the storage tank is capable of containing the monomethylamine at ambient temperature. It does not rely exclusively on the proper operation of the refrigeration system to ensure that the vapor pressure does not exceed the design pressure of the storage tank.

---

1 PHAST® (Process Hazard Analysis Software Tool) is a comprehensive hazard analysis software tool broadly used in the chemical and petroleum industries developed and marketed by DNV Software is the commercial software house of DNV (Det Norske Veritas).

2 American Industrial Hygiene Association (April 20, 1988). *Emergency Response Planning Guidelines: Chlorine*. Akron, OH: American Industrial Hygiene Association. Referenced in Carrithers, et al. (1996).

Table 11.1: Effect of Refrigeration on Distance to ERPG-3 Concentration for a 5.1 cm. (2 inch) Monomethylamine Pipe Rupture

Monomethylamine Storage Temperature (°C)	Distance to ERPG-3 (500 ppm) Concentration, km (miles)
10	1.9 (1.2)
3	1.1 (0.7)
-6	0.6 (0.4)

11.3.3. Elimination of a Chlorine Water Treatment System

Chlorine is often used for water treatment in a chemical manufacturing facility. Because the chlorine handling facilities are often in the utilities area, they may not be recognized as presenting a significant potential hazard.

One ton cylinders of chlorine were used for disinfecting water at a chemical manufacturing facility, with liquid chlorine piped through 1.3 cm. (1/2 inch) pipe to an injection nozzle, where it was mixed with the water being treated. Modeling of the consequences of a rupture of the 1.3 cm (1/2 inch) liquid chlorine pipe indicated that the distance to the ERPG-3 chlorine concentration of 20 ppm (AIHA, 1988) could be as far as about 1.6 km (1 mile), depending on weather conditions. An alternative chlorination system was considered that uses a chlorine vaporizer immediately adjacent to the cylinders. In this system, the long outdoor transfer line from the chlorine cylinder storage feed area to the water treatment area contained chlorine as a gas, instead of as a liquid, significantly reducing the inventory in the transfer pipe. This alternative reduced the distance to the ERPG-3 concentration of 20 ppm chlorine to about 0.8 km (0.5 mile) for the release scenario of rupture of the chlorine transfer line.

For this installation, alternative water disinfecting systems were also investigated. It was found to be feasible to use sodium hypochlorite treatment, and this essentially eliminated the hazard of a chlorine vapor cloud entirely.

11.3.4. Reduction of Chlorine Transfer Line Size

A manufacturing process included a chlorination process using liquid chlorine. The original facility used a 5.1 cm (2 inch) transfer line from the chlorine storage facility to the manufacturing building. A hazards review questioned the line size, and it was determined that it could be reduced to 2.5 cm (1 inch) without impacting the manufacturing process. Table 11.2 shows the impact of this reduction on the hazard zone resulting from the potential failure of the transfer pipe. Several typical weather conditions are considered. For purposes of this example, the hazard zone was defined as the distance to the ERPG-3 concentration for chlorine, 20 ppm. For all weather conditions, the distance to the ERPG-3 concentration was reduced by a factor of two to three.

11.3.5. Substitution of Aqueous Ammonia For Anhydrous Ammonia

Dilution of a hazardous material can be an important strategy for improving the inherent safety of a chemical process or storage facility. It can reduce the storage or vapor pressure of a hazardous material and reduce the atmospheric concentration of hazardous vapor from a spill.

Table 11.2: Effect of Reduction of Line Size on Hazard Zone from Potential Failure of a Chlorine Transfer Line

Chlorine Transfer Line Diameter	**Distance in kilometers (miles) to ERPG-3 Concentration for Chlorine (20 ppm) (Accident Scenario - Line Rupture)**		
Weather Conditions	D Stability 5.5 km/h (3.4 mph) Wind Speed	D Stability 18 km/h (11.2 mph) Wind Speed	F Stability 5.5 km/h (3.4 mph) Wind Speed
5.1 cm (2 inch)	5.5 (3.4)	2.1 (1.3)	6.8 (4.2)
2.5 cm (1 inch)	1.9 (1.2)	0.65 (0.4)	2.7 (1.7)

Approximately 227,000 kg (500,000 pounds) of anhydrous ammonia were stored in a large pressurized storage tank. The tank was rated for 6.2 atm absolute (75 psig, 90 psia) working pressure, and it had a pressure relief valve set for 6.0 atm absolute (72 psig, 87 psia). This value compares to the vapor pressure of ammonia of about 6.4 atm absolute (93 psig, 108 psia) at 16 °C (60 °F). The ammonia was kept under refrigeration to maintain the storage tank pressure below the tank pressure rating and relief valve set point. Occasionally, the refrigeration system would fail, and the storage tank pressure would slowly increase as the ammonia temperature rose. There was a potential to open the relief valve if the refrigeration system could not be returned to service before the relief valve set point was reached.

The system was reviewed, and the team determined that 28% aqueous ammonia could be substituted for anhydrous ammonia in all of the processes and products where it was currently used with minimal impact—including economic considerations—on these manufacturing processes.

To illustrate the benefit of substituting aqueous for anhydrous ammonia, the consequences of two release scenarios were modeled using PHAST®:

- Release of the entire contents of the ammonia storage tank over a 10 minute period. This scenario is equivalent to the "worst-case" scenario and is a requirement of the EPA Risk Management Program (RMP) regulations (EPA, 1996).[3]
- Failure of a 5.1 cm (2 inch) diameter ammonia transfer pipe between the storage tank and the ammonia transfer pump. This scenario might be considered an appropriate "alternative release scenario" as required by the EPA Risk Management Program regulations (EPA, 1996).

The results of the analysis are summarized in Figure 11.5 for both potential release scenarios. The use of aqueous ammonia greatly reduces the downwind ammonia concentration in the resulting vapor cloud.

The aqueous ammonia system also does not require a refrigeration system, which greatly simplifies the storage facility. The storage tank is now capable of containing the ammonia under all ambient temperature conditions, eliminating the reliance on proper operation of the refrigeration system to maintain a storage tank pressure less than the relief valve set pressure.

This case study also illustrates the importance of periodic review of all facilities that includes a search for inherently safer design options. There had, at one time, been a requirement for anhydrous ammonia at this site. It was logical, and safer, to use a single ammonia storage facility for anhydrous ammonia, rather than two separate systems. However, the process that required anhydrous ammonia had been shut down, and it was now possible to convert the other processes to the use of aqueous ammonia, resulting in an inherently safer system.

---

[3] Environmental Protection Agency (EPA) (January 22, 1996). *Off-site Consequence Analysis Guidance* (Draft). Washington, D.C.: U.S. Environmental Protection Agency; referenced by Carrithers, et al.

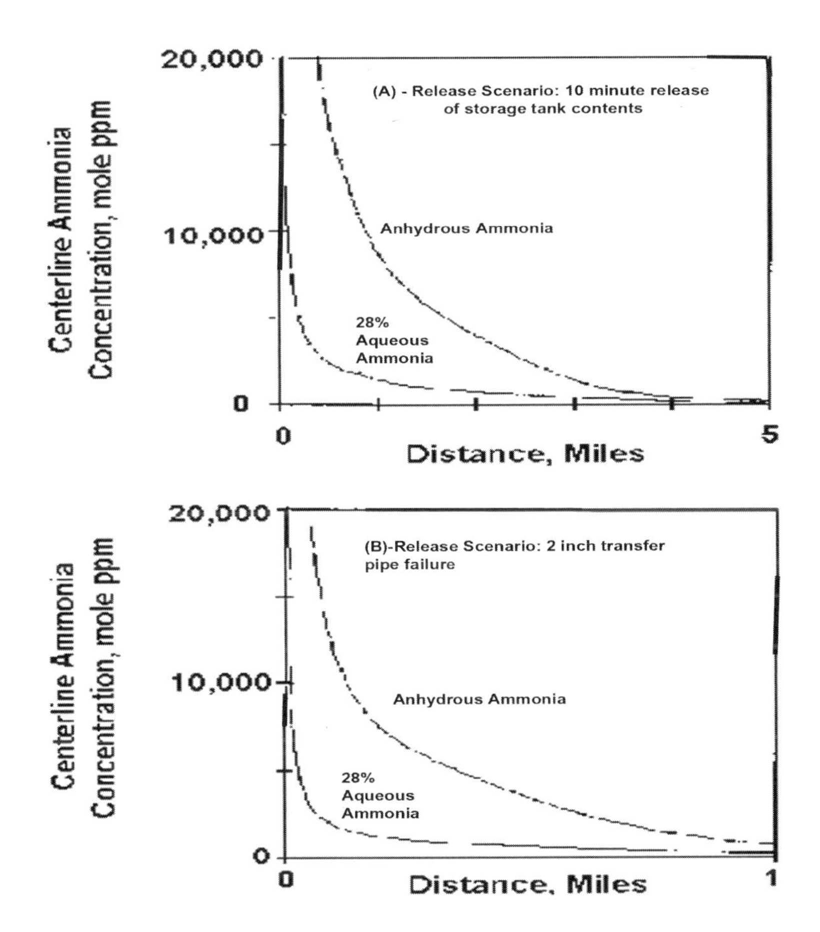

Figure 11.5: Comparison of centerline vapor cloud concentration as a function of distance from the release for anhydrous and 28% aqueous ammonia storage for two release scenarios
(Weather – D. Stability, 3.4 mph wind speed)

11.3.6 Limitation of Magnitude of Deviations for Aqueous Ammonia

The hazard of a substance can be reduced by limiting the possible magnitude of process deviations. By designing equipment to be more tolerant of error (or more robust), the frequency of a release can be reduced (CCPS, 1993).

For the reasons listed in the previous example, an aqueous ammonia system was installed to supply ammonia for reaction in a process that could tolerate the water solution. The system was initially installed using an available surplus tank of

1.7 atm absolute (10 psig, 25 psia) design pressure, and fitted with a 1.7 atm absolute (10 psig) relief valve. The system injected anhydrous ammonia in ratio with water into an eductor in a circulating heat exchanger loop (Figure 11.6). For the purpose of this example, the desired concentration was about 30 wt%, and the normal operating temperature was about 38 °C (100 °F) or less.

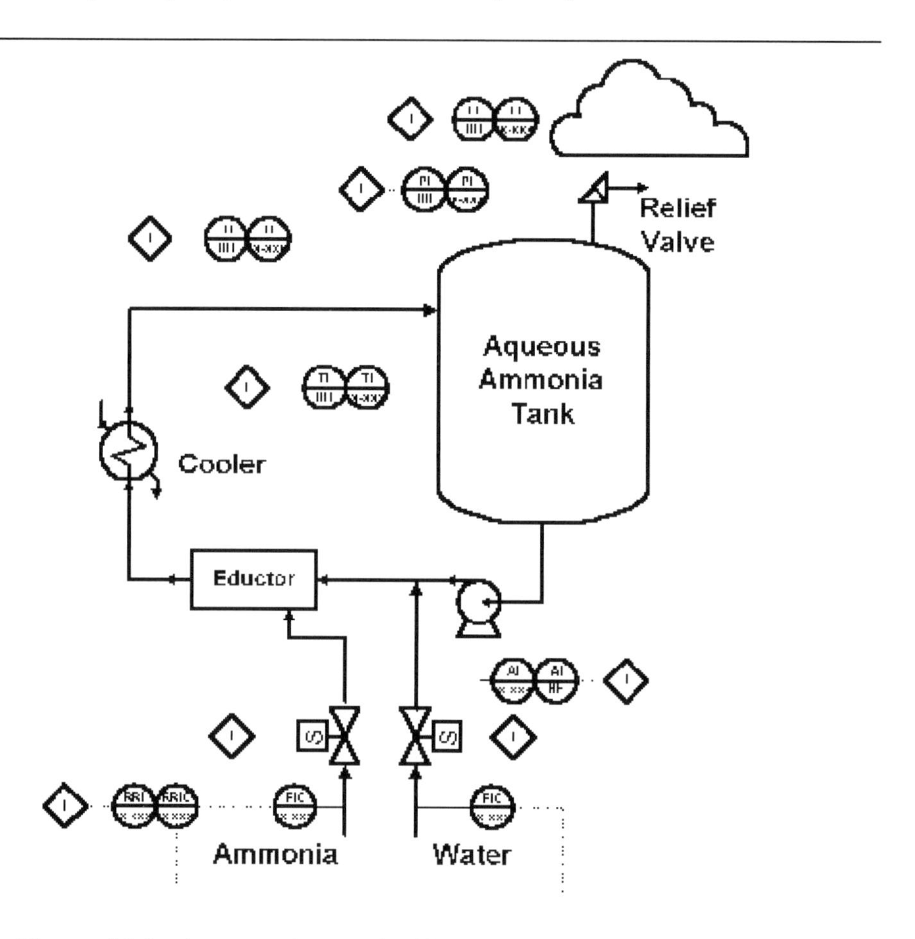

Figure 11.6: Aqueous Ammonia: Limitations of Magnitude of Deviations

Table 11.3 gives the total vapor pressure of aqueous ammonia for conditions near the design point. The tank design pressure will be exceeded and the relief valve will open if the temperature rises, or if the concentration increases, even by small increments, as shown by the shaded area and the black border on the right side.

Table 11.3: Total vapor pressure of aqueous ammonia
(adapted from Perry, 1984, page 3-73)[4]

Temperature	Wt % NH_3 in solution			
°F	19%	24%	29%	34%
90	7.41 psia	11.4 psia	17.2 psia	25.5 psia
100	9.3 psia	14.2 psia	21.3 psia	31.2 psia
110	11.6 psia	17.6 psia	26.1 psia	37.8 psia
120	14.4 psia	21.5 psia	31.7 psia	45.6 psia
130	17.7 psia	26.2 psia	38.2 psia	54.6 psia
140	21.5 psia	31.5 psia	45.7 psia	65.8 psia

10 psig Tank Threshold of Release

50 psig Tank Threshold of Release

Initiating causes for a pressure surge leading to a release include:

- loss of flow in the circulation loop
- upset in the ratio control
- loss of coolant to the heat exchanger
- high temperature of coolant in the heat exchanger
- upsets in the ratio control

During plant operation, several pressure surges from these causes opened the relief valve. The resulting cloud required evacuation of adjacent process units. The emergency response was to use a fire monitor to knock down the cloud while the unit corrected the control or heat transfer problem and waited for the tank pressure to come down. To prevent releases, an SIS was implemented that isolated the make-up ammonia when there was high temperature, high pressure, improper feed ratio, high level, or loss of the circulating pump. The SIS had to be designed to

[4] From *Perry's Chemical Engineer's Handbook, 6th Edition.* page 3-73. See Reference Section for full citation.

have a fast response time, since temperature and pressure rose quickly during a upset.

A new tank of 4.4 atm absolute (50 psig, 65 psia) design pressure was installed to make the plant less sensitive to upset. The concentration would have to reach about 34% at a temperature of 60 °C (140 °F) to cause a release, shown by the heavy black border in the lower right corner of Table 11.3. The new tank essentially eliminated releases from the relief valve. The SIS did not have to act as quickly, and, perhaps, some shutdown initiators could be eliminated.

This example illustrates the fact that the cheapest equipment—even if it is free—may not always be the most cost-effective option, especially when the economics of the consequences of releases and the cost of SISs are considered.

11.3.7. A Vessel Entry Example

Nitrogen is often piped to process vessels for process reasons, as well as to inert equipment handling flammable materials. When it is necessary to enter such vessels for inspection or maintenance operations, one of the concerns is ensuring that the nitrogen is properly disconnected or otherwise isolated from the vessel. Figure 11.7 schematically shows an inherently safer approach for being sure that the nitrogen has been disconnected *prior* to vessel entry. All nitrogen lines to the vessel are fed through a single pipe, which passes through a hose or removable section of pipe directly across the vessel manway. It is not possible to open the vessel manway without first removing the nitrogen hose or removable pipe section. Figure 11.8 shows how this has been implemented for two vessels.

Of course, this design does not ensure that the atmosphere inside the vessel is safe for entry, and all permit and vessel entry procedures are still needed. There are many other mechanisms that could lead to a hazardous atmosphere inside the vessel. For instance, it may contain hazardous materials, the vessel may not have been properly purged prior to entry, and other process pipe connections may not be as reliable. However, by making it very difficult to avoid disconnecting the nitrogen supply to the vessel when opening the manway, this design does provide an inherently more reliable mechanism for ensuring that one hazard to personnel involved in a vessel entry operation has been eliminated.

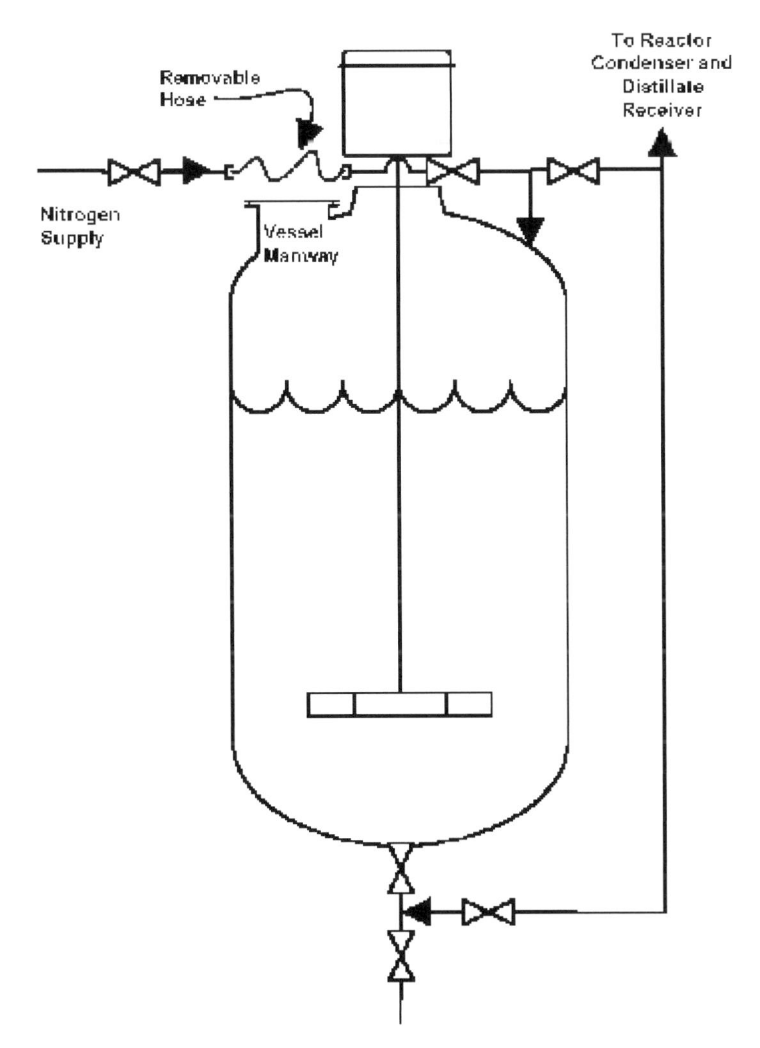

Figure 11.7: Nitrogen supply line routed over the vessel manway. The nitrogen must be disconnected to open the manway.

Figures 11.8a and 11.8b: Two examples of nitrogen supply lines piped to vessels across the manway. The manway cannot be opened without disconnecting the nitrogen.

11.4. PROCESS ROUTE SELECTION – EARLY R&D EXAMPLE

Inherent safety has been institutionalized into procedures that researchers must follow when developing a process chemistry and/or a process design. Researchers are required to review hazards and document a process hazards analysis (PHA) for each experimental set-up and/or significant change in that set-up. A checklist is a required part of that PHA effort and inherent safety (IS) questions are included in the instructions for completing the checklist.

The format (template) for technical reports on product, chemistry, and process development includes a section on IS, as does the report format for applying for permission to seek a patent. The principal instruction on the template for each is similar to the following:

> *If the process or chemistry being addressed is aimed at the implementation of a new manufacturing process, or improvements/changes to an existing manufacturing process, then a discussion of the anticipated hazard/risk level at the commercial scale must be included. This discussion should be from an IS perspective and include consideration of the quantity of hazardous materials involved, and the severity of process conditions. The use of a standard "index" sheet—a form that dictates IS be considered—is required.*

The index sheet is a chart that gives five levels of definition (from low to high) for toxicity, flammability, and reactivity (i.e., the material factor), for quantity (i.e., the quantity factor), and for reaction severity, pressure, temperature, corrosiveness/erosiveness, dust content, operability, and experience (i.e., the process factor). The researcher is asked to assign a "level of severity" to each factor, and to sum them.

The higher the resulting number, the more hazardous the chemistry or process is, from an IS standpoint. Several alternative chemistries and/or processes must be proposed and the "index" sheet used for them as well. If competing chemistries and/or processes exhibit a lower total "level of severity," the researcher is obligated to defend his choice. No chemistry that exhibits "severe" factors in all categories is accepted.

11.5. EXAMPLE OF AN INHERENTLY SAFER STUDY OF A STEAM PRODUCTION FACILITY

The following comes from a paper written and presented by Karen Study at a 2005 safety conference sponsored by the Mary Kay O'Connor Process Safety Center in College Station, Texas.[5]

Choosing an inherently safer alternative may seem straightforward. However, sometimes what initially seems to be the most obvious IS alternative may not actually provide the best overall risk reduction. In this case study, an "inherently safer" alternative was selected and later discarded due to issues uncovered during the detailed design phase. The option ultimately chosen was inherently safer than both the original design and the "inherently safer" alternative.

Facility Description: The unit produces a large amount of steam using a multiple burner boiler with natural gas and a low BTU off-gas as its fuel sources. The boiler waste gas (flue gas) is sent to an elevated stack where it is discharged to the environment. This flue gas is mainly nitrogen and water, with oxygen and carbon dioxide. As with all boilers, there is also NO_x present in the flue gas. A team was formed to assess different NO_x reduction options. After evaluating several options to achieve the required NO_x emission reduction targets, the design team chose to install a Selective Catalytic Reactor (SCR).

Initial Design Proposal (Liquid Anhydrous Ammonia): To supply ammonia to the SCR, the design team chose to tap into an existing liquid anhydrous ammonia piping header that supplied a nearby processing unit. Piping was minimized as much as possible, to ~600 feet of 2 inch pipe. A vaporizer skid, which used steam to vaporize the liquid ammonia prior to injecting into the SCR, was to be installed near the boiler. See Figure 11.9 for a high level overview of this option.

---

[5] See References for full citation.

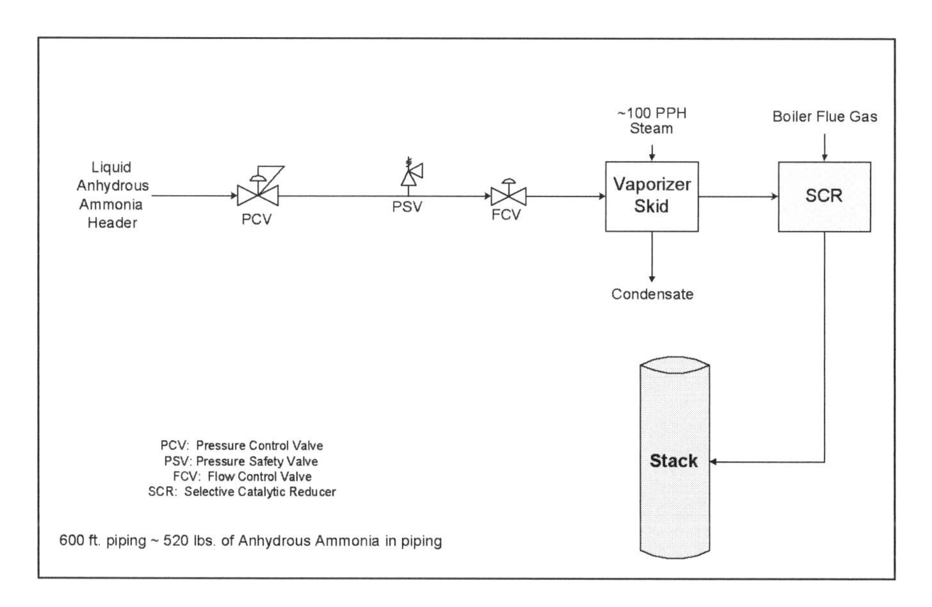

Figure 11.9: Initial ammonia supply proposal:
Liquid anhydrous ammonia supply *(Study, 2005)*

After the option was selected, the process safety group was consulted to provide input. Due to concerns regarding incrementally increasing risks associated with the current liquid anhydrous ammonia piping system, the safety group recommended using aqueous ammonia available from a nearby processing unit. This seemed to be a straightforward alternative (see Figure 11.9 above for anhydrous/aqueous ammonia comparison).

Aqueous Ammonia Design Proposal: An aqueous ammonia user in the plant could deliver it to the boiler facility via a connection downstream from the storage tank. Since this tank was much further from the boiler than the anhydrous ammonia header, the length of piping required was much greater. The tank also required new positive displacement pumps to supply the aqueous ammonia. And, a temporary supply alternative had to be built into the design since periodic shutdowns were required at the aqueous ammonia tank. To accommodate these periodic supply needs, additional connections and provisions were made for tank truck deliveries of aqueous ammonia. See Figure 11.10 for an overview of the aqueous supply option.

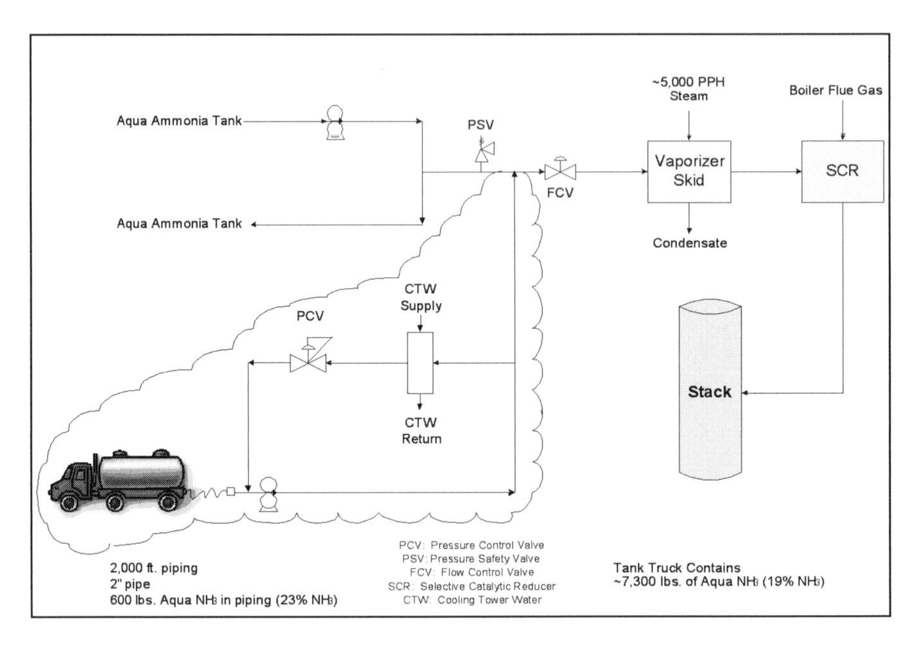

Figure 11.10: Aqueous Ammonia Supply Proposal

The design team conducted a Hazard and Operability analysis that raised several concerns regarding the tank truck delivery system and associated operations:

- the risk of spills and operator errors for the tank truck portion of the delivery system was higher for this option than an anhydrous ammonia system
- higher capital, operating and maintenance costs
- the reliability of the addition of pumps in the system.

So, the project was recycled back to the option selection phase.

Final Round of Option Selection: During the new round of option selection, the use of anhydrous ammonia vapor was evaluated. This option had not been fully evaluated previously based on early assumptions that:

- the available anhydrous ammonia vapor was insufficient for the SCR.
- aqueous ammonia was a safer alternative.
- the vapor stream would be more difficult to control than the liquid stream.

A nearby processing unit that vaporized anhydrous ammonia from the plant header prior to use was evaluated and, based on an evaluation of the downtime of their system, it was determined that anhydrous ammonia vapor could be supplied to the SCR at a sufficiently high on-stream rate, without a secondary supply system.

Flow control issues could be addressed by using redundant instruments that had a good operating history in similar service. To prevent condensation in the transfer piping, the ammonia pressure was dropped to 25 psig at the tie-in point to the header. This made the risks associated with the use of anhydrous ammonia vapor less than if high-pressure vapor was used. Redundant instrumentation for control valves and letdown regulators reduced downtime for the system. A small amount of low-pressure steam was added to the ammonia prior to injection into the SCR as a diluent to more evenly disperse the ammonia in the catalyst bed. The anhydrous ammonia vapor option is depicted in Figure 11.11. (Redundant instrumentation is not shown.)

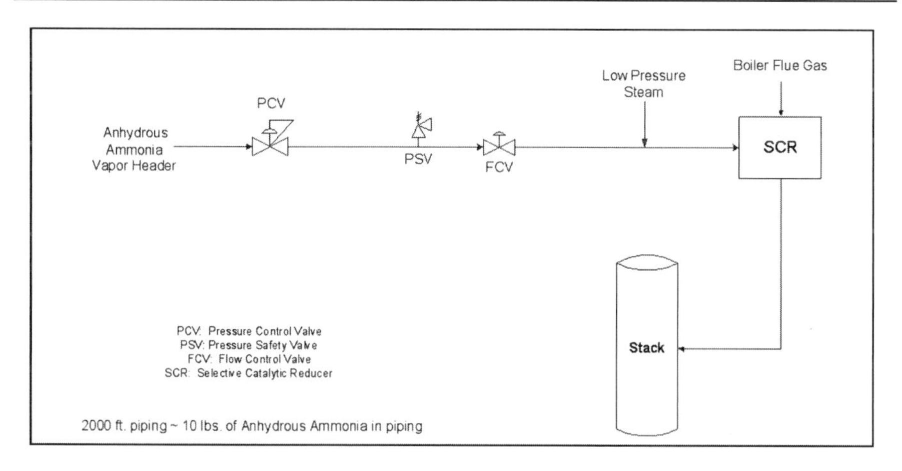

Figure 11.11: Anhydrous ammonia vapor option

Consequence Analysis: For comparison, the designers modeled a complete break in the piping supplying the SCR, resulting in an almost instantaneous release of the entire pipe contents. The consequence analysis for liquid anhydrous ammonia estimated a maximum distance of 1,170 feet to the IDLH concentration under the worst-case weather conditions (F atmospheric stability, 1.5 m/s wind speed). For the aqueous ammonia piping system using the same leak scenario and weather conditions, the impact distance for IDLH level results was 930 feet. The ammonia vapor scenario impact distance was 125 feet.

Another issue with the aqueous ammonia system is the temporary tank truck needed when the primary source is not available. Connecting and disconnecting hoses to the truck introduces new release scenarios involving smaller amounts of ammonia in close proximity to production operators. Plus, the large tank truck inventory greatly increases risks during a catastrophic failure scenario.

Modeling aside, one significant difference among all three options is simply the mass of material inside the transfer piping and associated equipment. The anhydrous ammonia vapor system contained an order of magnitude less ammonia than the other piping systems, and two orders of magnitude less than the aqueous ammonia tank truck. See Table 11.4 for a comparison of the options based on mass differences.

Table 11.4: Mass Comparison for Ammonia Transfer Options

Option	Piping Length (ft)	Volume (ft^3)	NH_3 Mass (lbs)*
Anhydrous Ammonia Liquid	600	14	520
Aqueous Ammonia (23 wt% NH_3)	2,000	47	600
Aqueous Ammonia Tank Truck (19 wt% NH_3)	N/A	652	7,300
Anhydrous Ammonia Vapor	2,000	47	10

**Density of ammonia and water at 80 °F; 37 lb/ft^3 and 62 lb/ft^3 respectively*

Conclusion and Action: The design team ultimately chose to use anhydrous ammonia vapor to supply the SCR. Receiving ammonia in the anhydrous vapor form was determined to be an economically viable option, and the safety analysis indicated that the vapor form of anhydrous ammonia was safer than design options incorporating either liquid anhydrous or aqueous ammonia.

In comparing both the anhydrous vapor and aqueous liquid options to the liquid anhydrous option, using the vapor decreases risks along the piping run without introducing new risks to the system. Based on modeling analysis performed for the piping run, the liquid aqueous ammonia option did have a slightly lower risk compared to the liquid anhydrous ammonia system. However, the aqueous system introduced additional risks by the need to have a temporary supply station. Such a station would have increased risk to operating personnel due to hose handling while hooking up and disconnecting the tank truck. Not only did the anhydrous vapor option not require such a secondary supply, but the piping system contained an order of magnitude less ammonia than the other systems considered.

Judging both the aqueous and anhydrous vapor options from an economic standpoint, the vapor system also had lower capital costs, as well as lower projected operating costs, since very little steam was required prior to injection into the SCR. The vapor system had lower maintenance costs since no pumps were required, and was predicted to be more reliable due to the simple design and absence of moving parts.

Conclusion: This project illustrates the principle that decision-making between inherently safer designs involves evaluating several different metrics, including volume of hazardous material, distances affected by a release, frequency of release (for example, during those connections for aqueous), risk (including consequence severity and frequency), and life cycle cost. The option selected depends upon the metrics used in the decision-making, and the weighting factors among those metrics. There is not a single metric that can be considered as "correct" for selecting one inherently safer design over another option. It always involves trade-offs.

11.6 CASE STUDY: BHOPAL

The following case study is adapted, with permission, from an article called "How to Make Inherent Safety Practice a Reality," written by Faisal I. Khan, and Paul R. Amyotte, and published in the February 2003 issue of The Canadian Journal of Chemical Engineering. *(Volume 81, pp. 2-16)*

Several thousand kilograms of a toxic gas, rich in methyl isocyanate (MIC), were released into the atmosphere from the Union Carbide India, Ltd. pesticide facility. MIC was the intermediate product of carbaryl (a pesticide) manufacturing and was stored in three horizontal stainless steel vessels. The MIC was usually stored under refrigerated conditions, but in the days before the incident, the refrigeration unit had been shut down. Water entered an MIC storage tank, creating an exothermic reaction which heated the MIC beyond its boiling point. The concrete mounds above the tank cracked, releasing MIC vapor through a relief valve. The scrubber and flare systems designed to remove and destroy the MIC were not in operation on the day of the incident. The Bhopal facility relied on engineered and procedural safeguards. Khan and Amyotte analyzed this incident to identify inherent safety considerations that might have been prevented or mitigated the tragedy.

Minimization: MIC was an intermediate in carbaryl production. The facility could have stored less, as evidenced by the fact that MIC stocks had been reduced by 75% within a year of the incident. Furthermore, by changing the release diameter—the parameter reflecting chemical inventory—from a standard 50-mm orifice to 30 mm, the CEI-hazard distance[6] would be reduced by 28%. If the MIC inventory (or release rate) had been minimized, the incident potential would have been reduced (2nd order IS measure).

Substitution: Carbaryl was manufactured by reacting phosgene and methylamine to produce MIC, which was then reacted with alpha-napthol. An alternative process reacts the phosgene and methylamine in a different order to avoid the production of MIC. In this alternate process route, phosgene is reacted with alpha-napthol, and then the intermediate—a less hazardous chloroformate—is reacted with methylamine (Figure 11.12). Application of the substitution principle at the process route selection stage could have played a role in averting the incident consequences (first order IS measure).

Moderation: Moderation of the storage conditions (temperature) would have been enabled by the refrigeration system had it been operating. Rather than being stored at 0°C or lower, as standard procedures required, the MIC was actually at ambient temperature — obviously much closer to its boiling point of 39.1 °C. With the contaminant presence leading to an exothermic reaction and elevated MIC temperatures and vapor generation, it is not clear how pressure might have been effectively moderated. However, a 90% reduction in the operating pressure involved would have resulted in a 60% decrease in the CEI-hazard distance.

Simplification: The Bhopal facility relied on end-of-pipe monitoring and control systems attached to the storage vessels, the reliability of which required they be in good working order. Reliance on such systems can create a dual problem: the safeguards may not be available when needed, and their existence may provide a false sense of security for process operators who may ignore initial warning signs, such as a pressure increase. Many of these issues are brought into focus when one analyzes a system through use of the simplification principle.

---

[6] Calculated using Dow's Fire & Explosion Hazard Classification Guide and Dow's Chemical Exposure Index Guide.

Methyl Isocyanate Route:

$$CH_3NH_2 + COCl_2 \rightarrow CH_3CNO + 2HCl$$

Methylamine *Phosgene* *Methyl Isocyanate* *Hydrochloric Acid*

$$CH_3CNO + C_{10}H_8O \rightarrow C_{12}H_{11}NO_2$$

Methyl Isocyanate *Napthol* *Carbaryl*

Non-Methyl Isocyanate Route:

$$C_{10}H_8O + COCl_2 \rightarrow C_{11}H_7ClO_2 + HCl$$

Napthol *Phosgene* *Napthol Chloroformate* *Hydrochloric Acid*

$$C_{11}H_7ClO_2 + CH_3NH_2 \rightarrow C_{12}H_{11}NO_2 + HCl$$

Napthol Chloroformate *Methylamine* *Carbaryl* *Hydrochloric Acid*

Figure 11.12. Alternative route for carbaryl production *(based on Crowl and Louvar, 2002).*

11. 7 SUMMARIES IN BRIEF: EXAMPLES BY IS STRATEGY

Implementation of inherently safer design principles over the process life cycle has been increasingly practiced by the chemical processing industry. Khan and Amyotte published an extensive list of examples in their article, "How to Make Inherent Safety Practice a Reality" (2003, abbreviated here as KA). Overton and King in their article, "Inherently Safer Technology: An Evolutionary Thing," describe inherent safety applications by The Dow Chemical Company (2006, abbreviated here as OK). This section will present worked examples and case studies which represent the principles and concepts discussed in this book. They are illustrative of the many approaches that companies may take to achieve inherently safer processes.

Minimize: Use smaller quantities of hazardous substances.

- Although the reaction of nitric acid and glycerin is quite rapid, early industrial implementation of this chemistry used large batch reactors. Current technology has enabled efficient contacting of the reactants, making the actual reaction process continuous and rapid. These continuous reactors are orders of magnitude smaller than the older technology batch reactors, thus greatly reducing the potential hazard from this extremely hazardous reaction (KA).

- A continuous process has been developed for manufacturing phosgene on demand, thus eliminating the need for storage of liquid phosgene. Various important issues, such as quality control, understanding of transient reactor operation, and process control, were successfully resolved by a fundamental understanding of the chemical reaction. This enabled the design of a system that met all of the user's requirements in an inherently safer way (KA).
- It is understood that mixing and gas-liquid phase mass transfer controls the rate and efficiency of chlorination reactions. Replacing a stirred tank reactor with a loop reactor, specifically designed to optimize mixing and gas-liquid phase transfer has been shown to significantly reduce reactor size, processing time, and chlorine usage (KA).
- A 50-L loop reactor replaced a 5000-L batch reactor in a polymerization process (KA).
- MDI plants at Dow's LaPorte, Texas, manufacturing site began a campaign in 1986 to reduce phosgene inventory through improved process control and in situ production. Figure 11.13 shows that implementing a series of major capital projects reduced phosgene inventory by 50 % over 10 years while increasing MDI production—just one example of the impact of an evolutionary strategy for inherently safer technologies (OK).

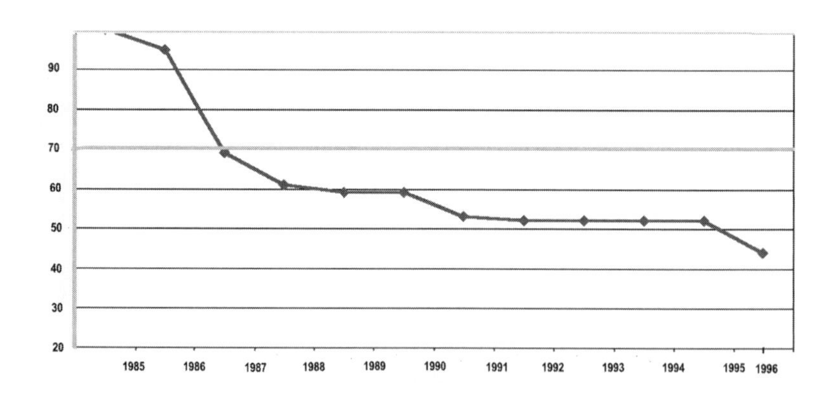

Figure 11.13: Dow LaPorte Phosgene Inventory Reduction

- Dow's Union Carbide subsidiary redesigned the acrolein derivatives unit in Taft, Louisiana, during the mid-1980s to dramatically reduce acrolein inventory. In addition, by partnering with a key customer, another operating unit was added to convert acrolein to a less hazardous derivative at the Dow site prior to shipment to the customer site, thus reducing transportation risks (OK).
- Dow also uses continuous reaction processes instead of batch reaction processes to produce certain types of polyethylene and polypropylene plastic resins. This technology reduces the inventory of flammable material in these continuous reaction processes (OK).
- Direct chlorination using 95% chlorine vapor direct from chlorine cells rather than purified chlorine eliminates the need for liquid chlorine storage (Figure 11.14) (OK).

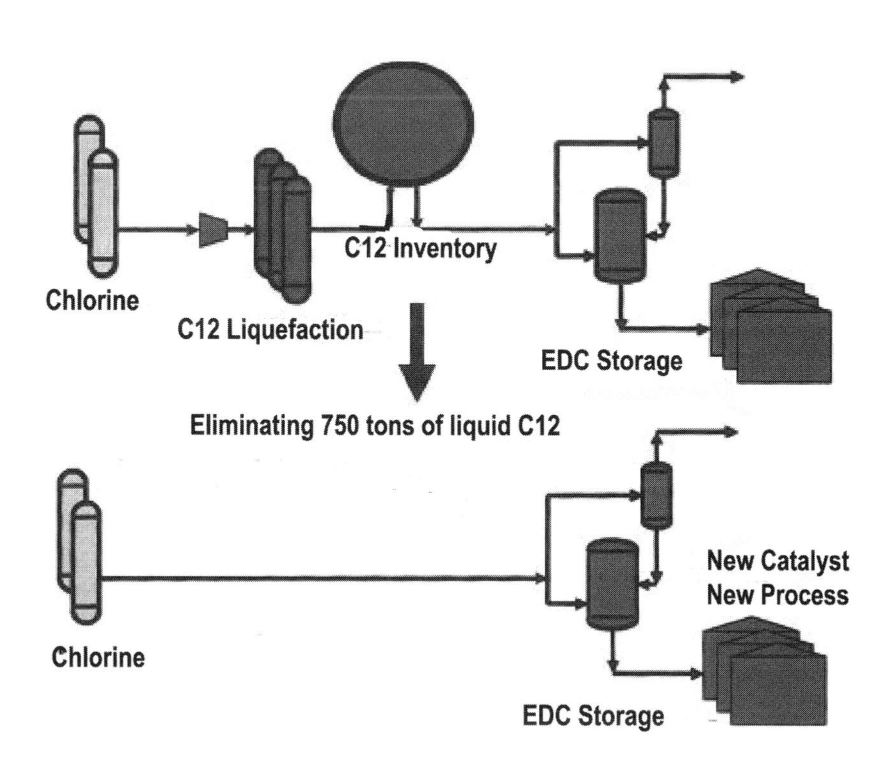

Figure 11.14: Use of Chlorine Vapor for Liquid Chlorine

- Dow's licensed EO METEOR$_{TM}$ technology significantly reduces the portion of the process using concentrated EO (Figure 11.15).

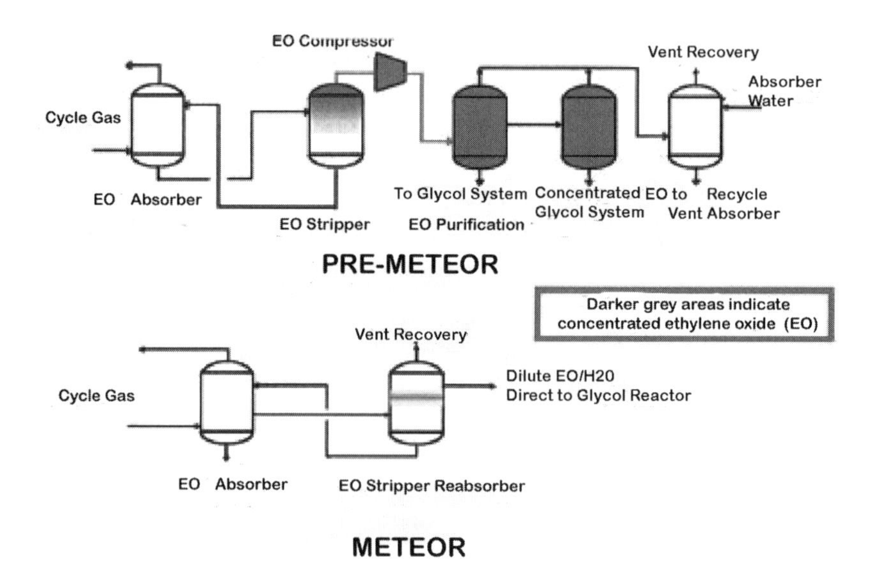

Figure 11.15: Streamlined METEOR$_{TM}$ Ethylene Oxide Recovery System

---

Substitute: Replace a material with a less hazardous substance.

- Manufacture of acrylic esters by oxidation of propylene to produce acrylic acid, followed by esterification to manufacture the various esters, is inherently safer than the older Reppe process, which used acetylene, carbon monoxide, and nickel carbonyl (KA).
- Disodium iminodiacetate (DSIDA), an intermediate used for agricultural chemicals, was traditionally manufactured using a process requiring ammonia, formaldehyde, hydrogen cyanide, and hydrogen chloride. Monsanto developed a new process for DSIDA that eliminates the use of hydrogen cyanide and formaldehyde, has a higher yield, is simpler, and produces a product of sufficiently high purity that purification is not required (Franczyk, 1997). Monsanto won a Presidential Green Chemistry Challenge Award for this process in 1996 (KA).
- Many cleaning and de-greasing operations have eliminated the use of organic solvents by substituting water-based systems (KA).

- A polymerization process conducted using a gradual addition batch process required a large quantity of organic solvent to keep the system viscosity low enough for effective mixing and heat transfer. In the event of an uncontrolled (runaway) polymerization, a large quantity of flammable and toxic material would be ejected through the reactor rupture disk. Instead of relying on an expensive and elaborate emergency relief discharge system to control the potential hazard, the basic process chemistry was re-considered. It was found that it was possible to make the product using a suspension polymerization process in water. Because of the significantly lower amount of solvent required in the new process, most of the material released in a runaway reaction would be water, with just a small amount of solvent and unreacted monomer. Reaction runaway was also less likely because of the higher heat capacity of water, allowing it to absorb more of the heat of reaction during a process upset (KA).
- Dow had been using benzene as an azeotroping agent in one of its oxide derivatives units, but replaced the benzene with a less hazardous chemical (OK).
- Dow replaced chlorine gas with sodium hypochlorite (bleach) as a water purification chemical at selected non-chlorine producing sites, and replaced Varsol in a process with another, less hazardous solvent (OK).

Moderate: Use less hazardous conditions, a less hazardous form of a material, or facilities which minimize the impact of a release to hazardous material or energy.

- In the 1930s, ammonia plants typically operated at pressures up to 600 bar. Over the years, improved understanding of the chemistry has resulted in a downward trend in the operating pressure of ammonia plants; in the 1980s, plants operating in the range of 100-150 bar were common. The newer, low-pressure plants are inherently safer, cheaper, and more efficient than their high-pressure counterparts (KA).
- Dow's newest process for producing diphenyl oxide uses a catalyst to operate at temperatures and pressures that are significantly lower than the processes that it replaced (OK).
- Storing highly toxic or flammable materials (for example, chlorine, LNG) as a refrigerated liquid at atmospheric pressure, rather than at atmospheric temperature under pressure reduces the potential consequences of an accidental spill (OK).

- Using an aqueous hydrogen chloride (HCl) solution, rather than anhydrous HCl in selected processes also reduces the potential consequences of a spill (OK).
- One of Dow's facilities near a populated area replaced the use of aqueous 36% HCl with 17% HCl. Although this change did double the number of deliveries, and with that, increased transportation risks, and increased risk associated with connecting/disconnecting fill lines during deliveries, given that the partial pressure has reduced by a factor of 1,000 (i.e., reduced toxic hazard in the event of a spill), it was viewed as an overall reduction in total inherent risk (OK).
- Multiple processes within Dow use small diameter piping or restricting orifices to limit the flow rate in the event of an accidental release or operational error, making it almost impossible for a reaction to occur too quickly (OK).
- In some cases, Dow uses process equipment designed to contain the maximum foreseeable pressures from the process, eliminating the need for pressure relief (OK).
- Using high wall dikes and other types of secondary containment to severely limit the surface area available for evaporation of toxic liquid spills minimizes the impact, as shown in Figure 11.16 (OK).

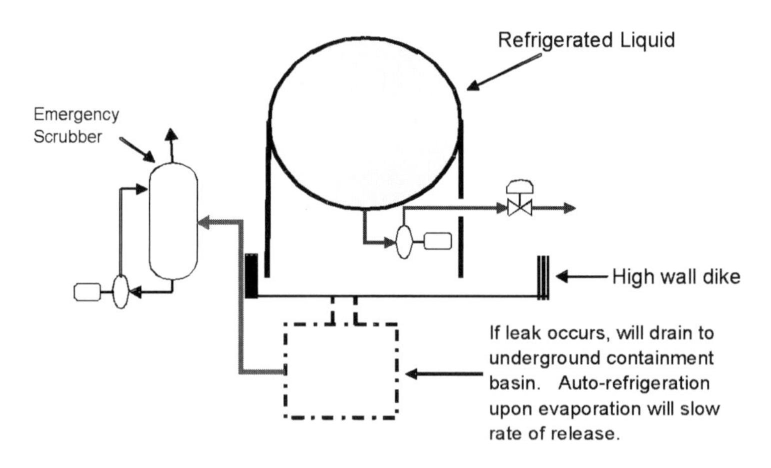

Figure 11.16: Dike and Secondary Containment System

Simplify: Design facilities that eliminate unnecessary complexity, make operating errors less likely, and are forgiving of errors that are made.

- An understanding of basic process chemistry allows process chemists and engineers to design simpler reaction systems and facilities. Processes where most or all of the unit operations are designed to take place in one vessel/reactor permit the generation of a hazardous intermediate in the vessel where it will be used, eliminating the need to store it, or to move it around the plant in piping systems. The inventory of hazardous material is thus limited to a maximum of one batch. This example illustrates the beneficial overlap that can occur with inherent safety principles, in this case, simplification and minimization (KA).
- Simplification sometimes involves a tradeoff between the complexity of an overall plant and complexity within one particular piece of equipment. For example, a reactive distillation process for producing methyl acetate requires only three columns and the associated support equipment. The older process required a reactor, an extractor, and eight other columns, along with the associated support equipment. The new process is simpler, safer, and more economical, but the successful operation of the reactive distillation component is itself more complex and knowledge-intensive (KA).
- Redesigned latex reactor cleaning equipment eliminates the potential for incorrect installation that could result in unwanted reactivity or cross contamination (OK).
- Dow eliminated the use of hoses in several hazardous services in favor of hard piped connections (OK).

11.8 ADDITIONAL LITERATURE GIVING EXAMPLES OF INHERENTLY SAFER OPERATIONS

- Amyotte, P.R., Goraya, A.U., Hendershot, D.C., and Khan, F.I. (2007). Incorporation of Inherent Safety principles in process safety management. *Process Safety Progress, 26* (4), 333-346.
- CCPS (1993). "Inherently Safer Plants." *Guidelines for Engineering Design for Process Safety* (Chapter 2). New York: American Institute of Chemical Engineers.
- Commission of the European Community (1997). INSIDE Project and INSET Toolkit. Available for download at http://www.aeat-safety-and-risk.com/html/inset.html.

- Kletz, T.A. (1998). *Process Plants - A Handbook for Inherently Safer Design*, London, UK: Taylor and Francis.
- "Layer of Protection Analysis and Inherently Safer Processes." *Process Safety Progress*, *18*, (4), 214-220, Winter 1999.
- 'Inherently Safer Approaches to Plant Design', DP Mansfield, ,
- Mansfield, D.P. and Cassidy, K. (1994). Inherently safer approaches to plant design: The benefits of an inherently safer approach and how this can be built into the design process. Presentation at IChemE Hazards XII—European Advances in Process Safety, April 19-21, 1994, UMIST, Manchester, UK. AEA/CS/HSE/R1016, HSE Books/HMSO, August 1994, ISBN 0853564159.
- Sanders, R.E. (2003). "Designs that lacked inherent safety: case histories." *Journal of Hazardous Materials, 104*, 1-3, 149-161.

12.

Future Initiatives

Applying inherently safer design principles to new and existing processes will improve the overall safety of the chemical process industry. The greatest gains in inherent safety will come from the invention of new chemistry and new processing technologies that eliminate the need for hazardous materials and operations. The key to spurring future invention in inherent safety is establishing a broad appreciation of its potential benefits within each company, and among the entire research and chemical engineering community, including academia. This awareness is especially important among chemists and others working on the earliest development of process chemistry. Initiatives such as those presented below would lead to a quicker application of these concepts.

12.1 INCORPORATING INHERENTLY SAFER DESIGN INTO PROCESS SAFETY MANAGEMENT

Inherently safer design within each company should be incorporated into process safety management programs to make its application a "way of life." Seamlessly integrating IS concepts into all facets of process safety will encourage the development of inherently safer processes and procedures. To do so, the company must identify the best way to go about incorporating these concepts. Should inherent safety be considered separately and distinctly, or as part of the process hazards analysis programs? What are the advantages and disadvantages of each method of application?

Incident investigations are a part of process safety management. In investigations, lessons are learned as to how inherently safer design could have prevented or mitigated the results. How can these learnings be disseminated such that future incidents in similar processes are avoided?

There are barriers impeding progress in applying inherently safer design. How might they be best identified and overcome?

12.2 ENCOURAGING INVENTION WITHIN THE CHEMICAL AND CHEMICAL ENGINEERING COMMUNITY

Publicizing the virtues of inherent safety beyond the *process safety* community and into the *broader chemistry and chemical engineering* community is essential to spur innovation. The authors of this book encourage readers to look for opportunities to "beat the drum" for inherent safety at every opportunity as they interact with this broader audience. Awareness can be raised in a variety of ways. Books by Trevor Kletz (1984, 1991b), CCPS (1993a), and Englund (1993) have proven to be successful vehicles. The IChemE and IPSG Inherently Safer Processing training program (IChemE and IPSG, 1995) is another successful format for promoting this topic. Other means need to be explored and exploited.

12.3 INCLUDING INHERENT SAFETY INTO THE EDUCATION OF CHEMISTS AND CHEMICAL ENGINEERS

Teaching inherently safer design concepts in undergraduate chemistry, chemical engineering, and related disciplines will be a great benefit as students move into industry after graduation.

12.4 DEVELOPING INHERENTLY SAFER DESIGN DATABASES AND LIBRARIES

The industry needs to develop inherently safer design databases that are readily available, cataloged, cross-referenced and indexed. These might take the form of libraries of information. Several examples of needed databases are:

- A continually updated database that describes and catalogs inherently safer design successes and failures.
- A collection of databases of chemicals, and of functional groups, ranked relative to their reactivity, stability, toxicity, and flammability categories. This would assist in the evaluation of the potential benefits of substituting one, somewhat safer, chemical for another.

- A database of the hazards associated with different types of equipment and unit operations, including the applicability of inherently safer design in each. As innovative solutions to hazards in equipment and process operations are discovered, these could be shared with others through this database, reducing risk in similar equipment and processes. A summary of design approaches for a number of common types of chemical process equipment is published in CCPS (1997).

12.5 DEVELOPING TOOLS TO APPLY INHERENTLY SAFER DESIGN

Inherent safety is still a young and evolving area of process risk assessment and decision-making. New tools and analytic approaches are needed to help process risk assessors and decision-makers identify inherent safety opportunities and evaluate them against other competing risks and opportunities.

12.5.1 The Broad View and Life Cycle Cost of Alternatives

One of the recognized barriers to the successful application of inherently safer design is the lack of appreciation of the benefits that can be derived from viewing a process broadly, rather than narrowly. Employing a cradle-to-grave and a feed-end-to-product-end view will lead to the development of processes which are as inherently safe as possible. Examples of myopic design could include the following.

- A chemist may think only of optimizing the route of synthesis to avoid a runaway reaction hazard and not consider the safety and design implications of flammable reaction by-products.
- A designer fails to identify and consider the environmental land use implications of dismantling a process at the end of its useful life.

At any given time, individuals are more concerned with a particular unit operation, or with a particular life cycle stage of the process. A broader approach, however, would require an analysis of the impact of the process on:

- upstream and downstream operations
- ancillary operations, such as waste stream management and the handling of byproducts
- later stages in the process life cycle

Tools to apply such a broader view of a process would pay inherently safer dividends. One tool might include instructions on how to estimate the life cycle cost of proposed alternative solutions. Such a tool is not presently fully developed and available in the public domain. Training to assist in estimating life cycle cost is also needed.

12.5.2 Benefits of Reliability Analysis

Reliability relies on quality hardware, correct installation and commissioning, and prudent inspection and test programs. The frequency and duration of failure are minimized, thus reducing the likelihood of such systems being unavailable when challenged by process hazards. Guidance in assessing such reliability could lead to simplifying SISs (CCPS, 2007; Green and Dowell, 1995 and 1996).

Reliability analysis of both the overall processes, and individual unit operations within those processes, is needed. High process reliability may reduce the overall need for large in-process inventories by minimizing the need for process intermediate storage facilities, or storage and blending facilities for "off-spec" product. It may also maximize the ability to produce product "as needed," thereby reducing on-plant product storage requirements. Highly reliable management of raw material purchasing and delivery systems may reduce the need for on-plant raw material storage.

12.5.3 Potential Energy

It is obvious that, all other things being equal, a process operated at high temperature and high pressure, containing exothermically reactive materials, and containing a large inventory, is inherently less safe than one at ambient temperature and pressure, containing a small inventory of thermally-stable materials. A measure of the magnitude of this difference in safety is the difference in the potential energy in the process. By comparing the potential energies of alternative process strategies, the lowest energy alternative can be identified. Methods and guidance are needed to estimate the total potential energy in different types of situations encountered in processes.

12.5.4 A Table of Distances

Defining appropriate spacing of unit operations within a process, and appropriate distancing—from one process to another; from employees' nonessential to day-to-day process operation; and from the process to the public—is an inherently safer approach. A definition of "appropriate spacing" would assist in evaluating the process location alternatives. This definition may take the form of a table of distances as a function of the type of hazard, inventory quantity, and other factors.

12.5.5 Quantitative Measures of Inherent Safety

Several available quantitative risk indices can be used to measure the degree of application of inherent safety in a process. These include the Dow Chemical Exposure Index (Dow/AIChE, 1994a), the Dow Fire and Explosion Index (Dow/AIChE, 1994b), the Mond Index (ICI, 1985; Tyler, 1985), and the Prototype Index of Inherent Safety by Edwards, et al. (1996). Refinement of these quantitative measurement techniques and convergence to a single set of accepted indices would be beneficial. Accepted metrics would aid in comparing alternatives and in quantifying the degree of process improvement that the various applications of inherently safer design bring.

12.5.6 Other Suggestions

- Increase emphasis on training and promoting the benefits of IS in companies, universities, and through continuing education to practicing engineers and chemists.
- Develop tools to aid the engineer in conducting IS reviews through the life of a process.
- Make IS tools such as the INSET Toolkit (Mansfield, et. al., 1996) available to a broad community of engineers involved in process development and design.
- Increase efforts to promote IS to process development chemists, who work on processes early in the life cycle and have the greatest leverage for reducing or eliminating hazards.
- Promote IS to smaller companies and organizations, where the concept is less well known and understood.
- Develop and expand on the concepts of life cycle costing to better understand the potential economic advantages of inherently safer design.
- Continue research on process intensification through new and innovative technologies. The industry also needs to identify ways of demonstrating and promoting new and inherently safer technologies, perhaps through multi-company consortia, such as the Center for Waste Reduction Technology.
- Publicize successes in implementing inherently safer processes, through symposia and literature articles. The requirements of the EPA RMP regulations may provide additional incentive to companies to demonstrate that they are committed to reducing the potential hazard zones in the area of a plant.

- Many significant advances are possible, but they, too, will require research, development and implementation over a long time period.
- Develop methods to measure various inherent safety process options, an essential first step to widespread implementation.
- Develop a method to measure inherent safety using fuzzy logic mathematics—something that is now being researched.
- Government programs now support concepts research and development, such as green chemistry, solvent substitution, waste reduction, and sustainable growth, all of which are related to inherent safety. A similar approach involving industry, government and academia can enhance inherently safer chemical processes discovery, development and implementation.

Appendix A

Inherently Safer Technology Checklist

The following procedure is used to answer each ISD Evaluation Checklist item:

1. Determine the applicability of the item to the process being reviewed. Indicate "Y" or "N" in the appropriate column to indicate applicability.
2. If an item is applicable to the process under review, determine whether there are any ISD opportunities or applications available that could potentially reduce the risk (consequences and/or likelihood) of an accidental or intentional release. If an ISD opportunity is identified, it should be described in the Opportunities/Applications Column. If no new ISD opportunities are identified, this is noted in the Opportunities/ Applications column with supporting information, including general references to existing safeguards and ISD already implemented, where applicable.
3. If an ISD opportunity is identified, a screening evaluation of the alternative should be performed by the review team and the results summarized in the Feasibility column. If the ISD alternative is determined to be infeasible (due to cost, technology limitations, security, operability, safety, or other factors), this decision is noted as such along with supporting information, including general references to existing safeguards and ISD already implemented, where applicable. The current status of the opportunity (e.g., "to be evaluated further," "evaluation in-progress," "implementation in-progress," etc.) is documented in the "Current Status" column.
4. If an ISD opportunity is deemed potentially feasible, a recommendation for further evaluation or implementation is made by the team and entered in the "Recommendations" column.

While proceeding through the ISD review checklist, existing safeguards are reviewed for each ISD alternative determined to be infeasible. In some cases, the checklist suggests Layers of Protection (Safeguards), which can be active, passive, or procedural. Credible safeguards designed to prevent, detect, or mitigate a hazard may be as effective as implementing ISD alternatives. For security issues, countermeasures can be detection, deterrence, delay, or response. Definitions and examples of safeguards reviewed include:

- Active - Using controls, alarms, safety instrumented systems (SISs), and mitigation systems to detect and respond to process deviations from normal operation. Examples: High level, temperature, and flow SISs, atmospheric sensors with automatic shutdowns, and pressure relief systems. Active safeguards require external inputs to function, such as electrical power.
- Passive - Minimize the hazard by process and equipment design features which reduce either the frequency or consequences of the hazard without the active functioning of any device. Examples: dikes, containment buildings. Passive safeguards do not require external inputs to function.
- Procedural - Using human response in the form of operating procedures, administrative checks, and emergency response. Examples: emergency shutdown or emergency response procedures, operator checks and corrective actions.

In general, supporting information for rejection of inherently safer design alternatives should include the presence and adequacy of existing safeguards combined with cost, operability, or other issues that reduce the feasibility of implementing the inherently safer options.

SOURCES FOR THIS APPENDIX

1. AcuTech Consulting Group (2007). *ISD Checklist.*
2. Contra Costa County, CA, Industrial Safety Ordinance (September 12, 2002) *IST Evaluation Checklist 1-06, Revision 0.*
3. Hendershot, D.C. (2000). Process minimization: Making plants safer. *Chemical Engineering Progress*, 30 (1), 35-40.
4. Center for Chemical Process Safety (1998). *Guidelines for Design Solutions for Process Equipment Failures*. New York: CCPS/AIChE.
5. Stankiewicz, A. and Moulijn, J.A. (Eds.) (2004). *Re-Engineering the Chemical Processing Plant: Process Intensification.* New York: Marcel Dekker.
6. Kletz, T. A. (1998) *Process Plants: A Handbook for Inherently Safer Design.* Philadelphia: Taylor & Francis, 1998.

No.	Inherently Safer Design Alternatives	Applicable (Y/N)?	Opportunities/Applications	Feasibility	Current Status	Recommendation
1.0	**SUBSTITUTE**					
1.1	Is this (hazardous) process/product necessary?					
1.2	Is it possible to completely eliminate hazardous raw materials, process intermediates, or by-products by using an alternative process or chemistry?					
1.3	Is it possible to completely eliminate in-process solvents and flammable heat transfer media by changing chemistry or processing conditions?					
1.4	Is an alternate process available for this product that eliminates or substantially reduces the need for hazardous raw materials or production of hazardous intermediates?					

No.	Inherently Safer Design Alternatives	Applicable (Y/N)?	Opportunities/Applications	Feasibility	Current Status	Recommendation
1.5	Is it possible to substitute less hazardous raw materials?					
	• Noncombustible for flammable					
	• Less volatile					
	• Less reactive					
	• More stable					
	• Less toxic					
	• Low pressure steam rather than flammable heat transfer fluid (i.e., operated above flash point)					
1.6	Is it possible to substitute less hazardous final product solvents?					

No.	Inherently Safer Design Alternatives	Applicable (Y/N)?	Opportunities/Applications	Feasibility	Current Status	Recommendation
1.7	Is it possible to use a nonflammable refrigerant instead of a flammable one (or minimize inventory)?					
1.8	Are there any other alternatives for substituting or eliminating the use of hazardous materials in this process?					
2.0	**MINIMIZE**					
2.1	Inventory Reduction					
2.1.1	Can hazardous raw materials inventory be reduced?					
	• Just-in-time deliveries based on production needs					
	• Supplier management including strategic alliance					

No.	Inherently Safer Design Alternatives	Applicable (Y/N)?	Opportunities/Applications	Feasibility	Current Status	Recommendation
	• On-site generation of hazardous material (including in situ) from less hazardous raw materials					
	• Hazardous raw material inventory management system based on production forecast					
2.1.2	Can (hazardous) in-process storage and inventory be reduced?					
	• Direct coupling of process elements					
	• Eliminating or reducing size of in-process storage					

No.	Inherently Safer Design Alternatives	Applicable (Y/N)?	Opportunities/Applications	Feasibility	Current Status	Recommendation
	• Designing process equipment involving hazardous material with the smallest feasible inventory (see also Section 2.2)					
2.1.3	Can hazardous finished product inventory be reduced?					
	• Improving production scheduling/sales forecasting					
	• Improving communication with transporters/material handlers					
	• Hazardous finished product inventory management system based on sales forecast					

No.	Inherently Safer Design Alternatives	Applicable (Y/N)?	Opportunities/Applications	Feasibility	Current Status	Recommendation
2.2	Process Intensification Considerations					
2.2.1	Can alternate equipment with reduced hazardous material inventory requirement be used?					
	• Centrifugal extractors in place of extraction columns					
	• Flash dryers in place of tray dryers					
	• Continuous reactors in place of batch					
	• Plug flow or loop reactors in place of continuous stirred tank reactors					
	• Continuous in-line mixers (e.g., static mixer) in place of mixing vessels or reactors					

No.	Inherently Safer Design Alternatives	Applicable (Y/N)?	Opportunities/Applications	Feasibility	Current Status	Recommendation
	• Intensive mixers to minimize size of mixing vessel of reactor					
	• High heat-transfer reactors (e.g., microreactor, HEX reactor)					
	• Spinning-disk reactor (especially for high heat-flux or viscous liquids)					
	• Compact heat exchangers (higher heat transfer area per unit volume, e.g., spiral, plate & frame, plate-fin) in place of shell-and-tube					
	• More hazardous material on the tubeside in shell-and-tube exchangers					

No.	Inherently Safer Design Alternatives	Applicable (Y/N)?	Opportunities/Applications	Feasibility	Current Status	Recommendation
	• Use water or other non-flammable heat transfer medium, a vapor-phase medium, or a medium below its boiling point					
	• Wiped film stills in place of continuous still pots (distillation columns)					
	• Combine unit operations (such as reactive distillation or extraction in place of separate reactor with multi-column fractionation train or extractor; installing internal reboilers or heat exchangers) to reduce overall system volume					

No.	Inherently Safer Design Alternatives	Applicable (Y/N)?	Opportunities/Applications	Feasibility	Current Status	Recommendation
	• Use of acceleration fields (e.g., rotating packed bed for gas/liquid or liquid/liquid contacting for absorption, stripping, distillation, extraction, etc.)					
	• Alternate energy sources (such as lasers, UV light, microwaves, or ultrasound) to control reaction or direct heat to the unit operation					
2.2.2	Has the length of hazardous material piping runs been minimized?					
2.2.3	Has hazardous material piping been designed for minimum pipe diameter?					

No.	Inherently Safer Design Alternatives	Applicable (Y/N)?	Opportunities/Applications	Feasibility	Current Status	Recommendation
2.2.4	Can pipeline inventory be reduced by using the hazardous material as a gas rather than a liquid?					
2.2.5	Can process conditions be changed to reduce production of hazardous waste or by-products?					
2.3	Are there any other alternatives for minimizing the inventory of hazardous materials in this process?					
3.0	**MODERATE**					
3.1	Is it possible to limit the supply pressure of (hazardous) raw materials to less than the maximum allowable working pressure of the vessels to which they are delivered?					

No.	Inherently Safer Design Alternatives	Applicable (Y/N)?	Opportunities/Applications	Feasibility	Current Status	Recommendation
3.2	Is it possible to make reaction conditions (for hazardous reactants or products) (temperature, pressure) less severe by using a catalyst, or a better catalyst (e.g., structured or monolithic vs. packed-bed)?					
3.3	Can the process be operated at less severe conditions (for hazardous reactants or products) by considering:					
	• Improved thermodynamics or kinetics to reduce operating temperatures or pressures					
	• Changes in reaction phase (e.g., liquid/liquid, gas/liquid, or gas/gas)					

No.	Inherently Safer Design Alternatives	Applicable (Y/N)?	Opportunities/Applications	Feasibility	Current Status	Recommendation
	• Changes in the order in which raw materials are added					
	• Raw material recycle to compensate for reduced yield or conversion					
	• Operating at lower pressure to limit potential release rate					
	• Operating at lower temperature to prevent runaway reactions or material failure					
3.4	Is it possible to use less concentrated hazardous raw materials to reduce the hazard potential?					
	• Aqueous ammonia and/or HCl instead of anhydrous					

No.	Inherently Safer Design Alternatives	Applicable (Y/N)?	Opportunities/Applications	Feasibility	Current Status	Recommendation
	• Sulfuric acid instead of oleum					
	• Dilute nitric acid instead of concentrated fuming nitric acid					
	• Wet benzoyl peroxide instead of dry					
3.5	Is it possible to use larger particle size/reduced dust forming solids to minimize potential for dust explosions?					
3.6	Are all process materials (e.g., heating/cooling media) compatible with process materials in event of inadvertent contamination (e.g., due to a tank coil or heat exchanger tube failure)?					

No.	Inherently Safer Design Alternatives	Applicable (Y/N)?	Opportunities/Applications	Feasibility	Current Status	Recommendation
3.7	Is it possible to add an ingredient to volatile hazardous materials that will reduce its vapor pressure?					
3.8	For equipment containing materials that become unstable at elevated temperature or freeze at low temperature, is it possible to use heating/cooling media which limit the maximum and minimum temperatures attainable (i.e., self-limiting electric heat tracing or hot water at atmospheric pressure)?					
3.9	Can process conditions be changed to avoid handling flammable liquids above their flash points?					

No.	Inherently Safer Design Alternatives	Applicable (Y/N)?	Opportunities/Applications	Feasibility	Current Status	Recommendation
3.10	Is equipment designed to totally contain the materials that might be present inside at ambient temperature or the maximum attainable process temperature (i.e., higher maximum allowable working temperature to accommodate loss of cooling, simplified reliance on external systems like refrigeration to control temperature such that vapor pressure is less than equipment design pressure)?					
3.11	For processes handling flammable materials, is it possible to design the layout to minimize the number and size of confined areas and to limit the potential for serious overpressure in the event of a loss of containment and subsequent ignition?					

No.	Inherently Safer Design Alternatives	Applicable (Y/N)?	Opportunities/Applications	Feasibility	Current Status	Recommendation
3.12	Can process units (for hazardous materials) be designed to limit the magnitude of process deviations?					
	• Selecting pumps with maximum capacity lower than safe rate of addition for the process					
	• For gravity-fed systems, limiting maximum feed rate to be within safe limits by pipe size or fixed orifice					
	• Minimum flow recirculation line for pumps/compressors (with orifice to control flow) to ensure minimum flow in event of deadheading					

No.	Inherently Safer Design Alternatives	Applicable (Y/N)?	Opportunities/Applications	Feasibility	Current Status	Recommendation
3.13	Can hazardous material liquid spills be prevented from entering drainage system/sewer (if potential for fire or hazardous reaction exists, e.g., water reactive material)?					
3.14	For flammable materials, can spills be directed away from the storage vessel to reduce the risk of a boiling liquid expanding vapor explosion (BLEVE) in the event of a fire?					
3.15	Can passive designs, such as the following, be implemented?					
	• Secondary containment (e.g., dikes, curbing, buildings, enclosures)					

No.	Inherently Safer Design Alternatives	Applicable (Y/N)?	Opportunities/Applications	Feasibility	Current Status	Recommendation
	• Use of properly vented blowdown tank for dumping of runaway reaction mass					
	• Permanent bonding and grounding systems for process equipment, tanks and vessels					
	• Use of gas inerting systems for handling flammables and explosive dusts (e.g., nitrogen, CO2)					
	• Use of diplegs with anti-siphon openings for feed to flammable liquid storage tanks					
	• Fireproofing insulation vs. fixed/portable fire protection					

No.	Inherently Safer Design Alternatives	Applicable (Y/N)?	Opportunities/Applications	Feasibility	Current Status	Recommendation
3.16	Can gases be transported and stored at low or atmospheric pressure on a high capacity adsorbent instead of using pressurized gas cylinders?					
3.17	Are there any other alternatives for moderating the use of hazardous materials in this process?					
4.0	**SIMPLIFY**					
4.1	Can equipment be designed such that it is difficult or impossible to create a potential hazardous situation due to an operating or maintenance error?					
	• Easy access and operability of valves to prevent inadvertent errors					

No.	Inherently Safer Design Alternatives	Applicable (Y/N)?	Opportunities/Applications	Feasibility	Current Status	Recommendation
	• Elimination of all unnecessary cross-connections					
	• Use of dedicated hoses and compatible couplings for reactants where hose connections are used					
	• Designing temperature-limited heat transfer equipment to prevent exceeding maximum process or equipment design temperatures					
	• Use of corrosion resistant materials for process equipment, piping and components					

No.	Inherently Safer Design Alternatives	Applicable (Y/N)?	Opportunities/Applications	Feasibility	Current Status	Recommendation
	• Operating at higher temperature to avoid cryogenic effects such as embrittlement failures					
	• Using alternative agitation methods (e.g., external circulation using sealless pump which eliminates potential releases due to agitator seal failures)					
	• Use of mixing feed nozzle instead of agitator for vessel mixing					
	• Using underground or shielded tanks					
	• Specifying fail-safe operation on utility failure (e.g., air, power)					

No.	Inherently Safer Design Alternatives	Applicable (Y/N)?	Opportunities/Applications	Feasibility	Current Status	Recommendation
	• Allocating redundant inputs and outputs to separate modules of the programmable electronic system to minimize common cause failures					
	• Provide continuous pilots (independent, reliable source) for burner management systems					
	• Using refrigerated storage vs. pressurized storage					
	• Using independent power buses for redundant equipment to minimize consequences of partial power failures					

No.	Inherently Safer Design Alternatives	Applicable (Y/N)?	Opportunities/Applications	Feasibility	Current Status	Recommendation
	• Minimizing equipment wall area to minimize corrosion/fire exposure					
	• Minimizing connections, paths and number of flanges in hazardous processes					
	• Avoiding use of threaded connections in hazardous service					
	• Using double-walled pipe					
	• Minimizing number of bends in piping (potential erosion points)					
	• Using expansion loops in piping rather than bellows for thermal expansion					

No.	Inherently Safer Design Alternatives	Applicable (Y/N)?	Opportunities/Applications	Feasibility	Current Status	Recommendation
	• Designing equipment isolation mechanisms for maintenance in the process					
	• Limiting manual operations such as filter cleaning, manual sampling, hose handling for loading/unloading operations, etc.					
	• Designing vessels for full vacuum to eliminate risk of vessel collapse					
	• Designing both shell- and-tube side of heat exchangers to contain the maximum attainable pressure, eliminating the need for pressure relief (may still be needed to meet fire safety requirements)					

No.	Inherently Safer Design Alternatives	Applicable (Y/N)?	Opportunities/Applications	Feasibility	Current Status	Recommendation
	• Designing/selecting equipment which makes incorrect assembly impossible					
	• Using equipment that clearly identifies status:					
	• Check valves with easy to identify direction of flow					
	• Gate valves with rising spindles to clearly indicate open or closed position					
	• Spectacle (or figure 8) blinds instead of slip plates					
	• Manual quarter-turn block valves with handles that clearly indicate position					

No.	Inherently Safer Design Alternatives	Applicable (Y/N)?	Opportunities/Applications	Feasibility	Current Status	Recommendation
	• For automated block valves, display actual valve position in addition to the output to the valve					
	• Designing equipment with an MAWP to contain the maximum pressure generated without reliance on pressure relief systems, even if the "worst credible event" occurs?					
	• Use open vent or overflow line to secondary containment for overpressure, overfill and vacuum protection					
	• Eliminate utility connections above pressure rating of vessel					

No.	Inherently Safer Design Alternatives	Applicable (Y/N)?	Opportunities/Applications	Feasibility	Current Status	Recommendation
	• Carrying out several process steps in separate processing vessels rather than a single multi-purpose vessel (to reduce the complexity and number of raw materials, utilities, and auxiliary equipment connected to a specific vessel)					
4.2	Can passive leak-limiting technology be used to limit potential loss of containment?					
	• Blowout resistant gaskets (e.g., spiral wound)					
	• Increasing wall strength of piping and equipment					
	• Maximize use of all-welded pipe					

No.	Inherently Safer Design Alternatives	Applicable (Y/N)?	Opportunities/Applications	Feasibility	Current Status	Recommendation
	• Using fewer pipe seams and joints					
	• Providing extra corrosion/erosion allowance (e.g., Sch. 80 vs. 40)					
	• Reducing or eliminating vibration (e.g., through vibration dampening or equipment balancing)					
	• Minimizing the use of open-ended (bleed or vent), quick-opening valves (for example, quarter-turn ball or plug valves)					
	• Eliminating open-ended (bleed or vent), quick-opening valves (for example, quarter-turn ball or plug valves) in hazardous service					

No.	Inherently Safer Design Alternatives	Applicable (Y/N)?	Opportunities/Applications	Feasibility	Current Status	Recommendation
	• Using incompatible hose connections to prevent mis-connection (e.g., air/ nitrogen, raw materials)					
	• Use of round valve handles for open-ended quarter-turn valves to minimize potential for bumping open					
	• Improving valve seating reliability (e.g., using system pressure to seal valve seats where possible, using valve seat geometry, valve operations, and flow to eliminate or reduce seat damage)					
	• Eliminating unnecessary expansion joints, hoses, and rupture disks					

No.	Inherently Safer Design Alternatives	Applicable (Y/N)?	Opportunities/Applications	Feasibility	Current Status	Recommendation
	• Use of articulated arms instead of hoses for loading/unloading of hazardous materials					
	• Eliminating unnecessary sight glasses/glass rotameters; use high-pressure/armored sight glasses as needed					
	• Eliminate use of glass, plastic or other brittle material as material of construction					
	• Use of seal-less pumps (e.g., canned, magnetic drive)					

No.	Inherently Safer Design Alternatives	Applicable (Y/N)?	Opportunities/Applications	Feasibility	Current Status	Recommendation
	• Minimizing the number of different gaskets, nuts, bolts, etc. used to reduce potential for error					
4.3	Has attention to control system human factors[1] been addressed through:					
	• Simplified control displays					
	• Limited instrumentation complexity					
	• Clearly displayed information about normal and abnormal process conditions					
	• Logical arrangement of controls and displays that match operator expectations					

[1] See *Human Factors Methods for Improving Performance in the Process Industries* (CCPS 2007) for a comprehensive HF checklist.

No.	Inherently Safer Design Alternatives	Applicable (Y/N)?	Opportunities/Applications	Feasibility	Current Status	Recommendation
	• Clearly displayed information about normal and abnormal process conditions					
	• Logical arrangement of controls and displays that match operator expectations					
	• Separate displays that present similar information in a consistent manner					
	• Safety alarms that are easily distinguished from process alarms					
	• Correction of nuisance alarms and elimination of redundant alarms as soon as practical to help prevent complacency					

No.	Inherently Safer Design Alternatives	Applicable (Y/N)?	Opportunities/Applications	Feasibility	Current Status	Recommendation
	• Control system displays that give adequate feedback for all operational actions					
	• Layout of control system displays that are logical, consistent, and effective					
	• Controls that are distinguishable, accessible, and easy to use					
	• Controls which meet standard expectations (color, direction of movement, etc.)					
	• Control arrangements which logically follow the normal sequence of operation					

No.	Inherently Safer Design Alternatives	Applicable (Y/N)?	Opportunities/Applications	Feasibility	Current Status	Recommendation
	• Operating procedure format and language which operators believe are easy to follow and understand and that include necessary information					
4.4	Are there any other alternatives for simplifying operations involving hazardous materials in this process?					
5.0	**LOCATION/SITING/TRANSPORTATION**					
5.1	Can the plant be located to minimize the need for transportation of hazardous materials? (e.g., co-located with supplier/customer, on-site production of hazardous raw materials)					

No.	Inherently Safer Design Alternatives	Applicable (Y/N)?	Opportunities/Applications	Feasibility	Current Status	Recommendation
5.2	Can hazardous process units be located to eliminate or minimize:					
	• Adverse effects from adjacent hazardous installations					
	• Off-site impacts					
	• On-site impacts on employees and other plant facilities including control rooms, fire protection systems, emergency response and communication facilities, and maintenance and administrative facilities					

No.	Inherently Safer Design Alternatives	Applicable (Y/N)?	Opportunities/Applications	Feasibility	Current Status	Recommendation
5.3	Can a multi-step process, where the steps are done at separate sites, be divided up differently to eliminate the need to transport hazardous materials?					
5.4	Can materials be transported:					
	• In a less hazardous form (e.g., refrigerated liquid vs. pressurized)					
	• In a safer transport method (e.g., via pipeline, top- vs. bottom-unloaded, rail vs. truck)					
	• Along a safer route (e.g., avoiding high risk areas such as high population areas, tunnels, or high-accident-rate sections of roadway)?					

Appendix B

Inherent Safety Analysis Approaches

Inherent safety can be analyzed in a number of ways, but, in all case, the intent is to formalize the consideration of inherent safety, rather than to include it by circumstance. By formally including inherent safety, in either a direct or indirect way, facilities can fully realize the potential benefits of inherent safety. In addition, all IS considerations will be fully documented.

Three analysis methods can be used to evaluate implementation of IS:

1. Inherent Safety Analysis: Checklist Process Hazard Analysis (PHA)
2. Inherent Safety Analysis: Independent Process Hazard Analysis (PHA)
3. Inherent Safety Analysis: Integral to Process Hazard Analysis (PHA)

Method 1 employs a checklist containing a number of practical inherent safety considerations organized around the four strategies of minimization, substitution, moderation, and simplification. The advantage of this approach is that it is very direct and asks pointed questions that have proven to be valuable in reducing hazards at past locations. The disadvantage is that, as with any checklist, it may be limiting in that other ideas may surface if the team was asked to more creatively determine applications for the inherent safety strategies given a safety objective. (Note that the checklist that appears here is only a representative subset to illustrate its use. See Appendix A for the complete checklist.)

For the second method, the team is asked to avoid a particular hazard at a designated part of the process. In this case, the team reviews a problem, determines which of the inherently safer strategies may apply, and then brainstorms possible ways the hazard can be reduced or eliminated.

The third method integrates ISD into every PHA study the facility may conduct, including "What-If?," HAZOP, FMEA or other similar methodology. The concept is both to include questions (for "What-If?") or guidewords (for HAZOP) to introduce ISD to the discussion, and then to use the four strategies mentioned above as a possible means to mitigate each hazard identified. The input from those discussions are then evaluated alongside the other layers of protection strategies that may be used.

Each method is explained below. In each case, the analyst reviews the potential for applying inherent safety either at the process level or at the node level. A node is defined the same way as in Process Hazards Analysis studies, or existing studies. It is possible to do the analysis at the higher level required for PHA study if the process is relatively simple and the ISD opportunities are limited. If the study is conducted at a detailed node level, it may result in additional considerations being given to smaller but important details. For example, at a macro level the hazardous chemical in the process cannot be substituted, but on a micro level there may be opportunities to do so in given areas or with certain types of equipment.

In all cases it is recommended to use a risk-ranking scheme that defines likelihood and consequences on a scale such as the one shown on Tables B.1-B.3. Inherent safety should be considered in light of risks as with other risk management strategies.

Table B.1. Risk Matrix (R)

Likelihood					
	Very High (4)	Medium 2	High 3	Very High 4	Very High 4
	High (3)	Low 2	Medium 3	High 4	Very High 4
	Medium (2)	Low 2	Medium 3	Medium 3	High 4
	Low (1)	Low 1	Low 2	Low 3	Medium 3
		Low (1)	Medium (2)	High (3)	Very High (4)
		Severity			

Table B.2. Severity (S)

Category	Low (1)	Medium (2)	Moderate (3)	High (4)
Health & safety impacts	Minor injury or health effect	Moderate injury or health effect	Major injury or health effect; offsite public impacts	Fatality offsite, multiple onsite injuries or fatalities,
Asset damage (replacement cost)	$	$$	$$$	$$$$
Business interruption (days unavailable or $)	$	$$	$$$	$$$$
Environmental impact (remediation damages)	$	$$	$$$	$$$$

Table B.3. Likelihood (L)

Likelihood	Short descriptor	Description
1	Low	Not expected to occur in life of facility
2	Medium	Possible to occur in life of facility
3	High	Possible to occur in range of 1 year to 10 years
4	Very High	Possible to occur at least once a year

B.1. INHERENT SAFETY ANALYSIS–CHECKLIST PROCESS HAZARD ANALYSIS (PHA)

Table B.4 offers an example of a checklist approach. The analyst asks the questions from the checklist (potential opportunities) and the team documents the potential consequences of any issue that may be applicable to the process or node under study. Considering the four ISD strategies, the team documents the potential recommendations that may address the concern ranked in the following order:

- First order ISD
- Second order ISD
- Layers of Protection

Table B.4 Case Summary: Inherent Safety Analysis – Checklist Process Hazard Analysis (PHA)

Location: Orange, New Jersey						**Risk Ranking**			**Unit:** Hydrofluoric Acid Alkylation Unit	**Analysis Date:** April 1, 2008
PFD No.: 1234-5678										
Node: Isobutane Storage										
Design Conditions/Parameters: Storage of isobutene in five bullets and two process vessels near the unit										
#	**QUESTION**	**POTENTIAL OPPORTUNITIES**	**FEASIBILITY**	**CONSEQUENCES**	**EXISTING SAFEGUARDS**	**S**	**L**	**R**	**RECOMMENDATIONS**	**COMMENTS/ STATUS**
1	Reduce hazardous raw materials inventory	Lower storage tank volume or eliminate some storage if possible.	Lowering tank volumes is already done. There may be one tank that could be eliminated.	Potential release from storage and exposure to south plant from unconfined vapor cloud explosion.	1. Administrative controls limit fill level of the five tanks.	4	1	3	1. Eliminate one of five flammable storage bullets to reduce potential releases from storage.[1]	In review.
2	Reducing in-process storage and inventory	Interim storage adds to inventory and could be eliminated.	Will require engineering analysis to evaluate.	Potential leak, fire and explosion.	1. High level alarms 2. Flammable gas detectors	4	1	3	2. Consider eliminating interim storage and providing a continuous flow operation.[2]	In review

[1] This uses the concept of ***Minimization*** to avoid a hazard. Originally, feed could be taken from any of five tanks, yet the process could be managed with only three tanks. A hazardous condition was reduced so this represents a **Second order Inherent Safety** change.

[2] Also uses the concept of ***Minimization***. A number of unit tanks could possibly be eliminated, thus eliminating a hazardous condition. This represents a **First order Inherent Safety** change.

Table B.4 Case Summary: Inherent Safety Analysis – Checklist Process Hazard Analysis (PHA)

Location: Orange, New Jersey **PFD No.**: 1234-5678 **Node**: Isobutane Storage **Design Conditions/Parameters:** Storage of isobutene in five bullets and two process vessels near the unit					**Risk Ranking**			**Unit**: Hydrofluoric Acid Alkylation Unit	**Analysis Date:** April 1, 2008
#	**QUESTION**	**POTENTIAL OPPORTUNITIES**	**FEASIBILITY**	**CONSEQUENCES**	**EXISTING SAFEGUARDS**	**S**	**L**	**R**	**RECOMMENDATIONS**	**COMMENTS/ STATUS**
3	Reducing finished product inventory	Not applicable (NA)[3]								
4	Reduce hazardous material by using alternate equipment		No alternatives available or feasible.[4]							
5	Minimize length of hazardous material piping runs	There are many piping runs that are no longer used that might be decommissioned.	Need to consider in engineering and operations evaluation but appears to be feasible.	Potential larger release		4	1	3	3. Consider moving the location of the planned storage vessels to a location not closer than 250 feet, but closer than the 1500 feet location presently planned.[5]	

[3] Some questions are not relevant to the particular process.

[4] Some questions may result in a conclusion that there are no alternatives or feasible inherent safety applications.

[5] See Footnote 3

Table B.4 Case Summary: Inherent Safety Analysis – Checklist Process Hazard Analysis (PHA)

Location: Orange, New Jersey						**Risk Ranking**			**Unit**: Hydrofluoric Acid Alkylation Unit	**Analysis Date**: April 1, 2008
PFD No.: 1234-5678										
Node: Isobutane Storage										
Design Conditions/Parameters: Storage of isobutene in five bullets and two process vessels near the unit										
#	**QUESTION**	**POTENTIAL OPPORTUNITIES**	**FEASIBILITY**	**CONSEQUENCES**	**EXISTING SAFEGUARDS**	**S**	**L**	**R**	**RECOMMENDATIONS**	**COMMENTS/ STATUS**
6	Smallest diameter piping	The line planned for the feed is oversized and could contain a larger inventory than is necessary.	Possible following engineering evaluation to reduce diameter.	Potential larger release		3	2	3	4. Reduce the planned feed line size from 6 to 4 inch.[6]	
7	Eliminate hazardous raw materials, process intermediates, or by-products by using an alternative process or chemistry		No alternative for isobutane in the process[7]							

[6] See Footnote 3.

[7] See Footnote 4.

Table B.4 Case Summary: Inherent Safety Analysis – Checklist Process Hazard Analysis (PHA)

Location: Orange, New Jersey						**Risk Ranking**			**Unit**: Hydrofluoric Acid Alkylation Unit	**Analysis Date:** April 1, 2008
PFD No.: 1234-5678										
Node: Isobutane Storage										
Design Conditions/Parameters: Storage of isobutene in five bullets and two process vessels near the unit										
#	**QUESTION**	**POTENTIAL OPPORTUNITIES**	**FEASIBILITY**	**CONSEQUENCES**	**EXISTING SAFEGUARDS**	**S**	**L**	**R**	**RECOMMENDATIONS**	**COMMENTS/ STATUS**
8	Eliminate in-process solvents and flammable heat transfer media.	Substitute the flammable solvent.	A nonflammable solvent may be commercially available.	Possible source of flammable release	Existing fire prevention controls and fire suppression systems.	3	3	4	5. Consider substituting the solvent used with a non-flammable solvent.[8]	

[8] This uses the concept of ***Substitution*** to avoid a hazard. A nonflammable alternative solvent was substituted for a planned flammable solvent. A hazardous condition was eliminated so this represents a **First order Inherent Safety** change.

B.2. INHERENT SAFETY ANALYSIS - INDEPENDENT PROCESS HAZARD ANALYSIS (PHA)

Table B.5 is an example of an ISD approach, which is similar to a typical PHA, but focuses exclusively on inherent safety.

The analyst considers a hazard, such as runaway reaction caused by water reactivity in a reactor, and sets a safety objective, such as to "Minimize potential for runaway reaction in the feed to the reactor." The team then documents each potential cause of the hazard being realized, and reviews the consequences, existing safeguards, and potential means of eliminating it or reducing its risk through ISD strategies.

Considering the four ISD strategies, the team documents potential recommendations that may address the concern using the order of first order ISD, Second Order ISD, followed by Layers of Protection. Each strategy is considered and ideas that are feasible, practical, and best address the hazard are generated. The approach acknowledges that other risk management strategies besides ISD may be more effective.

Table B.5: Inherent Safety Analysis - Independent Process Hazard Analysis (PHA)

Node: 1. Feed system to reactor

Objective: 1. Minimize potential for runaway reaction in the feed to the reactor

CAUSES	CONSEQUENCES	EXISTING SAFEGUARDS	S	L	R	OPPORTUNITIES	FEASIBILITY	RECOMMENDATIONS	COMMENT//STATUS
1. High water content in feed tank due to settlement or water carryover from upstream process	1. Potential for operator error to leave water online or valve not fully closed, or failure of the valve, allowing leak of water into the feed line. Excess water in the reactor may cause shorter run life due to catalyst fouling; this has a possible safety hazard in more startups and shutdowns over the life of the process. Worst credible case: excessive water may cause a runaway reaction.	1. Control of unit operation to meet feed and operator monitoring of process conditions.	4	4	4	Evaluate way to positively eliminate water from entering the reactor rather than controls.	It may be feasible to switch to a 'clean' tank without the potential for water with minor piping changes.	1. Change from feeding from Tank 1 to only Tank 3 since Tank 1 has high water settlement potential. Tank 1 has water in upstream units that cannot be completely avoided whereas Tank 3 is clean feedstock.[1]	

[1] This uses the concept of ***Substitution*** to avoid a hazard. Previously the feed was taken from a tank with water contaminants that had to be controlled. The alternative tank did have this inherent condition. A hazardous condition was avoided altogether so this represents a **First order Inherent Safety** change.

Table B.5: Inherent Safety Analysis - Independent Process Hazard Analysis (PHA)

Node: 1. Feed system to reactor

Objective: 1. Minimize potential for runaway reaction in the feed to the reactor

CAUSES	CONSEQUENCES	EXISTING SAFEGUARDS	S	L	R	OPPORTUNITIES	FEASIBILITY	RECOMMENDATIONS	COMMENT//STATUS
2. Water into the feed from wrong valve opened in one of the water wash cross connections		1. Proper procedures for water washing	4	2	4	Evaluate ways to eliminate water contamination risk from human error	Operating procedures can be improved.	2. Improve operating procedures for water washing to ensure operators check the valve closure and water flow following a water wash.[2]	
		2. Operator training 3. Temperature instrumentation to monitor reaction rate.					There is an excess number of cross connections so some can be eliminated.	3. Reduce the number of water cross connections from the feed to the reactor from 3 to 1.[3]	

[2] This uses the concept of ***Applying Procedural Safeguards*** to avoid a hazard. Existing operating procedures had not explicitly called for checking on these aspects. The change in operating procedure may reduce the likelihood of water being left on mistakenly. The inherent condition of water usage and lead potential still exists, but the likelihood, and therefore the risk, may have been reduced.

[3] Note: This uses the concept of ***Minimization*** to avoid a hazard. The water wash was previously done from multiple connections. The alternative approach of eliminating unnecessary connections reduces the likelihood of the incident by eliminating equipment, representing a **Second order Inherent Safety** change.

B.3. INHERENT SAFETY ANALYSIS – INTEGRAL TO PROCESS HAZARD ANALYSIS (PHA)

Table B.6 is an example of a third ISD approach. Essentially, it is the same as the second method presented above. But, they differ in the way in which the hazard is identified. In this example, a HAZOP method utilizes deviations from design intent while in the previous example the hazard was recognized and directly addressed by ISD. The way in which ISD is considered remains the same. The ISD hierarchy of analysis of First order ISD, then Second order ISD, and finally Layers of Protection applies no matter what the analysis method unless the scope is limited to only identify ISD potential recommendations.

Figure B.6: Inherent Safety Analysis – Integral to Process Hazard Analysis (PHA)

Node: 1. Feed system to reactor									
Intent: 1. Feed to the process									
Guideword: As Well As			Parameter: Flow					Deviation: Contamination	
CAUSES	CONSEQUENCES	EXISTING SAFEGUARDS	S	L	R	OPPORTUNITIES	FEASIBILITY	RECOMMENDATIONS	COMMENT/STATUS
1. Settlement or water carryover from upstream process.	1. Excess water in the feed and then the reactor may cause shorter run life due to catalyst fouling; this poses a possible safety hazard due to more frequent start-ups and shut-downs over the life of the process. Worst credible case: excessive water may cause a runaway reaction.	1. Control of unit operation to meet feed and operator monitoring of process conditions.	4	4	4	Evaluate ways to stop water from entering the reactor rather than controls.	It may be feasible to switch to a 'clean' tank without the potential for water with minor piping changes.	1. Change feeding from Tank 1 to only Tank 3. Tank 1 has high water settlement potential, and has water in upstream units that cannot be completely avoided. Tank 3 is clean feedstock.[1]	

[1] Note: This uses the concept of ***Substitution*** to avoid a hazard. The feed was previously taken from a tank that had water contaminants that had to be controlled. The alternative tank did not have this inherent condition. A hazardous condition was avoided altogether so this represents a **First order Inherent Safety** change.

Figure B.6: Inherent Safety Analysis – Integral to Process Hazard Analysis (PHA)

Node: 1. Feed system to reactor

Intent: 1. Feed to the process

Guideword: As Well As			**Parameter: Flow**					**Deviation: Contamination**	
CAUSES	**CONSEQUENCES**	**EXISTING SAFEGUARDS**	**S**	**L**	**R**	**OPPORTUNITIES**	**FEASIBILITY**	**RECOMMENDATIONS**	**COMMENT/STATUS**
2. Potential for operator error to leave water online or valve not fully closed, or failure of the valve allowing leak of water into the feed line	1. Excess water in the reactor may cause shorter run life due to catalyst fouling; this has a possible safety hazard in more startups and shutdowns over the life of the process. Worst credible case: excessive water may cause a runaway reaction.	1. Proper procedures for water washing	4	2	4	Evaluate ways to eliminate water contamination risk from human error	Operating procedures can be improved.	2. Improve operating procedures for water washing to ensure operators check the valve closure and water flow following a water wash.[2]	
		2. Operator training					There is an excess number of cross connections so some can be eliminated.	3. Reduce the number of water cross connections to the feed to the reactor from 3 to 1.[3]	

[2] Note: This uses the concept of ***Applying Procedural Safeguards*** to avoid a hazard. The operating procedure wasn't explicit on checking these aspects which may reduce the likelihood of the water being mistakenly left on. The inherent condition of water usage and lead potential still exists but the likelihood, and therefore the risk, may have been reduced.

Figure B.6: Inherent Safety Analysis – Integral to Process Hazard Analysis (PHA)

Node: 1. Feed system to reactor									
Intent: 1. Feed to the process									
Guideword: As Well As			Parameter: Flow					Deviation: Contamination	
CAUSES	CONSEQUENCES	EXISTING SAFEGUARDS	S	L	R	OPPORTUNITIES	FEASIBILITY	RECOMMENDATIONS	COMMENT/STATUS
		3. Temperature instrumentation to monitor reaction rate							

[3] This uses the concept of ***Minimization*** to avoid a hazard. The water wash was previously done from a multiple connections, while the alternative approach eliminates unnecessary connections to reduce the likelihood of an incident. Reducing the likelihood of the hazardous condition through elimination of equipment makes this a **Second order Inherent Safety** change.

Appendix C

Applying Inherent Safety to Risk Based Process Safety

The following table identifies opportunities within the Risk Based Process Safety (RBPS) strategies and implementation tactics described in *Guidelines for Risk Based Process Safety* (CCPS 2007b).

RBPS Book Location	ISD Applicable
Chapter 1. Introduction 1.7 Application of the RBPS Guidelines	The RBPS elements are meant to apply to the entire process life cycle. Some elements may not be active in early life cycle stages (i.e., during conceptual design there is little need for developing operating procedures). Other elements may be active, but the information available in early stages may not be very detailed; therefore, the work performed in that element would be more preliminary (i.e., hazard identification and risk analysis). On the other hand, for some elements, the early life cycle stages provide a unique opportunity to minimize risk (i.e., identifying and incorporating inherently safer process characteristics). In later stages (i.e., decommissioning), some element work activities may not be as important or may no longer be needed (i.e., maintenance), while others may be necessary, but perhaps using simpler approaches (i.e., hazard reviews of decommissioning activities using checklists).

RBPS Book Location	ISD Applicable
Chapter 5. Process Safety Competency 5.2.1 Maintain a Dependable Practice	Almost all companies profess to be learning organizations that aspire to a high degree of competency. However, those that are successful in this pursuit intentionally foster learning by establishing objectives and making plans to achieve the objectives. Normally, one or more of several conditions are necessary for an organization to invest in process safety competency: • A business case describes the expected benefits, and the level of resources that must be invested to achieve those benefits. • The organization inherently values technology, and places particular value on enhancing its process safety competency. • There is a widely-held organizational norm that decisions should be based on knowledge that is supported by facts, and any significant improvement in the body of knowledge will lead to better decisions, reducing risk and improving performance.
Chapter 9. Hazard Identification and Risk Analysis (HIRA) 9.2.1 Maintain a Dependable Practice	***Clearly define the analytical scope of the Hazard Identification and Risk Analysis (HIRA) system and assure adequate coverage.*** A HIRA system should address all the types of process risks that management wants to control. For example, some companies may consider the possibility of a steam burn to be a process hazard, while others may exclude it because they do not consider steam to be a hazardous chemical within the scope of the HIRA system. Companies should define what types of processes and/or materials are subject to HIRAs and what types of consequences (i.e., worker fatality, public evacuation, process shut-down) are of interest. In addition, the system should address when HIRAs should be performed over a project's life cycle. Some companies may require a series of analyses over a project's life cycle, while others are content with an engineering design review. Each required study should have a defined objective, such as identifying inherently safer design options, that is consistent with the process safety information available for the review. Evolved systems should consider documenting the technical basis for any exemptions/exclusions from the HIRA program.

RBPS Book Location	ISD Applicable
Chapter 9. Hazard Identification and Risk Analysis 9.2.1 Maintain a Dependable Practice (cont.)	***Make consistent risk judgments***. Management must communicate its risk tolerance to the risk assessment teams so they can make risk judgments. This guidance may be in a simple, qualitative form, such as an instruction that the facility must conform to recognized and generally accepted good engineering practices (RAGAGEP), or that risks must be reduced as low as reasonably practicable (ALARP). Management may fund the development of corporate process safety standards and require adherence to them as well. However, neither of these approaches provides guidance for all situations. Thus, some organizations develop a risk matrix, such as Figure 9.3 to communicate their tolerance for any scenario falling within a specific range of severity and likelihood. These categories may be defined either qualitatively or quantitatively. Beyond that, some companies choose (or regulators impose) some absolute risk criterion that events of this magnitude shall not exceed a specified likelihood (i.e., the likelihood of a worker fatality at the facility shall not exceed 10^{-4} events/y).

Risk	Serious danger in immediate area	Serious danger inside battery limits	Serious danger site wide	Serious danger off-site
More than once per year	Action required unless risk ALARP	Action required at first opportunity	Immediate action required	Immediate action required
Once every few years	Action required unless risk ALARP	Action required unless risk ALARP	Action required at first opportunity	Immediate action required
Once in the plant lifetime	No action required	Action required unless risk ALARP	Action required unless risk ALARP	Action required at first opportunity
Not expected in plant lifetime	No action required	No action required	Action required unless risk ALARP	Action required unless risk ALARP

Figure 9.3. Example Risk Matrix

In addition to specific risk tolerance criteria, the company should also specify its preference for the types of risk control measures employed. For example, a company may prefer inherently safer design alternatives to those requiring hazard controls, passive to active controls, or engineered to administrative controls.

RBPS Book Location	ISD Applicable
Chapter 9. Hazard Identification and Risk Analysis 9.2.3 Assess Risks and Make Risk-Based Decisions	Once hazards have been identified and the risks associated with them have been analyzed, the acceptability of the risk must be judged. Some companies may judge a risk acceptable if the system conforms to a minimum standard (i.e., regulation, code) while other companies may require that risks be reduced as low as reasonably practicable (ALARP). Some companies may judge the risk to be unacceptable under any circumstances and require that the process be relocated or abandoned unless an inherently safer alternative can be found. In order for companies/facilities to adopt and implement appropriate risk judgment protocols, the following essential features should be considered: ***Apply the risk tolerance criteria.*** Management must communicate its expectation as to how the risk tolerance criteria will be applied in each risk analysis over the life cycle of the project. If a review only identifies hazards, then the risk tolerance criteria must be applied at that level. For example, Technology A has the potential for off-site effects and Technology B does not. If management (or the community) will not tolerate any off-site risk, then Technology B must be chosen, regardless of the advantages that Technology A may offer. Later in the life cycle, the risk tolerance criteria can be applied to each scenario identified. For example, when evaluating accident scenarios related to offloading, a team may determine that a hose may burst and cause a large material release, but there are means to isolate the release and mitigate the consequences. On a qualitative basis, the unloading system may conform to applicable codes and be judged acceptable. However, companies with more stringent criteria may require a layer of protection analysis (LOPA) of the scenario to determine if it meets their risk tolerance, or they may require a detailed quantitative risk analysis (QRA). In all cases, the risk tolerance criteria should guide the risk analysis team as to when a recommendation is required, when a recommendation is optional, and when a recommendation would be superfluous. The management guidance should also acknowledge that some considerations might override the normal risk tolerance criteria. For example, if a design does not meet a code or policy requirement, corrective action is required and risk judgment is irrelevant. Or if a “tolerable” risk can be further reduced at minimal cost, the desire to reduce risk as low as reasonably achievable may warrant consideration of the recommendation.

RBPS Book Location	ISD Applicable
Chapter 9. Hazard Identification and Risk Analysis 9.2.3 Assess Risks and Make Risk-Based Decisions	***Select appropriate risk control measures.*** Management should develop a system to ensure that the most appropriate risk control measures are selected, considering its risk tolerance criteria. In particular, engineering design tradeoffs should consider the true life cycle costs (and not just the capital costs) of active and administrative controls versus inherently safer and passive controls. Otherwise, there is a tendency to select controls with lower initial capital costs that may have much higher life cycle costs because of their operating and maintenance expenses. Another difficulty is that the implementation of any one recommendation to reduce risk may also decrease (or increase) the risk in other areas. Thus, the priority of other proposed risk controls may decrease (or increase) accordingly. If all the recommendations from any one risk analysis are adopted with no consideration of these interactive effects, resources may be misdirected to unnecessary controls for some risks while the controls for other risks are inadequate.
Chapter 14. Training and Performance Assurance 14.4.2 Identify What Training is Needed	***Select training methods based on hazards and potential consequences*** - Training tools range from simple self-study workbooks to complex, full-scale drills. It is grossly inefficient to apply complex tools when simple ones would suffice, but simple tools cannot effectively prepare workers to cope with some complex situations. One way to make the training process more efficient is to perform an initial screening based on hazards and potential consequences. If, for example, the screening does not identify any severe hazards (e.g., runaway reactions), or if the potential consequences are inherently limited (e.g.,, due to small quantities of reactants), then a simple lecture or demonstration may be sufficient to ensure that risks are tolerable. This prevents wasting resources over-training for minor risks. Conversely, a screening that identifies serious hazards and/or major consequences clearly indicates the need for rigorous simulations and exercises. This prevents wasting resources on ineffective training, which will require extensive remedial effort.

Glossary

Active – Capable of functioning, causing action or change; participating in a state of action. For example, using controls, alarms, safety instrumented systems, and mitigation systems to detect and respond to process deviations from normal operation; engineering controls.

Administrative controls –Procedural mechanisms, such as lockout/tagout procedures, for directing and/or checking human performance on plant tasks; *procedures*.

Alternative Methods –Optional courses of action to achieve an objective from which a decision maker is expected to choose. In the context of inherent safety, such options would reflect one or more of the *inherently safer design strategies*.

Analytical Hierarchy Process (AHP) –a mathematical decision making technique that allows consideration of both qualitative and quantitative aspects of decisions by reducing complex decisions to a series of one-on-one comparisons, then synthesizing the results.

Attenuation –Using materials under less hazardous conditions; *moderation*. May be accomplished by strategies that are either physical (lower temperatures, dilution) or chemical (development of a reaction chemistry which operates at less severe conditions).

Cost-benefit analysis/cost-performance analysis –Methods for assessing the value of a project by comparing its costs to measures of its performance, or more generally to the value of the benefits it produces. The analysis requires accurate cost data, as well as measures of performance in appropriate units and overall benefits. Cost- performance measurement is narrower in that it deals only with measures of performance as the basis for comparison.

Decision analysis – Evaluating complex alternatives in terms of values and uncertainty to illustrate how the defined alternatives differ from one another, and then generating suggestions for new and improved alternatives. Numbers quantify subjective values and uncertainties in order to better illuminate the decision situation. Numerical results then may be translated back into words in order to generate qualitative insight.

Ergonomics –the science of designing machines, products, and systems to maximize the safety, comfort, and efficiency of the people who use them; applying the study of posture and movement to the design of objects, systems and the environment to make them compatible with the needs, abilities and limitations of people; *human factors*.

Feasible –Able to be accomplished or possible; appropriate or suitable; likely *(Source: Webster's, 3rd edition, 2005).*

Hazard – A chemical or physical condition that has the potential for causing damage to people, property or the environment. A hazard is intrinsic to the material or to its conditions of storage or use.

With respect to chemicals, "hazard" may include toxicity (acute or chronic), flammability, corrosivity or reactivity.

Human factors – Selecting materials or equipment that can better tolerate human error in handling; making a process or piece of equipment easier to understand, easier to function as intended or more difficult to function improperly; *ergonomics*.

Inherent – Existing as an essential element or feature; intrinsic. In the case of *inherent safety*, a condition in which the hazards associated with a material or operation used in the process have been reduced or eliminated in a permanent and inseparable manner.

Inherent Safety –A concept or an approach to safety that focuses on eliminating or reducing the hazards associated with a set of conditions.

Inherently Safer Design Strategies –

- *Minimize* – reducing the quantity of material or energy contained in a manufacturing process or plant (also called *Intensification*).
- *Substitute* – replacing a material with a less hazardous substance; the replacement of a hazardous material or process with an alternative that reduces or eliminates the hazard.
- *Moderate* – using materials under less hazardous conditions; using less hazardous conditions, a less hazardous form of a material, or facilities which minimize the impact of a release of hazardous material or energy (also called *Attenuation* and *Limitation of Effects*).

- *Simplify* – designing facilities which eliminate unnecessary complexity and make operating errors less likely, and which are forgiving of errors which are made (also called *Error Tolerance*).

Inherently Safer – A condition in which the hazards associated with the materials and operations used in the process have been reduced or eliminated, and this reduction or elimination is permanent and inseparable.

Inherently Safer Design – A way of thinking about the design of chemical processes and plants that focuses on the elimination or reduction of hazards, rather than on their management and control.

Inherently Safer Process – One which reduces or eliminates the hazards associated with materials and operations in a manner that is permanent and inseparable from the process itself.

Inherently Safer System – A system that relies on the chemistry and physics (the quantity, properties and conditions of use of the process materials) rather than on control systems, alarms, safety instrumented systems, and procedures to prevent incidents.

Inherently Safer Technology –A piece of equipment, process or operation designed to reduce hazard or prevent incidents.

Kepner-Tregoe Decision Analysis –A weighted scoring method for identifying and ranking all factors critical to a decision, based on explicit objectives and limits on conscious and unconscious biases.

Less Hazardous Form –A chemical, process or operation that presents a lower intrinsic *hazard* to people, property or the environment such as a less toxic chemical, a less intensive process or a more simple operation.

Minimization –The reduction of the quantity of material or energy contained in a manufacturing process or plant (also called *Intensification*)

Moderation –Use of materials under less hazardous conditions; use of a less hazardous form of a material, or use of materials in a manner that minimizes the impact of a release of hazardous material or energy (also called *Attenuation* and *Limitation of Effects*).

Multi-attribute utility analysis – A quantifiable method for evaluating and comparing alternatives to assist in decision making about complex alternatives, based on the assumption that the apparent desirability of a particular alternative depends on how its attributes are viewed. Useful in material selection decisions.

Passive – Not active, but acted upon. In the case of *inherent safety*, minimizing the hazard by process and equipment design features that reduce either the frequency or consequence of a hazard, without the active functioning of any device.

Payoff matrix analysis – A technique from game theory that shows the payoff each player will receive at the outcome of a game in which each player's payoff is dependent on the combined actions of all players.

Procedural – An established series of steps; a prescribed or traditional set of forms, or requirements to be followed. In the context of *inherent safety*, the use of operating procedures, administrative checks, emergency response, and other management approaches to prevent incidents, or to minimize the effects of an incident; *administrative controls*.

Process Intensification – *Minimization*; also often used more specifically to describe new technologies which reduce the size of unit operations equipment, particularly reactors.

Process Safety Management (PSM) –The application of management principles to ensure the safety of chemical process facilities.

Process Risk Management Strategies:

- *Inherent* – Eliminating the hazard by using materials and process conditions which are non-hazardous; i.e., substituting water for a flammable solvent.
- *Passive* – Minimizing the hazard by process and equipment design features that reduce either the frequency or consequence of the hazard without the active functioning of any device; i.e., the use of equipment rated for higher pressure.
- *Active* – Using controls, alarms, and safety instrumented systems to detect and correct process deviations; i.e., a pump which is shut-off by a high level switch in the downstream tank when the tank is 90% full. These systems are commonly referred to as engineering controls.
- *Procedural* – Using operating procedures, administrative checks, emergency response, and other management approaches to prevent incidents, or to minimize the effects of an incident; i.e., hot work

procedures and permits. These approaches are commonly referred to as administrative controls.

Reactive Distillation –A technique for combining a number of process operations in a single device. While reactive distillation can eliminate equipment and simplify a process, there may be inherent safety conflicts resulting from this strategy. Combining a number of process operations into a single device increases the complexity of that device, but it also reduces the number of vessels or other pieces of equipment required for the process.

Risk –A measure of economic loss, human injury, or environmental damage in terms of both the incident likelihood and the magnitude of the loss, injury, or damage (CCPS, 2000, 1995b). A measure of potential economic loss, human injury, or environmental damage (cost) in terms of the probability of the loss, injury, or damage over a period of time, normally a year.

With respect to chemicals, risk is a function of the *hazard* of the chemical and the *probability* of exposure to a specified endpoint.

Safer Chemical – A chemical that meets the same requirements but exhibits less of one or more hazardous properties than another chemical; a chemical that is less toxic, less flammable, less caustic, less reactive, etc.; a less hazardous chemical.

Safety – Tolerable risk in comparison to the benefit of the activity, considering who receives the benefit as compared to who bears the risk.

Safety Management Systems – Elements including management of change, process hazards analysis, mechanical integrity, and others.

Simplification – The design of facilities or processes in a manner to reduce or eliminate unnecessary complexity and make operating errors less likely.

Substitute – To replace a material with a less hazardous substance, or with an alternative that reduces or eliminates the hazard.

References

CHAPTER REFERENCES

AEA Technology (July, 1997). *The INSET Toolkit* Warrington, Cheshire, UK/ (available from AEA Technology at http://www.aeat-safety-and-risk.com/html/inset.html in February 2005).

Agreda, V. H., Partin, L.R. and Heise, W.H. (1990). High-purity methyl acetate via reactive distillation. *Chemical Engineering Progress, 86* (2), 40-46.

Akay, G., and Azzopardi, B.J. (1995). In *Proceedings of the First International Conference on Science, Engineering and Technology of Intensive Processing, September 18-20, 1995*. Nottingham, U. K.: University of Nottingham.

Allen, D. (1992). The role of catalysis in industrial waste reduction. In D.T. Sawyer and A.E. Martell (Eds.). *Industrial Environmental Chemistry* (pp. 89-98). New York: Plenum Press.

Althaus, V. E., and Mahalingam, S. (1992). Inherently safer process designs. In W.F. Early, V.H. Edwards, and E.A. Waltz (Eds.). *South Texas Section AIChE Process Plant Safety Symposium, February 18-19, 1992* (pp. 546-555). Houston, TX: American Institute of Chemical Engineers South Texas Section.

The American College Dictionary (1967). New York: Random House.

American Congress of Government Industrial Hygienists (ACGIH) (2001), Threshold Limit Values for Chemical Substances and Physical Agents and Biological Exposure Indices. Cincinnati, OH: ACGIH.

American Industrial Hygiene Association (AIHA) (1988). *Emergency Response Planning Guidelines: Chlorine*, Akron, OH: American Industrial Hygiene Association.

American Institute of Chemical Engineers (AIChE) (1995). *CHEMPAT: A Program to Assist Hazard Evaluation and Management*. New York: American Institute of Chemical Engineers (Publication Z-1).

American National Standards Institute/Instrument Society of America (ANSI/ISA) (2004). *Application of Safety Instrumented Systems for the*

Process Industries. Research Triangle Park, NC: Instrument Society of America (ISA-S84.01-2004).

American Petroleum Institute (API) (1998). *Welded Steel Tanks for Oil Storage* (API Standard 650). Washington, DC: API.

American Petroleum Institute (API) (2002). *Design And Construction of Large, Welded, Low-Pressure Storage Tanks* Washington, DC: API.

American Petroleum Institute (API) (2007). *Pressure-Relieving and Depressurizing Systems, 5th Edition*. Washington, DC: API.

American Society for Testing and Materials (ASTM International). (2006). *E2012-06 Standard Guide for the Preparation of a Binary Chemical Compatibility Chart*. West Conshohocken, PA: ASTM International.

American Society for Testing and Materials (ASTM International). American Society for Testing and Materials (ASTM) International, CHETAH: Chemical Thermodynamic & Energy Release Evaluation.

Amyotte (November 2003). et al. Reduction of dust explosion hazard by fuel substitution in power plants. *Trans IchemE* Institution of Chemical Engineers, *81*, *Part B*.

Amyotte, P.R. Goraya, A.U, Hendershot, D.C., and Khan, F.I. (2006). Incorporation of inherent safety principles in process safety management. In *Proceedings of 21st Annual International Conference–Process Safety Challenges in a Global Economy, World Dolphin Hotel, Orlando, Florida, April 23-27, 2006* (pp.175-207) New York: American Institute of Chemical Engineers.

Ankers, R. (1995). Introducing inherently safer concepts early in process development with PRORA. In *CCPS Inherently Safer Process Workshop, May 17, 1995, Chicago, IL*. New York: American Institute of Chemical Engineers.

Ashford, N.A. (1993). *The Encouragement of Technological Change for Preventing Chemical Accidents: Moving Firms From Secondary Prevention and Mitigation to Primary Prevention*. Cambridge, MA: Center for Technology, Policy and Industrial Development, Massachusetts Institute of Technology.

Auger, J.E. (1995). Build a proper PSM program from the ground up. *Chemical Engineering Progress, 91* (1), 47-53.

Bain, D.H. (1999) *Empire Express: Building the First Trans-continental Railroad*. New York: Viking.

Bea, R.G., Holdsworth, R.D., and Smith, C. (Eds.) (1997). *1996 International Workshop on Human Factors in Offshore Operations*. New York: American Bureau of Shipping.

Benson, R.S., and Ponton, J.W. (March 1993). Process miniaturization - A route to total environmental acceptability? *Trans. IchemE*, *71*, *Part A*, 160-168.

Berger, S.A., and Lantzy, R.J. (1996). Reducing Inherent Risk Through Consequence Modeling. In H. Cullingford (Ed.). *1996 Process Plant Safety Symposium*, *Volume 1, April 1-2, 1996, Houston, TX* (pp. 15-23). Houston, TX: South Texas Section of the American Institute of Chemical Engineers.

Bodor, N. (1995). Design of biologically safer chemicals. *Chemtech, 25* (10), 22-32.

Bollinger, R.E., Clark, D.G., Dowell, A.M., Ewbank, R.M., Hendershot, D.C., Lutz, W.K., et al. (1996). *Inherently Safer Chemical Processes: A Life Cycle Approach* (D. A. Crowl, Ed.). New York: American Institute of Chemical Engineers.

Boykin, R.F., Kazarians, M. and Freeman, R.A. (1986). Comparative fire risk study of PCB transformers. *Risk Analysis*, *6* (4), 477-488.

Bretherick, L. (1990). *Handbook of Reactive Chemical Hazards. 4th Edition.* London, UK: Butterworths.

Bretherick, L. (1995) Handbook of Reactive Chemical Hazards, 5th Edition. London, UK: Butterworths.

Burch, W.M. (1986). Process modifications and new chemicals. *Chemical Engineering Progress*, *82* (4) 5-8.

Carrithers, G.W., Dowell, A.M., and Hendershot, D.C. (1996). It's never too late for inherent safety. In *International Conference and Workshop on Process Safety Management and Inherently Safer Processes, October 8-11, 1996, Orlando, FL* (pp. 227-241). New York: American Institute of Chemical Engineers.

Catanach, J.S., and Hampton, S.W. (1992). Solvent and surfactant influence on flash points of pesticide formulations. *ASTM Spec. Tech. Publ. 11*, 149-57.

Center for Chemical Process Safety (CCPS) (1988a). *Guidelines for Safe Storage and Handling of High Toxic Hazard Materials*. New York: American Institute of Chemical Engineers.

Center for Chemical Process Safety (CCPS) (1988b). *Guidelines for Vapor Release Mitigation.* New York: American Institute of Chemical Engineers.

Center for Chemical Process Safety (CCPS) (1989a). *Guidelines for Chemical Process Quantitative Risk Analysis*. New York: American Institute of Chemical Engineers.

Center for Chemical Process Safety (CCPS) (1989b). *Guidelines for Technical Management of Chemical Process Safety*. New York: American Institute of Chemical Engineers.

Center for Chemical Process Safety (CCPS) (1992). *Guidelines for Hazard Evaluation Procedures, Second Edition With Worked Examples*. New York: American Institute of Chemical Engineers.

Center for Chemical Process Safety (CCPS) (1993a). *Guidelines for Engineering Design for Process Safety*. New York: American Institute of Chemical Engineers.

Center for Chemical Process Safety (CCPS) (1993b). *Guidelines for Safe Automation of Chemical Processes*. New York: American Institute of Chemical Engineers.

Center for Chemical Process Safety (CCPS) (1994a). *Guidelines for Preventing Human Error in Process Safety*. New York: American Institute of Chemical Engineers.

Center for Chemical Process Safety (CCPS) (1994b). *Guidelines for Evaluating the Characteristics of Vapor Cloud Explosions, Flash Fires, and BLEVES*. New York: American Institute of Chemical Engineers.

Center for Chemical Process Safety (CCPS) (1995a). *Tools for Making Acute Risk Decisions With Chemical Process Safety Applications*. New York: American Institute of Chemical Engineers.

Center for Chemical Process Safety (CCPS) (1995b). *Guidelines for Chemical Transportation Risk Analysis*. New York: American Institute of Chemical Engineers.

Center for Chemical Process Safety (CCPS) (1995c). *Guidelines for Writing Effective Operating and Maintenance Procedures*. New York: American Institute of Chemical Engineers.

Center for Chemical Process Safety (CCPS) (1995d). *Guidelines for Technical Planning for On-Site Emergencies*. New York: American Institute of Chemical Engineers.

Center for Chemical Process Safety (CCPS) (1995e). *Guidelines for Chemical Reactivity Evaluation and Application to Process Design*. New York: American Institute of Chemical Engineers.

Center for Chemical Process Safety (CCPS) (1995f). *Guidelines for Safe Storage and Handling of Reactive Materials*. New York: American Institute of Chemical Engineers.

Center for Chemical Process Safety (CCPS) (1996a). *Guidelines for Use Vapor Cloud Dispersion Models*. New York: American Institute of Chemical Engineers.

Center for Chemical Process Safety (CCPS) (1996b). *Guidelines for Evaluating Process Plant Buildings for External Explosions and Fires.* New York: American Institute of Chemical Engineers.

Center for Chemical Process Safety (CCPS) (1998). *Guidelines for Design Solutions to Process Equipment Failures.* New York: American Institute of Chemical Engineers.

Center for Chemical Process Safety (CCPS) (1999). *Avoiding Static Ignition Hazards in Chemical Operations.* New York: American Institute of Chemical Engineers.

Center for Chemical Process Safety (CCPS) (2000). *Guidelines for Chemical Process Quantitative Risk Analysis*. New York: American Institute of Chemical Engineers.

Center for Chemical Process Safety (CCPS) (2001). *Layers of Protection Analysis: Simplified Process Risk Assessment*. New York: American Institute of Chemical Engineers.

Center for Chemical Process Safety (CCPS) (2002). *Guidelines for Managing and Analyzing the Security Vulnerabilities of Fixed Chemical Sites.* New York: American Institute of Chemical Engineers.

Center for Chemical Process Safety (CCPS) (2003a). *Understanding Explosions.* New York: American Institute of Chemical Engineers.

Center for Chemical Process Safety (CCPS) (2003b). *Essential Practices for Managing Chemical Reactivity Hazards.* New York: American Institute of Chemical Engineers.

Center for Chemical Process Safety (CCPS) (2006). *Human Factors Methods for Improving Performance in the Process Industries.* New York: American Institute of Chemical Engineers.

Center for Chemical Process Safety (CCPS) (2007). *Guidelines for Safe and Reliable Instrumented Protective Systems.* New York: American Institute of Chemical Engineers.

Center for Chemical Process Safety (CCPS) (2008). *Guidelines for Chemical Transportation Risk Analysis*, *Second Edition.* New York: American Institute of Chemical Engineers.

Clayton, G.D. and Clayton, F.E. (Eds.). (1991). *Patty's Industrial Hygiene and Toxicology, Fourth Edition*. New York: John Wiley and Sons.

Collins, R.L. (1978). *Flying IFR*. New York: Delacorte Press/E. Friede.

Coward, H.F. and Jones, G.W. (1952). Limits of flammability of gases and vapors. *Bureau of Mines Bulletin 503*. Springfield, VA: National Technical Information Services.

Crowl, D.A., and Louvar, J.F. (1990, 2002). *Chemical Process Safety Fundamentals With Applications* (pp. 14-15). Englewood Cliffs, NJ: Prentice Hall.

CSChE (2002). *Process Safety Management, Third Edition*. Ottawa, ON: Canadian Society for Chemical Engineering.

Dale, S.E. (1987). Cost-effective design considerations for safer chemical plants. In J.L. Woodward (ed.). *Proceedings of the International Symposium on Preventing Major Chemical Accidents, February 3-5, 1987, Washington, D. C.* (pp. 3.79-3.99). New York: American Institute of Chemical Engineers.

Dartt, C.B., and Davis, M.E. (1994). Catalysis for environmentally benign processing. *Ind. Eng.Chem. Res. 33*, 2887-299.

Davies, C.A., Freedman, E., Frurip, D.J., Hertel, G.R., Seaton, W.H., and Treweek, D.N. (1990). *Chetah Version 4.4: The ASTM Chemical Thermodynamic and Energy Release Evaluation Program. 2nd Edition.* Philadelphia, PA: American Society for Testing and Materials.

Davis, G.A., Kincaid, L., Menke, D. , Griffith, B., Jones, S., Brown, K., and Goergen, M. (1994). *The Product Side of Pollution Prevention: Evaluating the Potential for Safe Substitutes*. Cincinnati, Ohio: Risk Reduction Engineering Laboratory, Office of Research and Development, U. S. Environmental Protection Agency.

The design of inherently safer plants (1988). *Chemical Engineering Progress, 84* (9), 21.

DeSimone, J.M., Maury, E.E., Guan, Z., Combes, J.R., Menceloglu, Y.Z., Clark, M.R., et al. (1994). Homogeneous and heterogeneous polymerizations in environmentally-responsible carbon dioxide. *Preprints of Papers Presented at the 208th ACS National Meeting, August 21-25, 1994, Washington, DC* (pp. 212-214). Center for Great Lakes Studies, University of Wisconsin-Milwaukee, Milwaukee, WI: Division of Environmental Chemistry, American Chemical Society.

Doherty, M., and Buzad, G. (27 August, 1992). Reactive distillation by design. *The Chemical Engineer*, s17-s19.

Dow Chemical Company (1994a). *Dow's Chemical Exposure Index Guide, 1st Edition*. New York: American Institute of Chemical Engineers.

Dow Chemical Company (1994b). *Dow's Fire and Explosion Index Hazard Classification Guide, 7th Edition*. New York: American Institute of Chemical Engineers.

Dowell, A.M. (2001). Regulations: Build a system or add layers? *Process Safety Progress, 20* (4).247-242.

Eckhoff R.K. (1997). *Dust Explosions in the Process Industries.* Oxford: Butterworth Heinemann.

Edwards, D.W. & Lawrence, D. (1993). Assessing the inherent safety of chemical process routes: Is there a relation between plant costs and inherent safety? *Trans. IChemE. 71*, Part B, 252-258.

Edwards, D.W., Lawrence, D., and Rushton, A.G. (1996). Quantifying the inherent safety of chemical process routes. In *5th World Congress of Chemical Engineering, July 14-18, 1996, San Diego, CA* (Paper 52d). New York: American Institute of Chemical Engineers.

Engineering Equipment and Materials Users Association (1993). *Alarm Systems—A Guide to Design, Management and Procurement* (Publication No. 191). London, UK: EEMUA.

Englund, S.M. (1990). Opportunities in the design of inherently safer chemical plants. *Advances in Chemical Engineering, 15*, 69-135.

Englund, S. M. (1991a). Design and operate plants for inherent safety - Part 1. *Chemical Engineering Progress, 87* (3), 85-91.

Englund, S.M. (1991b). Design and operate plants for inherent safety - Part 2. *Chemical Engineering Progress, 87* (5), 79-86.

Englund, S.M. (1993). Process and design options for inherently safer plants. In V. M. Fthenakis (ed.). *Prevention and Control of Accidental Releases of Hazardous Gases* (9-62). New York: Van Nostrand Reinhold.

European Commission-Joint Research Centre (1997). *Guidance on the Preparation of a Safety Report to meet the requirements of Council Directive 96/82/EC (Seveso II).* Luxembourg: Office for Official Publications of the European Communities, Institute for Systems Informatics and Safety, Major Accident Hazards Bureau.

Etowa, C.B., Amyotte, P.R., Pegg, M.J. and Khan, F.I. (2002) Quantification of inherent safety aspects of the Dow indices. *J Loss Prev, 15*, 477-487.

Evans, L., Frick, M.C., and Schwing, R.C. (1990). Is it safer to fly or drive? *Risk Analysis, 10,* 239–246.

Fauske, H.K. (2006). Managing Chemical Reactivity–Minimum Best Practice. *Process Safety Progress 25* (2), 120-129.

Federal Emergency Management Agency (FEMA), U. S. Department of Transportation (DOT), and U. S. Environmental Protection Agency (EPA) (ca. 1989). *Handbook of Chemical Hazard Analysis Procedures.* Washington, D.C.: FEMA Publications Office.

Flam, F. (14 October, 1994). Laser chemistry: The light choice. *Science 266*, 215-217.

Forsberg, C.W., Moses, D.L., Lewis, E.B., Gibson, R., Pearson, R., Reich, W.J., et al. (1989). *Proposed and Existing Passive and Inherent Safety-Related Structures, Systems, and Components (Building Blocks) for Advanced Light Water Reactors*. Oak Ridge, TN: Oak Ridge National Laboratory.

Franczyk, T. S. (1997). The Catalytic Dehydrogenation of Diethanolamin. In *Paper Preprints from the 213th ACS National Meeting, April 13-17,*

1997, San Francisco, CA. (342). Washington, DC: American Chemical Society Division of Environmental Chemistry.

Frank, W. L. (1995). Evaluation of a containment building for a liquid chlorine unloading facility. In *Proceedings of the 29th Annual Loss Prevention Symposium, July 30 - August 2, 1995, Boston, MA* (Paper 5b*).* E.D. Wixom and R.P. Benedetti (eds.). New York: American Institute of Chemical Engineers.

Freeman, R.A. (1996). Personal Communication to D.C. Hendershot, June 6, 1996.

French, R.W., Williams, D.D., and Wixom, E.D. (1996). Inherent safety, health and environmental (SHE) reviews. *Process Safety Progress 15* (1), 48-51.

Gay, D.M., and Leggett, D.J. (1993). Enhancing thermal hazard analysis awareness with compatibility charts. *Journal of Testing and Evaluation 21* (6), 477-80.

Gentile, M (2004) Development of a hierarchical fuzzy logic model for the evaluation of inherent safety. PhD Thesis. Texas A&M University, College Station, TX.

Gillett, J. (14 September, 1995). Validation Hazards. *The Chemical Engineer*, 26-28.

Goraya, A., Amyotte, P.R. and Khan, F.I. (2004). An inherent safety-based incident investigation methodology. *Process Safety Progress*, *23* (3), 197-205.

Govardhan, C.P., and Margolin, A.L. (4 September,1995). Extremozymes for industry: From nature and by design. *Chemistry & Industry*, 689-93.

Gowland, R.T. (1996a). Applying inherently safer concepts to a phosgene plant acquisition. *Process Safety Progress, 15* (1), 52-57.

Gowland, R.T. (1996b). Putting numbers on inherent safety. *Chemical Engineering* 103 (3), 82-86.

Green, D.L., and Dowell, A.M. (1995). How to design, verify, and validate emergency shutdown systems. *ISA Transactions 34*, 261-272.

Green, D.L., and Dowell, A.M. (1996). Cookbook safety shutdown system design. In H. Cullingford (Ed.). *1996 Process Plant Safety Symposium, Volume 1, April 1-2, 1996, Houston, TX* (pp.552-565). Houston, TX: South Texas Section of the American Institute of Chemical Engineers.

Gruhn, P. (1999). Lessons learned on safety instrumented system design (To engineer is human). In *AIChE 33rd Loss Prevention Symposium, Houston, TX, March 14-18, 1999.*

Gupta, J.P., Hendershot, D.C., and Mannan, M.S. (November, 2003). The real cost of process safety—A clear case for inherent safety. *Trans IChemE, Part B*, *81*.

Gupta, J. and Edwards D.W. (2002). Inherently safer design: Present and future. *Trans IChemE, Part B, 80,* 115.

Gupta, J. and Edwards D.W. (2003). A simple graphical method for measuring inherent safety. *Journal of Hazardous Materials, 104,* 15-30.

Gupta, J. and Hendershot, D.C. (2005). Inherently safer design. In S. Mannan (Ed.) in *Lees Loss Prevention in the Process Industries: Hazard Identification, Assessment & Control, 3rd Edition.* London, UK: Butterworth-Heinemann.

Hall, N. (7 October, 1994). Chemists clean up synthesis with one-pot reactions. *Science, 266,* 32-34.

Harris, N.C. (1987). Mitigation of accidental toxic gas releases. In J. L. Woodward (Ed.). *Proceedings of the International Symposium on Preventing Major Chemical Accidents, February 3-5, 1987, Washington, D.C.* (pp. 3.139-3.177). New York: American Institute of Chemical Engineers.

Heikkilä, A.-M., Hurme, M. & Järveläinen, M. (1996). Safety considerations in process synthesis. *Computers Chem. Engng, 20*(Suppl. A), S115-S120.

Heikkilä, A.-M. (1999). *Inherent safety in process plant design.* VTT Publications 384. Technical Research Centre of Finland, Espoo. (D Tech Thesis for the Helsinki University of Technology).

Hendershot, D.C. (1987). Safety considerations in the design of batch processing plants. In J. L. Woodward (Ed.). *Proceedings of the International Symposium on Preventing Major Chemical Accidents, February 3-5, 1987, Washington, D.C.* (pp. 3.2-3.16). New York: American Institute of Chemical Engineers.

Hendershot, D. C. (1991a). Design of inherently safer chemical processing facilities. Presented at the Texas Chemical Council Safety Seminar (Session D), June 11, 1991, Galveston, TX.

Hendershot, D. C. (1991b). The use of quantitative risk assessment in the continuing risk management of a chlorine handling facility. In B.J. Garrick, and W.C. Gekler (Eds.). *The Analysis, Communication, and Perception of Risk* (pp. 555-565). New York: Plenum Press.

Hendershot, D. C. (1995a). Conflicts and decisions in the search for inherently safer process options. *Process Safety Progress, 14* (1), 52-56.

Hendershot, D.C. (1995b). Some thoughts on the difference between inherent safety and safety. *Process Safety Progress 14* (4), 227-228.

Hendershot, D.C. (1996). Risk guidelines as a risk management tool. In *1996 Process Plant Safety Symposium, April 1-2, 1996, Houston, TX.* Houston, TX: South Texas Section of the American Institute of Chemical Engineers.

Hendershot, D.C. (1997). Safety through design in the chemical process industry: Inherently safer process design Presented at the Benchmarks for World Class Safety Through Design Symposium, sponsored by the Institute for Safety Through Design, National Safety Council, Bloomingdale, IL, August 19-20, 1997.

Hendershot, D.C. (2000). Process minimization: Making plants safer. *Chemical Engineering Progress*, 30 (1), 35-40.

Hendershot, D.C. (2003). "Inherently Safer Design." Course conducted for Mary Kay O'Connor Process Safety Center, April 15-16, 2003.

Hendershot, D.C. (2005a). An overview of inherently safer design. For presentation at *Risk Management:The Path Forward*: *20th Annual CCPS International Conference, April 11-13, 2005, Atlanta, Georgia.*

Hendershot, D.C. et al. (2005b) Implementing inherently safer design in an existing plant. Presented at the 7th Biennial Process Plant Safety Symposium, American Institute of Chemical Engineers 2005 Spring National Meeting, Atlanta, GA, April 10-14, 2005.

Hendershot, D.C. (2005c) Tell me why. For presentation at the *8th Annual International Symposium, Mary Kay O'Connor Process Safety Center, October 25-26, 2005, Texas A&M University, College Station, TX.*

Hendershot, D.C., et al. (2006a). Implementing inherently safer design in an existing plant. *Process Safety Progress, 25* (1), 52-57.

Hendershot, D.C. (2006b). An overview of inherently safer design. *Process Safety Progress*, *25* (2), 103-107.

Henderson, D.C, and Post, R.L. (2000). Inherent safety and reliability in plant design. In *Beyond Regulatory Compliance, Making Safety Second Nature*. Mary Kay O'Connor Process Safety Center 2000 Annual Symposium. College Station, TX: Texas A&M University.

Hochheiser, S. (1986). *Rohm and Haas History of a Chemical Company*. Philadelphia, PA: University of Pennsylvania Press.

Hopkins, A. (2005). *Safety, Culture and Risk: The Organizational Causes of Disasters*. Sydney: CCH.

Howard, W.B. (1996). Research needs for process safety technology. In H. Cullingford (Ed.). *1996 Process Plant Safety Symposium, Volume 1, April 1-2, 1996, Houston, TX* (pp.231-239). Houston, TX: South Texas Section of the American Institute of Chemical Engineers.

Imperial Chemical Industries (ICI) (1985). *The Mond Index, Second Edition*. Winnington, Northwich, Chesire, U. K.: Imperial Chemical Industries PLC.

INSET (2001) Toolkit Combined Version in Single Document, Volumes 1 and 2 – The Full Toolkit, The Inherent SHE In Design (INSIDE) Project.

International Association of Oil and Gas Producers (2005). *Human Factors - A Means of Improving HSE Performance.* London, UK: OGP.

The Institution of Chemical Engineers and The International Process Safety Group (1995). *Inherently Safer Process Design.* Rugby, England: The Institution of Chemical Engineers.

The Institution of Chemical Engineers (1991). New linear alkybenzene process. *The Chemical Engineer (UK), 488,* 16.

Instrument Society of America (ISA) (1985). *Graphic Symbols for Process Displays* (ISA-5.5-1985). Research Triangle Park, NC: Instrument Society of America.

Instrument Society of America (ISA) (1996). *Application of Safety Instrumented Systems for the Process Industries* (ISA-S84.01). Research Triangle Park, NC: Instrument Society of America.

International Association of Oil and Gas Producers (2005). *Human Factors - A Means of Improving HSE Performance.* London, UK: OGP.

ISO 9000 (1994). *Quality Management and Quality Assurance Standards - Guidelines for Selection and Use.* Geneva, Switzerland: International Organization for Standardization.

Jarabek, A.M., et. al. (1994). Mechanistic insights aid the search for CFC substitutes: Risk assessment of HCFC-123 as example. *Risk Analysis, 14* (3) 231-250.

Johnston, K.P. (17 March 1994). Safer solutions for chemists. *Nature 368*, 187-88.

Kauffman, G. B. (21 February 1994). CFCs, TEL and "Midge." *Chemistry and Industry*, 143.

Kelly, B.D. (2000). "Management of Change in Process Plants – A Participative Workshop," Calgary, AB, November 2000.

Kepner, C.H., and Tregoe, B.B. (1981). *The New Rational Manager.* Princeton, NJ: Princeton Research Press.

Khan, F.I. et al. (2001). Safety Weight Hazard Index (SWeHI): A new user-friendly tool for swift yet comprehensive hazard identification and safety evaluation in chemical process industries. *Trans IChemE, Part B, 79B* (2), 65.

Khan, F.I. (2003) et al. Evaluation of available indices for inherently safer design options. *Process Safety Progress, 22* (2), 83-97.

Khan F.I. and Amyotte, P.R. (2003). How to make inherent safety practice a reality. *The Canadian Journal of Chemical Engineering, 81*, 2-16.

Khan, F.I. and Amyotte, P.R. (2005). I2SI: A comprehensive quantitative tool for inherent safety and cost evaluation. *Journal of Loss Prevention in the Process Industries, 18*, 310-326.

Kharbanda, O.P., and Stallworthy, E.A. (1980 and 1988). *Safety in the Chemical Industry.* London: Heinemann Professional Publishing, Ltd.

Kletz, T.A. (6 May, 1978). What you don't have, can't leak. *Chemistry and Industry*, 287-292.

Kletz, T.A. (1984). *Cheaper, Safer Plants, or Wealth and Safety at Work.* Rugby, Warwickshire, England: The Institution of Chemical Engineers.

Kletz, T.A. (1985). Inherently safer plants. *Plant/Operations Progress 4* (3), 164-167.

Kletz, T.A. (ca. 1988). Seminar Presentation. Union Carbide Corporation.

Kletz, T.A. (1988). *Learning From Accidents in Industry.* London: Butterworths.

Kletz, T.A. (25 April, 1991a). Billiard balls and polo mints. *The Chemical Engineer, 495*, 21-22.

Kletz, T.A. (1991b). *Plant Design for Safety.* Rugby, Warwickshire, England: The Institution of Chemical Engineers.

Kletz, T.A. (1996). Inherently safer design - The growth of an idea. *Process Safety Progress, 15* (1), 5-8.

Kletz, T.A. (1998). *Process Plants: A Handbook for Inherently Safer Design.* Philadelphia, PA: Taylor & Francis.

Kletz, T.A. (1999). The constraints on inherently safer design and other innovations. *Process Safety Progress, 18* (1), 64

Kletz, T.A. (2003). Inherently safer design—Its scope and future. *Process Safety and Environmental Protection, 81*(B6), 401-405.

Kuchta, J.M. (1985). Investigation of fire and explosion accidents in the chemical, mining, and fuel-related industries - a manual. *Bureau of Mines Bulletin 680.* Springfield, VA: National Technical Information Services.

Lawrence, D. (1996). Quantifying inherent safety for the assessment of chemical process routes. PhD Thesis, Loughborough University, Loughborough, UK.

Lees, F.P. (1996). *Loss Prevention in the Process Industries, 2nd Edition.* Oxford, UK: Butterworth-Heinemann.

Leggett, D. (2002). Chemical reaction hazard identification and evaluation: Taking the first steps. Presented at AIChE Loss Prevention Symposium, New Orleans, LA, March 2002.

Lewis, B. and von Elbe, G. (1987). *Combustion Flames and Explosions of Gases, 3rd Edition.* Orlando, FL: Academic Press.

Lewis, D.J. (1979). The Mond Fire, Explosion and Toxicity Index applied to plant layout and spacing. In *13th Annual Loss Prevention Symposium, April 2-5, 1979, Houston, TX* (pp. 20-26). New York: American Institute of Chemical Engineers.

Lin, D., Mittelman, A., Halpin, V. and Cannon, D. (1994). *Inherently Safer Chemistry: A Guide to Current Industrial Processes to Address High Risk*

Chemicals. Washington, DC: Office of Pollution Prevention and Toxics, U. S. Environmental Protection Agency.

Lorenzo, D. K. (1990). *A Manager's Guide to Reducing Human Errors: Improving Human Performance in the Chemical Industry*. Washington, D.C.: Chemical Manufacturers Association.

Lutz, W. K. (1995a). Take chemistry and physics into consideration in all phases of chemical plant design. *Process Safety Progress 14* (3), 153-162.

Lutz, W. K. (1995b). Putting safety into chemical plant design. *Chemical Health and Safety 2* (6), 12-15.

Luyben, W.L. and Hendershot, D.C. (2004). Dynamic disadvantages of intensification in inherently safer process design. *Ind. Eng. Chem. Res.*, *43* (2), 2004.

Mandich, N.V.and Krulik, G.A. (1992). Substitution of nonhazardous for hazardous process chemicals in the printed circuit industry. *Me. Finish. 90* (11), 49-51.

Mannan, S. (2002). *White Paper-Challenges in Implementing Inherent Safety Principles in New and Existing Chemical Processes*. College Station, TX: Texas A&M University.

Mansfield, D.P. (1994). *Inherently Safer Approaches to Plant Design*. Warrington, Cheshire, U. K.: United Kingdom Atomic Energy Authority.

Mansfield, D.P. (1996). Viewpoints on implementing inherent safety. *Chemical Engineering, 103* (3), 78-80.

Manzer, L.E. (1993). Toward catalysis in the 21st century chemical industry. *Catalysis Today, 18*, 199-207.

Manzer, L.E. (1994). Chemistry and catalysis. In P.T. Anastas and C.A. Farris (Eds.). *Benign by Design: Alternative Synthetic Design for Pollution Prevention* (pp. 144-154). Washington, D. C.: American Chemical Society.

Marshall, V. C. (1990). The social acceptability of the chemical and process Industries. *Trans. IChemE, Part B, 68*, 83-93.

Marshall, V.C. (1992). The management of hazard and risk. *Applied Energy, 42*, 63-85.

McCarthy, A.J., Ditz, J.M. and Geren, P.M (1996). Inherently safer design review of a new alkylation process. *5th World Congress of Chemical Engineering, July 14-18, 1996, San Diego, CA (Paper 52b)*. New York: American Institute of Chemical Engineers.

McQuaid, J. (1991). Know your enemy: The science of accident prevention. *Trans. IchemE, Part B. 69*, 9-19.

Medard, L.A. (1989). *Accidental Explosions, Volume 2: Types of Explosive Substances*. West Sussex, England: Ellis Horwood Limited.

Misono, M., and Okuhara T. (1993). Solid superacid catalysts. *Chemtech*, *23* (11), 23-29.

Mizerek, P. (January/February 1996). Disinfection techniques for water and wastewater. *The National Environmental Journal*, 22-28.

Myers, P.; Mudan, K. and Hachmuth, H. (January 1992). The risks of HF and sulfuric acid alkylation. In *International Conference on Hazard Identification and Risk Analysis, Human Factors and Human Reliability in Process Safety*. New York: AIChE/CCPS, Health & Safety Executive, U.K. and European Federation of Chemical Engineering.

National Fire Protection Association (NFPA) (1994b). *Guide for Venting of Deflagrations*. NFPA 68, Quincy, MA: NFPA.

National Fire Protection Association (NFPA) (2002). *Explosion Prevention Systems*. NFPA 69, Quincy, MA: NFPA.

National Fire Protection Association (NFPA) (2002b). *Guide for Venting of Deflagrations*. NFPA 68, Quincy, MA: NFPA.

National Fire Protection Association (NFPA) (2003). *Flammable and Combustible Liquids Code*. NFPA 30, Quincy, MA: NFPA.

National Fire Protection Association (NFPA) (2004). *Recommended Practice for the Classification of Combustible Dusts and of Hazardous (Classified) Locations for Electrical Installations in Chemical Process Areas*, NFPA 499, Quincy, MA: NFPA.

National Fire Protection Association (NFPA) (2006). *Standard for the Prevention of Fire and Dust Explosions from the Manufacturing, Processing, and Handling of Combustible Particulate Solids* NFPA 654, Quincy, MA: NFPA.

National Institute for Occupational Safety and Health (NIOSH) (2005). *Pocket Guide to Chemical Hazards*, DHHS (NIOSH) Publication No. 2005-149.

National Oceanic and Atmospheric Administration (NOAA) (2006). *Chemical Reactivity Worksheet.* (http://response.restoration.noaa.gov).

Negron, R.M. (March/April 1994). Using ultraviolet disinfection in place of chlorination. *The National Environmental Journal*, 48-50.

Noronha, J.A., Merry, J.T., and Reid, W.C. (1982). Deflagration pressure containment for vessel safety design, *Plant/Operations Progress, 1* (1), 1-6.

New look fridges take off without HFCs (21 February 1994). *Chemistry and Industry*, 130.

Norman, D.A. (1988). *The Psychology of Everyday Things*. New York: Basic Books.

Norman, D.A. (1992). *Turn Signals are the Facial Expressions of Automobiles.* Reading, MA: Addison-Wesley Publishing Company.

Overton, T.A. and King G. (2006). Inherently safer technology: An evolutionary approach. *Process Safety Progress* 25 (2), 116-119.

Paint removers: New products eliminate old hazards (May 1991). *Consumer Reports*, 340-343.

Parshall, D.R. (April-June 1989). Suggestions for structuring the product design and manufacturing process to help create safe products and reduce litigation risk. *Hazard Prevention*, 12-17.

Perry, R.H. and Green, D. (1984). *Perry's Chemical Engineer's Handbook, 6th Edition.* New York: McGraw-Hill.

Petroski, H. (1985). *To Engineer Is Human: The Role of Failure in Successful Design.* New York: St. Martin's Press (reprinted in 1992 by Vintage Books, New York).

Petroski, H. (1995). *Engineers of Dreams: Great Bridge Builders and the Spanning of America.* New York: Alfred A. Knopf.

Pilz, V. (1995). Bayer's procedure for the design and operation of safe chemical plants. *Inherently Safer Process Design*, 4.54-4.65. Rugby, England: The Institution of Chemical Engineers.

Ponton, J. (1996). Some thoughts on the batch plant of the future. In *5th World Congress of Chemical Engineering, July 14-18, 1996, San Diego, CA* (Paper 52e). New York: American Institute of Chemical Engineers.

Puglionesi, P.S. and Craig, R.A. (1991). State-of-the-art techniques for chlorine supply release prevention. In *Environmental Analysis, Audits and Assessments: Papers from the 84th Annual Meeting and Exhibition of the Air and Waste Management Association, July 16-21, 1991, Vancouver, B.C.* (91-145.5). Pittsburgh, PA: Air and Waste Management Association.

Puranik, S.A., Hathi, K.K., and Sengupta, R. (1990). Prevention of hazards through technological alternatives. In *Safety and Loss Prevention in the Chemical and Oil Processing Industries, October 23-27, 1989, Singapore* (pp. 581-587). IChemE Symposium Series, no. 120. Rugby, Warwickshire, U. K.: The Institution of Chemical Engineers.

Purdy, G., and Wasilewski, M. (1995). Focused risk management for chlorine installations. *Mod. Chlor-Alkali Technol. 6,* 32-47.

Raghaven, K.V. (1992). Temperature runaway in fixed bed reactors: Online and offline checks for intrinsic safety. *Journal of Loss Prevention in the Process Industries, 5* (3), 153-159.

Rand, G.M., and Petrocelli, S.R. (1985). *Fundamentals of Aquatic Toxicology.* New York: Hemisphere Publishing.

Reason, J. (1990). *Human Error.* Cambridge, UK: Cambridge University Press.

Reid, R.A., and Christensen, D.C. (1994). Evaluate decision criteria systematically. *Chemical Engineering Progress*, *90* (7), 44-49.

Renshaw, F. M. (1990). A major accident prevention program. *Plant/Operations Progress, 9* (3), 194-197.

Reising, D.V. and Montgomery, T. (2005). Achieving effective alarm system performance: Results of ASM consortium benchmarking against the EEMUA guide for alarm systems. In *20th Annual International CCPS Conference, Risk Management: The Path Forward, Atlanta, GA, April 11-13 2005.*

Rogers, R.L., and Hallam, S. (1991). A chemical approach to inherent safety. In *IChemE Symposium Series,* No. 124, 235-41.

Rogers, R.L., Mansfield, D.P., Malmen, Y., Turney, R.D., and Verwoerd, M. (1995). The INSIDE Project: Integrating inherent safety in chemical process development and plant design. In G.A. Melhem and H.G. Fisher (Eds.). *International Symposium on Runaway Reactions and Pressure Relief Design, August 2-4, 1995, Boston, MA* (pp. 668-689). New York: American Institute of Chemical Engineers.

Rohm and Haas (2002). *2000 EHS and Sustainability Annual Report*, Revised December, 2002.

Rolt, L.T.C. (1960). *The Railway Revolution: George and Robert Stevenson (pg.147).* New York: St. Martin's Press.

Rothenberg, D.H., and Nimmo, I. (1996). The concept of abnormal situation management and mechanical reliability. In In H. Cullingford (Ed.). *1996 Process Plant Safety Symposium, Volume 1, April 1-2, 1996, Houston, TX* (pp.193-208). Houston, TX: South Texas Section of the American Institute of Chemical Engineers.

Sanders, R. E. (1993). *Management of Change in Chemical Plants Learning From Case Histories.* Oxford, UK: Butterworth-Heinemann.

Savage, P.E., Gopalan, S., Mizan, T.I., Martino, C.J., and Brock, E.E. (1995). Reactions at supercritical conditions: Applications and fundamentals. *AIChE Journal* 41 (7), 1723-1778.

Scheffler, N.E. (1996). Inherently safer latex plants. *Process Safety Progress, 15* (1), 11-17.

Scheffler, N.E., Green, L.S., and Frurip, D.J. (1993). Vapor suppression of chemicals using foams. *Process Safety Progress, 12* (3), 151-157.

Shah, S., Fischer, U. and Hungerbühler, K. (November 2003). A hierarchical approach for the evaluation of chemical process aspects from the perspective of inherent Safety. *Trans IChemE, 81*, Part B.

Sherringron, D.C. (7 January, 1991). Polymer supported systems: Towards clean chemistry? *Chemistry and Industry*, 15-19.

Siirola, J.J. (1995). An industrial perspective on process synthesis. In *AIChE Symposium Series*, 91, 222-233.

Sivak, M. and Flannagan, M.J. (2003). Flying and driving after the September 11 attacks. *Am Scientist 91*, 6–8.

Somerville, R.L. (1990). Reduce risks of handling liquified toxic gas. *Chemical Engineering Progress*, *86* (12), 64-68.

Sorensen, F. and Petersen, H.J.S. (1992). Substitution of organic solvents. *Staub - Reinhaltund Der Luft*, *52,* 113-18.

Stankiewicz, A. (2003). Reactive and hyrbid separations: Incentives, barriers, applications. In A. Stankiewicz and J.A. Moulijn (Eds.). *Re-engineering the Chemical Processing Plant: Process Intensification* (pp. 261-308). New York: CRC Press/Marcel Dekker, Inc.

Starks, C.M. (1987). Phase transfer catalysis: An overview. In C.M. Starks (Ed.). *Phase Transfer Catalysis: New Chemistry, Catalysts and Applications*, *September 8, 1985* (ACS Symposium Series No. 326). Washington, D.C.: American Chemical Society.

Starks, C.M., and Liotta, C. (1978). *Phase Transfer Catalysis Principles and Techniques*. New York: Academic Press.

Study, Karen, *In Search of an Inherently Safer Process Option***.** In *Beyond Regulatory Compliance, Making Safety Second Nature*. Mary Kay O'Connor Process Safety Center 2005 Annual Symposium. College Station, TX: Texas A&M University.

"Submarine!" (1992). *NOVA*, PBS Television Program. Boston, MA: WGBH Educational Foundation.

Sundell, M.J., and Nasman, J.H. (1993). Anchoring catalytic functionality on a polymer. *Chemtech*, *23* (12), 16-23.

Swain, A.D., and Guttmann, H.E. (1983). *Handbook of Human Reliability Analysis With Emphasis on Nuclear Power Plant Applications*. NUREG/CR-1278. Washington, D. C.: United States Nuclear Regulatory Commission.

Tickner, J. (3 October, 1994). The case for inherent safety. *Chemistry and Industry*, 796.

Tietze, L.F. (19 June, 1995). Domino reactions in organic synthesis. *Chemistry & Industry*, 453-457.

Turney, R.D. (1990). Designing plants for 1990 and Beyond: Procedures for the control of safety, health and environmental hazards in the design of chemical plants. *Trans. IChemE*, *Part B, 68*, 12-16.

Turney, R.D. (2001). Inherent safety: What can be done to increase the use of the concept?" In *Proceedings of 10th International Symposium on Loss Prevention and Safety Promotion in the Process Industries, Stockholm, Sweden, June 19-21, 2001* (pp. 519–528). Amsterdam, The Netherlands: Elsevier.

Tyler, B.J. (1985). Using the Mond Index to measure inherent hazards. *Plant/Operations Progress 4* (3), 172-75.

Tyler, B.J., Thomas, A.R., Doran, P., and Greig, T.R. (January/February 1996). A toxicity hazard index. *Chemical Health and Safety, 3*, 19-25.

Underwriter's Laboratories (UL) (2002). Standard for steel aboveground tanks for flammable and combustible liquids (UL-142). Northbrook, IL: UL.

U.S. Chemical Safety and Hazard Investigation Board (2002). *Investigative Report, Refinery Incident: Motiva Enterprises*. Washington, DC: Report No. 2001-05-I-DE.

U.S. Coast Guard (USCG) (1990). *Chemical Data Guide for Bulk Shipment by Water* (CIM 16616.6A). Washington, DC: USGC.

U.S. Coast Guard (USCG) (2005). *Compatibility of Cargoes, Chemical Compatibility Matrix* (46 CFR 150.170). Washington, DC: USGC.

U.S. Department of Energy (USDOE) (2005). *AEGLs, ERPGs, or Rev. 21 TEELs for Chemicals of Concern 2005* (DKC-05-0002). Washington, DC: USDOE.

U.S. Occupational Safety and Health Administration (OSHA) (1992). *Management of change element. Process Safety Management of Highly Hazardous Chemicals* 29CFR 1910.119(l).

Wade, D.E. (1987). Reduction of risks by reduction of toxic material inventory. In J.L. Woodward (Ed.). *Proceedings of the International Symposium on Preventing Major Chemical Accidents, February 3-5, 1987, Washington, D.*C (2.1-2.8.) New York: American Institute of Chemical Engineers.

Wallington, T.J., et al. (1994). The environmental impact of CFC replacements – HFCs and HCFCs. *Environmental Science and Technology*, *28* (7), 320-325.

Walsh, F., and Mills, G. (2 August, 1993). Electrochemical techniques for a cleaner environment. *Chemistry and Industry*, 576-579.

Wells. G.L., and Rose, L.M. (1986), *The Art of Chemical Process Design*, *266*. Amsterdam: Elsevier.

Wilday, A.J. (1991). The safe design of chemical plants with no need for pressure relief systems. In *IChemE Symposium Series No. 124*, 243-53.

Wilkinson, M., and Geddes, K. (December 1993). An award winning process. *Chemistry in Britain*, 1050-1052.

Wixom, E. D. (1995). Building inherent safety into corporate safety, health and environmental programs. In *CCPS Inherently Safer Process Workshop, May 17, 1995, Chicago, IL.*

Wolfe, T. (1979). *The Right Stuff.* New York: Farrar, Straus, and Giroux.

Yaws, C.L. (1999). *Chemical Properties Handbook*. New York: McGraw-Hill.

Yoshida, T. (1987). *Safety of Reactive Chemicals*. Amsterdam: Elsevier Science Publishers, B. V.

Yoshida, T., Wu, J., Hosoya, F., Hatano, H., Matsuzawa, T., and Wata, Y. (1991). Hazard evaluation of dibenzoylperoxide (BPO). *Proc. Int. Pyrotech. Semin.. 2* (17), 993-98.

Zabetakis, M. G. (1965). Flammability characteristics of combustible gases and vapors. *Bureau of Mines Bulletin 627*. Springfield, VA: National Technical Information Services.

Key Literature in Inherent Safety and Additional Reading

Agreda, V. H., Partin, L.R. and Heise, W.H. (1990). High-purity methyl acetate via reactive distillation. *Chemical Engineering Progress, 86* (2), 40-46.

Akay, G. and Azzopardi, B.J. (1995). In *Proceedings of the First International Conference on Science, Engineering and Technology of Intensive Processing, September 18-20, 1995*. Nottingham, U. K.: University of Nottingham.

Allen, D. (1992). The role of catalysis in industrial waste reduction. In D.T. Sawyer and A.E. Martell (Eds.). *Industrial Environmental Chemistry* (pp. 89-98). New York: Plenum Press.

Anastas, P.T. (1994). Benign by design chemistry: Pollution prevention through synthetic design. In *Preprints of Papers Presented at the 208th ACS National Meeting*, August 21-25, 1994, Washington, DC (pp. 204-205). Center for Great Lakes Studies, University of Wisconsin-Milwaukee, Milwaukee, WI: Division of Environmental Chemistry, American Chemical Society.

Arzoumanidis, G.G., and Karayannis, N.M. (1993). Fine-tuning polypropylene. *Chemtech*, 43-48.

Ashby, J. and Paton, D. (1993). The influence of chemical structure on the extent and sites of carcinogenesis for 522 rodent carcinogens and 55 different human carcinogen exposures. *Mutation Research, 286*, 3-74.

Ashford, N.A. (1993). *The Encouragement of Technological Change for Preventing Chemical Accidents: Moving Firms From Secondary Prevention and Mitigation to Primary Prevention*. Cambridge, MA: Center for Technology, Policy and Industrial Development, Massachusetts Institute of Technology.

Atherton, J.H. (March, 1993). Methods for study of reaction mechanisms in liquid/liquid and liquid/solid reaction systems and their relevance to the development of fine chemical processes. *Trans. IChemE, Part A, 71*, 111-118.

Bailey, J.D. (1992). Safer pesticide packaging and formulations for agricultural and residential applications. In *Reviews of Environmental Contamination and Toxicology* (pp. 17-27). New York: Springer-Verlag.

Balzani, V. (1994). "Greener Way to Solar Power." *New Scientist* 144, 1951 (12 November), 31-34.

Bashir, S., Chovan, T., Masri, B.J., Mukherjee, A., Pant, A., Sumit, S., et al. (1992). Thermal runaway limit of tubular reactors, defined at the inflection point of the temperature profile. *Ind. Eng. Chem. Res. 31* (9), 2164-2171.

Bell, A.T., Manzer, L.E., Chen, N.Y., Weekman, V.W., Hegedus, L.L., and Pereira, C.J. (1995). Protecting the environment through catalysis. *Chemical Engineering Progress*, *91* (2), 26-34.

Benson, R.S., and Ponton, J.W. (March 1993). Process miniaturization - A route to total environmental acceptability? *Trans. IchemE*, *71*, *Part A*, 160-168.

Berger, S.A. (1994). The pollution prevention hierarchy as an R&D management tool. In M. El-Halwagi and D.P. Petrides (Eds.). *Pollution Prevention Via Process and Product Modifications* (pp. 23-28). AIChE Symposium Series 303. New York: American Institute of Chemical Engineers.

Berger, S.A. (1995). "Estimating the environmental, safety, and health cost of processes in R&D." Presented at the American Industrial Hygienists Conference and Exhibition, May 26, 1995, Kansas City, MO.

Berger, S.A., Klawiter, I.L., Thomas, S.T., and Weber, V. (1994). "Estimate the environmental cost of new processes in R&D." Presented at the AIChE Spring National Meeting, April 1994, Atlanta, GA.

Berger, S.A., Cherry, J., Levengood, D., Weber, R., and Neman, D. (1996). A cost and time-efficient framework for inherent safety and pollution prevention during process development and engineering. In *International Conference and Workshop on Process Safety Management and Inherently Safer Processes*, *October 8-11, 1996, Orlando, FL* (pp. 416-428). New York: American Institute of Chemical Engineers.

Berger, S.A., and Lantzy, R.J. (1996). Reducing Inherent Risk Through Consequence Modeling. In H. Cullingford (Ed.). *1996 Process Plant Safety Symposium*, *Volume 1, April 1-2, 1996, Houston, TX* (pp. 15-23). Houston, TX: South Texas Section of the American Institute of Chemical Engineers.

Berglund, R.L. and Snyder, G.E. (December,1990). Waste minimization: The sooner, the better. *Chemtech*, 740-746.

Black, H. (1996). Supercritical carbon dioxide: the "greener" solvent. *Environmental Science and Technology* 30 (3), 124A-7A.

Blumenberg, B. (1992). Chemical reaction engineering in today's industrial environment. *Chemical Engineering Science, 47* (9-11,) 2149-2162.

Bodor, N. (October, 1995). Design of biologically safer chemicals. *Chemtech*, 22-32.

Borman, S. (November 30, 1992). Aromatic amine route is environmentally safer. *Chemical and Engineering News*, 26-27.

Bradley, D. (August 6, 1994). Solvents get the big squeeze. *New Scientist*, 32-35.

Bradley, D. (April 29, 1995). Incredible shrinking visions. *New Scientist*, 46-47.

Brennan, D.J. (May, 1993). Some challenges of cleaner production for process design. *Environmental Protection Bulletin* 024, 3-7.

Burch, W.M. (1986). Process modifications and new chemicals. *Chemical Engineering Progress*, *82* (4) 5-8.

Callari, J. (November, 1992). Environmental pressures force widespread change. *Plastics World*, 40-43.

Calvert, C. (November, 1992). Environmentally-friendly catalysis using non-toxic supported reagents. *Environmental Protection Bulletin* 021, 3-9.

Calvert, C. (1992). Environmentally friendly catalysts using non-toxic supported reagents. *IChemE North Western Branch Symposium Papers*, *3*, 4.1-4.16.

Carlsson, M., Habenicht, C., Kam, L.C., and Antal, M.J. (1994). Study of the sequential conversion of citric to itaconic to methacrylic acid in near-critical and supercritical water. *Ind. Eng. Chem. Res.*, *33*, 1989-1996.

Carrithers, G.W., Dowell, A.M., and Hendershot, D.C. (1996). It's never too late for inherent safety. In *International Conference and Workshop on Process Safety Management and Inherently Safer Processes, October 8-11, 1996, Orlando, FL* (pp. 227-241). New York: American Institute of Chemical Engineers.

Catanach, J.S., and Hampton, S.W. (1992). Solvent and surfactant influence on flash points of pesticide formulations. *ASTM Spec. Tech. Publ. 11*, 149-157.

Chementator: (August, 1993). A propylene process promises to slash operating costs. *Chemical* Engineering, 15.

Do away with solvents in polymer processing, 15.

Enzymes help make better phenolics, 17-18.

Chevalier, A.B.H. (1991). Passive safety features - A considered view of UK safety arguments. *Nuc. Energy*, *30* (2), 79-83.

Clark, D.G. (2007). Applying the limitation of effects ISP strategy when siting and designing facilities. In *Proceedings of the 41st Annual Loss Prevention Symposium, April 22-26, 2007, Houston, TX* (Paper 1C). New York: American Institute of Chemical Engineers.

Cleary, J. A. (June, 1993). Overview of the new sealless pump standard *Hydrocarbon Processing*, 49-51.

Clifford,T., and Bartle, K. (17 June, 1996). Chemical reactions in supercritical fluids. *Chemistry & Industry*, 449-452.

Coeyman, M., and R. Mullin (October 13, 1993). Raw materials suppliers key in on performance enhancement. *Chemical Week*, 34-35.

Coghlan, A. (19 November, 1994). Dry ice leaves dirty cleaners out in the cold. *New Scientist*, 28.

Costa, R., Recasens, F. and Velo, E. (1995). Inherent thermal safety of stirred-tank batch reactors: A prognosis tool based on pattern recognition of hazardous states. In G.A. Melhem and H.G. Fisher (Eds.). *International Symposium on Runaway Reactions and Pressure Relief Design, August 2-4, 1995, Boston, MA* (pp. 690-709). New York: American Institute of Chemical Engineers.

Cottam, A. N. (1991). Risk assessment and control in biotechnology.In *IChemE Symposium Series,* No. 124, 341-w348.

Crabtree, E.W., and El-Halwagi, M.M. (1994). Synthesis of environmentally acceptable reactions. In M. El-Halwagi, and D.P. Petrides (Eds.).. *Pollution Prevention Via Process and Product Modifications* (pp. 117-127). AIChE Symposium Series, 303. New York: American Institute of Chemical Engineers.

Cusumano, J A. (August, 1992). New technology and the environment. *Chemtech*, 482-89.

Dartt, C.B., and Davis, M.E. (1994). Catalysis for environmentally benign processing. *Ind. Eng.Chem. Res. 33*, 2887-299.

Davis, G.A., Kincaid, L., Menke, D., Griffith, B., Jones, S., Brown, K., and Goergen, M. (1994). *The Product Side of Pollution Prevention: Evaluating the Potential for Safe Substitutes*. Cincinnati, Ohio: Risk Reduction Engineering Laboratory, Office of Research and Development, U. S. Environmental Protection Agency.

The design of inherently safer plants (1988). *Chemical Engineering Progress*, *84* (9), 21.

DeSimone, J.M., Maury, E.E., Guan, Z., Combes, J.R., Menceloglu, Y.Z., Clark, M.R., et al. (1994). Homogeneous and heterogeneous polymerizations in environmentally-responsible carbon dioxide. In *Preprints of Papers Presented at the 208th ACS National Meeting, August 21-25, 1994, Washington, DC* (pp. 212-214). Center for Great Lakes Studies, University of Wisconsin-Milwaukee, Milwaukee, WI: Division of Environmental Chemistry, American Chemical Society.

DeVito, S. C. (November, 1996). Designing safer chemicals: Toxicological considerations. *Chemtech*, 34-47.

Doerr, W.W., and Hessian Jr., R.T. (1991). Control toxic emissions from batch operations. *Chemical Engineering Progress, 87* (9), 57-62.

Doherty, M., and Buzad, G. (27 August, 1992). Reactive distillation by design. *The Chemical Engineer*, s17-s19.

Donohue, M. D. and Geiger, J.L. (1994). Reduction of VOC emission during spray painting operations: A new process using supercritical carbon dioxide. In *Preprints of Papers Presented at the 208th ACS National Meeting, August 21-25, 1994, Washington, DC* (pp. 218-219). Center for Great Lakes Studies, University of Wisconsin-Milwaukee, Milwaukee, WI: Division of Environmental Chemistry, American Chemical Society.

Dowell, A.M. (1996). Vent systems: life cycle & inherently safer concepts. In *International Conference and Workshop on Process Safety Management and Inherently Safer Processes, October 8-11, 1996, Orlando, FL* (Workshop F: Case Studies on Inherent Safety: Cost Benefit Analysis; Life Cycle Cost).

Drexler, K. E. (1994). Molecular manufacturing for the environment. In *Preprints of Papers Presented at the 208th ACS National Meeting, August 21-25, 1994, Washington, DC* (pp. 263-265). Center for Great Lakes Studies, University of Wisconsin-Milwaukee, Milwaukee, WI: Division of Environmental Chemistry, American Chemical Society.

Drioli, E., and L. Giorno (1 January, 1996). Catalytic membrane reactors. *Chemistry and Industry*, 19-22.

Dutt, S. (1996). Safe design of direct steam injection heaters. In *The 5th World Congress of Chemical Engineering, July 14-18, 1996, San Diego, CA, Vol. II* (pp. 1107-1102). New York: American Institute of Chemical Engineers. (

Edwards, D.W. and Lawrence, D. (1993). Assessing the inherent safety of chemical process routes: Is there a relation between plant costs and inherent safety? *Trans. IChemE. 71*, Part B, 252-258.

Edwards, D.W. and Lawrence, D. (1995). Inherent safety assessment of chemical process routes. In *The 1995 ChemE Research Event: First European Conf. of Young Res. Chem. Eng*, July 14-18, 1996 (pp.62-64). Rugby, UK: Institution of Chemical Engineers.

Edwards, D.W., Lawrence, D., and Rushton, A.G. (1996). Quantifying the inherent safety of chemical process routes. In *5th World Congress of Chemical Engineering, July 14-18, 1996, San Diego, CA* (Paper 52d). New York: American Institute of Chemical Engineers.

Eierman, R. G. (1995). Improving Inherent Safety With Sealless Pumps. In E.D. Wixom and R.P. Benedetti (Eds.). *Proceedings of the 29th Annual Loss Prevention Symposium, July 31-August 2, 1995, Boston, MA* (Paper 1e). New York: American Institute of Chemical Engineers.

Emsley, J. (12 March, 1994). A cleaner way to make nylon. *New Scientist*, 15.

Englehardt, J. D. (1993). Pollution prevention technologies: A review and classification. *Journal of Hazardous Materials, 35*, 119-50.

Englund, S.M. (1990). Opportunities in the design of inherently safer chemical plants. *Advances in Chemical Engineering, 15*, 69-135.

Englund, S. M. (1990). "The design and operation of inherently safer chemical plants." Presented at the American Institute of Chemical Engineers 1990 Summer National Meeting, August 20, 1990, San Diego, CA, Session 43.

Englund, S. M. (1991a). Design and operate plants for inherent safety - Part 1. *Chemical Engineering Progress*, *87* (3), 85-91.

Englund, S.M. (1991b). Design and operate plants for inherent safety - Part 2. *Chemical Engineering Progress*, *87* (5), 79-86.

Englund, S.M. (1993). Process and design options for inherently safer plants. In V. M. Fthenakis (ed.). *Prevention and Control of Accidental Releases of Hazardous Gases* (9-62). New York: Van Nostrand Reinhold.

Englund, S. M. (1994). "Inherently safer plants—Practical applications." Presented at the American Institute of Chemical Engineers 1994 Summer National Meeting, August 14-17, 1994, Denver, CO, Paper No. 47b.

EPA urges chemical industry, universities to embrace "benign by design" production (August 27, 1993). *Chemical Regulation Reporter*, 989-990.

Finzel, W.A. (1991). Use low-VOC coatings. *Chemical Engineering Progress*, *87* (11), 50-53.

Flam, F. (9 September, 1994). EPA campaigns for safer chemicals. *Science, 265*, 1519.

Flam, F. (14 October, 1994). Laser chemistry: The light choice. *Science, 266*, 215-217.

Forsberg, C.W., Moses, D.L., Lewis, E.B., Gibson, R., Pearson, R., Reich, W.J., et al. (1989). *Proposed and Existing Passive and Inherent Safety-Related Structures, Systems, and Components (Building Blocks) for Advanced Light Water Reactors*. Oak Ridge, TN: Oak Ridge National Laboratory.

Forsberg, C.W. (1990). Passive and inherent safety technologies for light-water nuclear reactors. Presented at the America Institute of Chemical Engineers 1990 Summer National Meeting, August 19-22, 1990, San Diego, CA, Session 43.

French, R. W., Williams, D.D. and Wixom, E.D. (1995). Inherent safety, health and environmental (SHE) reviews. In E. D. Wixom and R. P. Benedetti (Eds.). *Proceedings of the 29th Annual Loss Prevention Symposium, July 31-August 2, 1995, Boston, MA* (Paper 1c). New York: American Institute of Chemical Engineers.

Friedlander, S.K. (1989). The implications of environmental issues for engineering R&D and education. *Chemical Engineering Progress*, *85* (11), 22-28.

Gerritsen, H.G., and Van't Land, C.M. (1988). Intrinsic continuous process safeguarding. *IChemE Symposium Series* No. 110, 107-115.

Gibbs, W.W. (November, 1994). Ounce of prevention. *Scientific American*, 103-105.

Goldschmidt, G., and Filskov, P. (1990). Substitution—A way to obtain protection against harmful substances at work. *Staub - Reinhaltund Der Luft*, *50*, 403-405.

Govardhan, C.P. and Margolin, A.L. (4 September,1995). Extremozymes for industry—From nature and by design. *Chemistry & Industry*, 689-693.

Gowland, R. T. (1995). Applying inherently safer concepts to an acquisition which handles phosgene. In E.D. Wixom and R.P. Benedetti (Eds.). *Proceedings of the 29th Annual Loss Prevention Symposium, July 31 - August 2, 1995, Boston, MA* (Paper No. 1d). New York: American Institute of Chemical Engineers.

Gowland, R.T. (1996). Putting numbers on inherent safety. *Chemical Engineering 103* (3), 82-86.

Green Chemistry Challenge (1995). *Chemical Health and Safety, 2* (6), 5.

Grewer, T., Klusacek, H., Loffler, U., Rogers, R.L. and Steinbach, J. (1989). Determination and assessment of the characteristic values for the evaluation of the thermal safety of chemical processes. *Journal of Loss Prevention in the Process Industries, 2*, 215-223.

Guettler, R. D., Jones, G.C., Posey, L. and Zare, R.N. (14 October, 1994). Partial control of an ion-molecule reaction by selection of the internal motion of the polyatomic reagent ion. *Science, 266*, 259-261.

Gygax, R.W. (1988). Chemical reaction engineering for safety. *Chemical Engineering Science, 43* (8), 1759-1771.

Haggin, J. (February 14,1994). Catalysis gains widening role in environmental protection. *Chemical and Engineering News*, 22-30.

Haggin, J. (April 18, 1994). Direct process converts methane to acetic acid. *Chemical and Engineering News*, 5.

Catalysis critical to benign process design. 22-25.

Haggin, J. (June 3, 1996). Chemical reaction engineering brings diversity of inputs to process designs. *Chemical and Engineering News*, 38-51.

Hall, N. (7 October, 1994). Chemists clean up synthesis with one-pot reactions. *Science, 266*, 32-34.

Hannon, J. (1992). Clean technology through process intensification. *IChemE North Western Branch Symposium Papers, 3*, 7.1-7.6.

Harris, C. (1993). Containment enclosures. In V. M. Fthenakis (Ed.). *Prevention and Control of Accidental Releases of Hazardous Gases* (pp. 404-410). New York: Van Nostrand Reinhold.

Hawksley, J. L., and. Preston, M.L (1996). Inherent SHE—20 Years of Evolution. In *International Conference and Workshop on Process Safety Management and Inherently Safer Processes, October 8-11, 1996,*

Orlando, FL (pp.183-196). New York: American Institute of Chemical Engineers.

Heil, J.A. (1995). *Inherent Safety Characteristics of Innovative Nuclear Reactors*. Petten, Netherlands: Netherlands Energy Res. Foundation.

Hendershot, D. C. (1987). Safety considerations in the design of batch processing plants. In J.L. Woodward (Ed.). *Proceedings of the International Symposium on Preventing Major Chemical Accidents, February 3-5, 1987, Washington, D.C.* (pp. 3.2-3.16). New York: American Institute of Chemical Engineers.

Hendershot, D.C. (1988). Risk reduction alternatives for hazardous material storage. In *Proc. 1988 Hazardous Materials Spills Conference, May 16-19, 1988, Chicago, IL* (pp. 611-618). New York: American Institute of Chemical Engineers.

Hendershot, D.C. (1988). Alternatives for reducing the risks of hazardous material storage facilities. *Environmental Progress 7* (3), 180-184.

Hendershot, D.C. (1991). “Design of inherently safer chemical processing facilities.” Presented at Texas Chemical Council Safety Seminar, June 11, 1991, Galveston, TX, Session D.

Hendershot, D.C. (1993). Inherently safer plants. In *Guidelines for Engineering Design for Process Safety* (pp. 5-52). New York: American Institute of Chemical Engineers.

Hendershot, D.C. (1994). Chemistry—The Key to Inherently Safer Manufacturing Processes. In *208th American Chemical Society National Meeting, August 21-25, 1994, Washington, DC* (Paper No. ENVR-135).

Hendershot, D.C. (1995). Conflicts and decisions in the search for inherently safer process options. *Process Safety Progress, 14* (1), 52-56.

Hendershot, D.C. (1995). Some thoughts on the difference between inherent safety and safety. *Process Safety Progress, 14* (4), 227-28.

Hendershot, D.C. (1996). Conflicts and decisions in the search for inherently safer process options. In G.F. Nalven (Ed.). *Plant Safety* (pp. 58-62). New York: American Institute of Chemical Engineers.

Hendershot, D.C. (1996). “The philosophy of inherently safer chemical process design.” Presented at the 89th Annual Meeting and Exhibition of the Air & Waste Management Association, June 23-28, 1996, Nashville, TN (Session 27).

Hendershot, D.C. (1996). Case studies of inherently safe design. In *The 5th World Congress of Chemical Engineering: Technologies Critical to a Changing World July 14-18, 1996, San Diego, CA* (pp. 98-99). New York: American Institute of Chemical Engineers.

Hendershot, D.C. (1996). How do you measure inherent safety early in process development? In *International Conference and Workshop on Process*

Safety Management and Inherently Safer Processes, *October 8-11, 1996, Orlando, FL* (Workshop F).

Hill, R.G. (July, 1995). Inherent safety. *Safety & Health News*, 4.

Hill, R.G. (April, 1995). Ongoing work—Inherent safety. *Safety and Health News*, 1.

Hodel, A.E. (March, 1993). Butyl acetate replaces toluene to remove phenol from water. *Chemical Processing*, 53-56.

Hugo, P., and Steinbach, J. (1986). A comparison of the limits of safe operation of a SBR and a CSTR. *Chemical Engineering Science, 41* (4), 1081-1087.

Humphrey, J L., and Seibert, A.F. (December, 1992). New horizons in distillation. *Chemical Engineering*, 86-98.

Illman, D.L. (September 6,1993). “Green” technology presents challenge to chemists. *Chemical and Engineering News*, 26-30.

Illman, D.L. (September 5, 1994). Environmentally benign chemistry aims for processes that don’t pollute. *Chemical and Engineering News*, 22-27.

Inherent Safety (July, 1995). *Safety & Health News*, 3.

Inherently Safe Technology Advances Despite Doubts (1994). *Chemical Process Safety Report, 5* (1), 1.

International Conference and Workshop on Process Safety Management and Inherently Safer Processes (1996). New York: American Institute of Chemical Engineers.

Irwin, A. (March, 1993). Changing the unchangeable. *Process Engineering*, 29-30.

Ishizaki, A., Tanaka, K, Takeshita, T., Kanemaru, T., Shimoji, T., and Kawano, T. (1993). Equipment and operation for fermentative PHB production using gaseous substrate to guarantee safety from explosion. *J. Chem. Eng. Jpn. 26* (2), 225-227.

Jacobson, A. and Willingham, G.L. (1994). Designing an Environmentally Safe Marine Antifoulant." In *Preprints of Papers Presented at the 208th ACS National Meeting*, *August 21-25, 1994, Washington, DC* (pp. 358-388). Center for Great Lakes Studies, University of Wisconsin-Milwaukee, Milwaukee, WI: Division of Environmental Chemistry, American Chemical Society.

Jarabek, A.M., et. al. (1994). Mechanistic insights aid the search for CFC substitutes: Risk Assessment of HCFC-123 as example. *Risk Analysis*, *14* (3) 231-250.

Jessop, P.G., Ikariya, T. and Noyori, R. (17 March, 1994). Homogeneous catalytic hydrogentation of supercritical carbon dioxide. *Nature,* 368, 231-233.

Joback, K.G. (1994). Solvent substitution for pollution prevention. In M. El-Halwagi, and D.P. Petrides (Eds.). *Pollution Prevention Via Process and Product Modifications* (pp. 98-103). AIChE Symposium Series, 303. New York: American Institute of Chemical Engineers.

Johnson, R.W., Unwin, S.D., and McSweeney, T.I. (1996). Inherent safety: how to measure it and why we need it. In *International Conference and Workshop on Process Safety Management and Inherently Safer Processes*, *October 8-11, 1996, Orlando, FL* (pp. 118-127). New York: American Institute of Chemical Engineers.

Johnston, K.P. (17 March 1994). Safer solutions for chemists. *Nature 368*, 187-88.

Jones, D.O. (1996). *Process Intensification of Batch, Exothermic Reactors.* Sudbury, Suffolk, UK: HSE Books.

Kauffman, G.B. (21 February,1994). CFCs, TEL and "Midge." *Chemistry and Industry*, 143.

Keeports, G.L. (1996). Major accident prevention program. In *International Conference and Workshop on Process Safety Management and Inherently Safer Processes*, *October 8-11, 1996, Orlando, FL* (Workshop E). New York: American Institute of Chemical Engineers.

Kelleher, T. and Fair, J.R. (1996). Distillation studies in a high-gravity contactor. *Ind. Eng. Chem. Res.*, *35*, 4646-4655.

Keller, A., Heinxle, E. and K. Hungerbuhler (1996). Development and assessment of inherently safe processes in the fine chemical industry." In *International Conference and Workshop on Process Safety Management and Inherently Safer Processes*, *October 8-11, 1996, Orlando, FL* (213-223). New York: American Institute of Chemical Engineers.

Kelley, K P. (April, 1992). Pollution prevention pays off. *Control*, 27-34.

Kharbanda, O.P. and Stallworthy, E.A. (1980 and 1988). *Safety in the Chemical Industry*. London: Heinemann Professional Publishing, Ltd.

Kirby, G.N. (1985). Explosion pressure shock resistance. *Chemical Engineering Progress*, *81* (11), 48-50.

Kletz, T.A. (6 May, 1978). What you don't have, can't leak. *Chemistry and Industry*, 287-292.

Kletz, T.A. (March, 1979). Is there a simpler solution? *The Chemical Engineer*, 161-64.

Kletz, T.A. (June, 1983). Inherently safer plant—The concept, its scope and benefits. *Loss Prevention Bulletin, 051*, 1-8.

Kletz, T.A. (1984). *Cheaper, Safer Plants, or Wealth and Safety at Work.* Rugby, Warwickshire, England: The Institution of Chemical Engineers.

Kletz, T.A. (1985). Inherently safer plants. *Plant/Operations Progress 4* (3), 164-167.

Kletz, T.A. (1990). Plants should be friendly. In *Safety and Loss Prevention in the Chemical and Oil Processing Industries, October 23-27, 1989, Singapore* (pp. 423-435). IChemE Symposium Series, no. 120. Rugby, Warwickshire, U. K.: The Institution of Chemical Engineers.

Kletz, T. A. (1991). Present trends in process safety. *Speculations in Science and Technology, Developments in Chemical Engineering, 15* (2), 83-90.

Kletz, T.A. (25 April, 1991). Billiard balls and polo mints. *The Chemical Engineer*, *495*, 21-22.

Kletz, T.A. (June 1991). Inherently safer plants: An update. *Plant/Operations Progress, 10* (2), 81-84.

Kletz, T.A. (1991). Inherently safer plants: Recent progress. *IChemE Symposium Series,* 124, 225-33.

Kletz, T.A. (1992). Titles can mislead. *Journal of Loss Prevention in the Process Industries*, *5* (5), 259.

Kletz, T.A. (1992). Managing risk in chemical manufacture. *Spec. Pub.- R. Soc. Chem*, 101, 92-105.

Kletz, T.A. (1993). What can we do to encourage the development of inherently safer designs. *Safety and Health News*, 3, 1.

Kletz, T.A. (1993). Accident prevention: Lessons Learned. In V. M. Fthenakis (ed.). *Prevention and Control of Accidental Releases of Hazardous Gases* (pp. 83-93). New York: Van Nostrand Reinhold.

Kletz, T.A. (1995). Inherently safer design: The growth of an idea. In E.D. Wixom and R.P. Benedetti (Eds.). *Proceedings of the 29th Annual Loss Prevention Symposium, July 31-August 2, 1995, Boston, MA (Paper 1a).* New York: American Institute of Chemical Engineers.

Kletz, T.A. (1996). Inherently safer design: Achievements and prospects. In *International Conference and Workshop on Process Safety Management and Inherently Safer Processes, October 8-11, 1996, Orlando, FL* (pp. 197-206). New York: American Institute of Chemical Engineers.

Koch, T. A., Krause, K.R. and Mehdizadeh, M. (1996). Improved safety through distributed manufacturing of hazardous chemicals. In *The 5th World Congress of Chemical Engineering, July 14-18, 1996, San Diego, CA, Vol. II* (pp. 1095-1098). New York: American Institute of Chemical Engineers.

Kollman, C.J. (1993). “Achieving environmental excellence in process development.” Presented at HazMat International ’93 Environmental Management and Technology Conference, June 9-11, 1993, Atlantic City, NJ (Session 20).

Komarek, R. (December, 1993). Briquets used to reduce environmental hazards. *Pollution Engineering*, 47-48.

Laing, I.G. (21 September, 1992). Waste minimisation: The role of process development. *Chemistry and Industry*, 682-686.

Laird, T. (1992). "Is your chemical process operating at its optimum?" Presented at Chemicals and the Environment Symposium, in Association With Chemical Specialities USA'92 Exhibition. Bramhall, Stockport England: Spring Innovations Limited.

Lammertink, E. (1992). "Safe storage of propylene oxide, a new approach." Presented at the International Process Safety Group Technical Meeting, April 29-May 1, 1992, Taormina, Italy (Paper 1.2)

Lawrence, D., and Edwards, D.W. (1994). Inherent safety assessment of chemical process routes by expert judgement. In *The 1994 IChemE Research Event* (886-888).

Lawrence, D., Edwards, D.W., and Rushton, A.G. (1993). Quantifying inherent safety. *Proc. Inst. Mech. Eng, 4,* 1-8.

Lazari, L., S. J. Burley, and S. T. S. Al-Hassani (1991). Containment of explosive blast in a spherical vessel with nozzles. *IChemE Symposium Series No. 124*, 135-153.

Lerou, J. J., and K. M. Ng (1996). Chemical reaction engineering: A multiscale approach to a multiobjective task. *Chemical Engineering Science, 51* (10), 1595-614.

Lim, P. K. (1994). Environmentally benign oil-water interfacial synthesis. *Preprints of Papers Presented at the 208th ACS National Meeting, August 21-25, 1994, Washington, DC* (pp. 238). Center for Great Lakes Studies, University of Wisconsin-Milwaukee, Milwaukee, WI: Division of Environmental Chemistry, American Chemical Society.

Lin, D., Mittelman, A., Halpin, V. and Cannon, D. (1994). *Inherently Safer Chemistry: A Guide to Current Industrial Processes to Address High Risk Chemicals*. Washington, DC: Office of Pollution Prevention and Toxics, U. S. Environmental Protection Agency.

Lin, M., and Ayusman, S. (14 April, 1994). Direct catalytic conversion of methane to acetic acid in aqueous medium. *Nature,* 238, 613-615.

Lodal, P. N. (1996). "Inherently Safer Processes Using the Low Cost Plant Approach." Presented at International Conference and Workshop on Process Safety Management and Inherently Safer Processes, October 8-11, 1996, Orlando, FL, Workshop E.

Lorentzen, G. (1995). The use of natural refrigerants: A complete solution to the CFC/HCFC predicament. *Int. J. Refrig. 18* (3), 190-197.

Lubineau, A. (19 February, 1996). Making a splash in synthesis: Water as a solvent. *Chemistry and Industry*, 123-26.

Lutz, W.K. (1995). Take chemistry and physics into consideration in all phases of chemical plant design. *Process Safety Progress 14* (3), 153-162.

Lutz, W.K. (1995). Putting safety into chemical plant design. *Chemical Health and Safety, 2* (6), 12-15.

Lutz, W.K. (1996). Advancing inherent safety into methodology. In *The 5th World Congress of Chemical Engineering, July 14-18, 1996, San Diego, CA, Vol. II* (pps.1015-1019). New York: American Institute of Chemical Engineers.

Lutz, W.K. (1996). "Advancing inherent safety into methodology." Presented at the International Conference and Workshop on Process Safety Management and Inherently Safer Processes, October 8-11, 1996, Orlando, FL, Workshop E.

Mandich, N.V.and Krulik, G.A. (1992). Substitution of nonhazardous for hazardous process chemicals in the printed circuit industry. *Me. Finish. 90* (11), 49-51.

Mansfield, D.P. (1994). *Inherently Safer Approaches to Plant Design.* Warrington, Cheshire, UK: United Kingdom Atomic Energy Authority.

Mansfield, D.P. (1996). Viewpoints on implementing inherent safety. *Chemical Engineering, 103* (3), 78-80.

Mansfield, D.P., and Cassidy, K. (1994). Inherently safer approaches to plant design. *IChemE Symposium Series*, *134,* 285-99.

Mansfield, D.P., Turney, R.D., Rogers, R.L., Verwoerd, M. and Bots, P. (1995). How to integrate inherent SHE in process development and plant design. Presented at *IChemE Conference, Major Hazards Onshore and Offshore*. UMIST'95, Manchester, UK.

Mansfield, D., Malmen, Y. and Suokas, E. (1996). "The Development of an Integrated Toolkit for Inherent SHE." *International Conference and Workshop on Process Safety Management and Inherently Safer Processes*, *October 8-11, 1996, Orlando, FL* (pp. 103-117). New York: American Institute of Chemical Engineers.

Manzer, L.E. (1993). Environmentally safer processes: Opportunities for catalysis and process R&D. *Chemistry for Sustainable Development*, *1*, 147-151.

Manzer, L.E. (1993). Toward catalysis in the 21st century chemical industry. *Catalysis Today, 18*, 199-207.

Manzer, L.E. (1994). Chemistry and catalysis. In P.T. Anastas and C.A. Farris (Eds.). *Benign by Design: Alternative Synthetic Design for Pollution Prevention* (pp. 144-154). Washington, D. C.: American Chemical Society.

Maroldo, S. G., and M. T. Vandersall (1993). Carbonaceous adsorbents for the prevention of bed fires. Presented at the Air & Waste Management Association, 86th Annual Meeting and Exhibition, June 13-18, 1993, Denver, CO (93-TP-31B.04).

Marshall, V.C. (1990). The social acceptability of the chemical and process industries. *Trans. IChemE, Part B*, *68*, 83-93.

Marshall, V.C. (1992). The management of hazard and risk. *Applied Energy*, *42*, 63-85.

Marshall, J., Mundt, A., Hult, M., McKealvy, T.C., Myers, P. and Sawyer, J. (1995). The relative risk of pressurized and refrigerated storage for six chemicals. *Process Safety Progress, 14* (3), 200-211.

McCarthy, A.J., Ditz, J.M. and Geren, P.M (1996). Inherently safer design review of a new alkylation process. *5th World Congress of Chemical Engineering, July 14-18, 1996, San Diego, CA (Paper 52b).* New York: American Institute of Chemical Engineers.

McQuaid, J. (1991). Know your enemy: The science of accident prevention. *Trans. IchemE, Part B. 69*, 9-19.

Mekitarian, A.E., and Keller, Jr., J.A (August, 1992). Product safety and modern grease formulations. *NLGI Spokesman*, 181-190.

Melhem, G.A. (1993). Hazard reduction benefits from reduced storage temperature of pressurized liquids. In V.M. Fthenakis (Ed.). *Prevention and Control of Accidental Releases of Hazardous Gases* (pps. 411-437). New York: Van Nostrand Reinhold.

Misono, M., and Okuhara T. (1993). Solid superacid catalysts. *Chemtech*, *23* (11), 23-29.

Mittelman, A., and Lin, D. (4 September,1995). Inherently safer chemistry. *Chemistry & Industry*, 694-696.

Mittelman, A., Lin, D., Cannon, D., and Ho, C K. (1994). Inherently Safe Chemistry Guide." *Preprints of Papers Presented at the 208th ACS National Meeting, August 21-25, 1994, Washington, DC* (pps. 271-272). Center for Great Lakes Studies, University of Wisconsin-Milwaukee, Milwaukee, WI: Division of Environmental Chemistry, American Chemical Society.

Mizerek, P. (January/February 1996). Disinfection techniques for water and wastewater. *The National Environmental Journal*, 22-28.

Moore, S., Samdani, S., Ondrey, G., and Parkinson, G. (March, 1994). New roles for supercritical fluids. *Chemical Engineering*, 32-35.

More Poke, Less Pounds (10 December, 1992). *The Chemical Engineer*, 18.

Mullin, R., and Plishner, E. (April 21, 1993). Big ticket changes in store for North American pulp and paper. *Chemical Week*, 32-34.

Nakazato, K. (1992). Anti-explosion system at recycling plants. *Inorganic Transformations and Ash Deposition During Combustion*, 915-920

Nasr, A. (1992). Prevent failures of mag drive pumps. *Chemical Engineering Progress*, *88* (11), 31-34.

Nasr, A. (July, 1994). Sealless pumps: Limitations and developments. *Hydrocarbon Processing*, 63-66.

Negron, R.M. (March/April 1994). Using ultraviolet disinfection in place of chlorination. *The National Environmental Journal*, 48-50.

Negron, R.M. (March/April 1994). Using ultraviolet disinfection in place of chlorination. *The National Environmental Journal*, 48-50.

New look fridges take off without HFCs (21 February 1994). *Chemistry and Industry*, 130.

New Refrigerant Gives More for Less (1994). *Environmental Protection Bulletin*, 032 (September), 33.

Newby, T., and Forth, D. (1991). Glandless pumps and valves—A technical update. *IChemE Symposium* Series No. 124, 119-134.

Newman, A. (1994). CFC phase-out moving quickly. *Environ. Sci. Technol.,* 28 (1), 35-37.

Newman, A. (1994). Designer chemistry. *Environ. Sci. Technol.* 28, 11, 463.

Ondrey, G. (March, 1995). Microreactor engineering: Birth of a new discipline? *Chemical Engineering*, 52.

Oosterkamp, W.J. (1993). Inherent and passive safety aspects in the design of advanced natural circulation boiling water reactors. In *Proc. ASME/JSME Nucl. Eng. Jt. Conf., Vol. 1* (pp. 713-728). New York: ASME.

Orrell, W., and Cryan, J. (August, 1987). Getting rid of the hazard. *The Chemical Engineer*, 14-15.

Paint Removers: New products eliminate old hazards (May 1991). *Consumer Reports*, 340-343.

Pan, H.Y., Minet, R.G., Benson, S.W. and Tsotsis, T.T. (1994). Process for converting hydrogen chloride to chlorine. *Ind. Eng. Chem. Res., 33*, 2996-3003.

Papayannokos, N.G., Koufopanos, C.A., and Karetsou, A. (1993). Studies on parametric sensitivity and safe operation criteria of batch processes. *Chem. Eng. Technol. 16*, 318-324.

Paquet Jr., D.A., and Ray, W.H. (1994). Tubular reactors for emulsion polymerization: I. Experimental investigation. *AIChE Journal, 40* (1), 73-87.

Paquet Jr., D. A., and Ray, W.H. (1994). Tubular reactors for emulsion polymerization: II. Model Comparisons With Experiments. *AIChE Journal, 40* (1), 88-96.

Park, D. E. (1996). "Inherently Safer Chemical Processes - A Life Cycle Approach." *International Conference and Workshop on Process Safety Management and Inherently Safer Processes, October 8-11, 1996, Orlando, FL* (pp. 316-319). New York: American Institute of Chemical Engineers.

Parthasarathy, R.V., and C.R. Martin (26 May, 1994). Synthesis of polymeric microcapsule arrays and their use for enzyme immobilization. *Nature, 369*, 298-301.

Paul, B.O. (February, 1994). Mag-drive pumps eliminate VOC emissions. *Chemical Processing*, 66-68.

Paul, E.L. (1988). Design of reaction systems for specialty organic chemicals. *Chemical Engineering Science, 43* (8), 1773-1782.

Pease, R. (10 June, 1995). Nanoworlds are made of this. *New Scientist*, 26-29.

Penteado, F.D., and. Ciric, A.R (1996). An MINLP approach for safe process plant layout. *Ind. Eng. Chem. Res.*, *35*, 1354-1361.

Perrin, D. A. (1996). Inherent safety in processes: Softer aspects of process safety. In *International Conference and Workshop on Process Safety Management and Inherently Safer Processes*, October 8-11, 1996, Orlando, FL (pp. 207-212). New York: American Institute of Chemical Engineers.

Perry, R.G. (1995). The paradox of inherent safety. Presented at the International Symposium on Runaway Reactions & Pressure Relief Design, August 2-4, 1995, Boston, MA.

Ponton, J. (1996). The disposable batch plant. In *5th World Congress of Chemical Engineering, July 14-18, 1996, San Diego, CA, Vol. II* (pp. 1119-1123). New York: American Institute of Chemical Engineers.

Popoff, F.P. (January 11, 1993). Full-cost accounting. *Chemical and Engineering News*, 8-10.

Poudrier, J.K. (September,1996). The chemical industry wins the green chemistry challenge. *Today's Chemist at Work*, 64-70.

Preston, M.L., and Turney, R.D. (1991). The process systems contribution to reliability engineering and risk assessment. In L. Puigjaner and A.L. Espuna, *Computer-Oriented Process Engineering* (pp.249-257). Amsterdam: Elsevier Science Publishers.

Preventing Pollution (March 23, 1995). *Right-to-Know Planning Guide,* 8 (14), 3.

Prugh, R.W. (1992). Hazardous fluid releases: Prevention and protection by design and operation. *Journal of Loss Prevention in the Process Industries, 5* (2), 67-72.

Puranik, S.A., Hathi, K.K., and Sengupta, R. (1990). Prevention of hazards through technological alternatives. In *Safety and Loss Prevention in the Chemical and Oil Processing Industries, October 23-27, 1989, Singapore* (pp. 581-587). IChemE Symposium Series, No. 120. Rugby, Warwickshire, U.K.: The Institution of Chemical Engineers.

Raghaven, K.V. (1992). Temperature runaway in fixed bed reactors: Online and offline checks for intrinsic safety. *Journal of Loss Prevention in the Process Industries, 5* (3), 153-159.

Report Updates Accident Patterns, Emphasizes Prevention (1997). *Chemical Process Safety Report* 7, 4 (February), 1.

Riezel, Y. (1996). Inherently safer process: an integral part of PSM implementation. In *International Conference and Workshop on Process Safety Management and Inherently Safer* Processes, October 8-11, 1996, Orlando, FL (pp. 320-328). New York: American Institute of Chemical Engineers..

Rogers, R.L., Hallam, and S. (1991). A chemical approach to inherent safety. *IChemE Symposium Series* No. 124, 235-241.

Rogers, R.L., Mansfield, D.P., Malmen, Y., Turney, R.D., and Verwoerd, M. (1995). The INSIDE Project: Integrating inherent safety in chemical process development and plant design. In G.A. Melhem and H.G. Fisher (Eds.). *International Symposium on Runaway Reactions and Pressure Relief Design, August 2-4, 1995, Boston, MA* (pp. 668-689). New York: American Institute of Chemical Engineers.

Rogers, W.G. (December, 1996). Working knowledge: Pop tops." *Scientific American*, 132.

Ror, B.O. (1992). Secondary containment examples in norsk hydro. Presented at the International Process Safety Group Technical Meeting, April 29 May 1, 1992, Taormina, Italy (Paper 1.1).

Rotman, D. (October 14, 1992). ICI claims breakthrough. *Chemical Week*, 24.

Rotman, D. (September 22, 1993). Chemists map greener synthesis pathways. *Chemical Week*, 56-57.

Rotman, D. (July 27, 1994). Supercritical CO2 points to greener polymer synthesis. *Chemical Week*, 12.

Rotman, D. (December 7, 1994). EPA, industry eye ways to safer chemistry. *Chemical Week*, 27.

Rotman, D. (June 14, 1995). California chemists make nanocatalyst. *Chemical Week*, 12.

Rotman, D., and Wood, A. (December 9, 1992). Successes emerge in search for cleaner processes. *Chemical Week*, 66.

Rushton, A.G., Edwards, D.W., and Lawrence, D. (May, 1994). Inherent safety and computer aided design. *Trans. IChemE 72, Part B*, 83-87.

Safe as a soda can (January, 1995). *ChemEcology*, 3-4.

Samdani, G.S. (October, 1994). "Miniaturization Reaches the CPI." *Chemical Engineering*, 41-43.

Savage, P.E., Gopalan, S., Mizan, T.I., Martino, C.J., and Brock, E.E. (1995). Reactions at supercritical conditions: Applications and fundamentals. *AIChE Journal, 41* (7), 1723-1778.

Scheffler, N.E. (1995). Inherently safer latex plants. In E.D. Wixom and R.P. Benedetti (Eds.), *Proceedings of the 29th Annual Loss Prevention Symposium, July 31 - August 2, 1995, Boston, MA.*. New York: American Institute of Chemical Engineers, Paper 1b

Schwartz, A.T. (March,1996). Sir Humphry Davy and the safety lamp. *Chemistry Review*, 12-15.

Seldon, R. A. (6 January, 1997). Catalysis and pollution prevention. *Chemistry & Industry*, 12-15.

Sherringron, D.C. (7 January, 1991). Polymer supported systems: Towards clean chemistry? *Chemistry and Industry*, 15-19.

Shinnar, R., Doyle, F.J., Budman, H.M., and Morari, M. (1992). Design considerations for tubular reactors with highly exothermic reactions. *AIChE Journal*, *38* (11), 1729-1743.

Shonnard, D.R., Herlevich, J., Parikh, P., Stark, D., Barna, B.A., Rogers, T.N., et al. (1996). Pollution assessment software as chemical industry process simulator enhancements. In *The 5th World Congress of Chemical Engineering, July 14-18, 1996, San Diego, CA, Vol. III* (pp. 329-337). New York: American Institute of Chemical Engineers.

Siirola, J.J. (1995). An industrial perspective on process synthesis. In *AIChE Symposium Series*, 91, 222-233.

Singh, J. (February 5, 1993). Assessing semi-batch reaction hazards. *The Chemical Engineer*, *537* (21), 23-25.

Singh, J. (1993). Safe disposal of reactive chemicals following emergency venting. *IChemE Symposium Series, 130,* 249-274.

Singh, M.M., Pike, R.M., and Szafran, Z. (1994). Pollution prevention in the organic and inorganic chemistry laboratory: Microscale approach. In *Preprints of Papers Presented at the 208th ACS National Meeting, August 21-25, 1994, Washington, DC* (pp.194-197). Center for Great Lakes Studies, University of Wisconsin-Milwaukee,

Snyder, P.G. (1996). Inherently safe(r) plant design. In H. Cullingford (Ed.). *1996 Process Plant Safety Symposium, Volume 1, April 1-2, 1996, Houston, TX* (pp.203-215). Houston, TX: South Texas Section of the American Institute of Chemical Engineers.

Snyder, P. G. (July, 1996). Is your plant inherently safer. *Hydrocarbon Processing*, 77-82.

Somerville, R. L. (1993). Control of liquefied toxic gas releases. In V. M. Fthenakis *Prevention and Control of Accidental Releases of Hazardous Gases* (pp. 438-447). New York: Van Nostrand Reinhold.

Sorensen, F. and Petersen, H.J.S. (1992). Substitution of organic solvents. *Staub - Reinhaltund Der Luft, 52*, 113-18.

Speechly, D., Thornton, R.E. and Woods, W.A. (1979). Principles of total containment system design. *North Western Branch Papers No. 2, Institution of Chemical Engineers*, 7.1-7.21.

Steensma, M., and Westererp, K.R. (1988). Thermally safe operation of a cooled semibatch reactor: Slow liquid-liquid reactions. *Chemical Engineering Science, 43* (8), 2125-2132.

Steensma, M., and Westererp, K.R (1990). Thermally safe operation of a cooled semibatch reactor: Slow reactions. *Ind. Eng. Chem. Res., 29*, 1259-1270.

Summary and Highlights: Industry/Department of Energy Workshop on Chemical Safety (1993). Washington, D.C.: U. S. Department of Energy, Assistant Secretary for Environment, Safety and Health, Office of Safety and Quality Assurance.

Sundell, M.J. and Nasman, J.H. (1993). Anchoring catalytic functionality on a polymer. *Chemtech, 23* (12), 16-23.

Tavener, S.J., and Clark, J.H. (6 January, 1997). Recent highlights in phase transfer catalysis. *Chemistry & Industry*, 22-24.

Techniques for evaluating chemical reaction hazards. (1993). In J.A. Barton and R.L. Rogers (Eds.). *Chemical Reaction Hazards-A Guide* (19-26). Rugby, Warwickshire, UK: Institution of Chemical Engineers.

Technology for a greener world (March 28, 1991). *The Chemical Engineer*, 13.

Ten Alps Publishing (15 May, 1995). A block of gas. *Chemistry and Industry*, 371.

Tickner, J. (3 October, 1994). The case for inherent safety. *Chemistry and Industry*, 796.

Tietze, L.F. (19 June, 1995). Domino reactions in organic synthesis. *Chemistry & Industry*, 453-457.

Tillman, J.W. (1991). *Achievements in Source Reduction and Recycling for Ten Industries in the United States*. Cincinnati, Ohio: U. S. Environmental Protection Agency.

Timberlake, D.L., and Govind, R. (1994). Expert system for solvent substitution. In *Preprints of Papers Presented at the 208th ACS National Meeting, August 21-25, 1994, Washington, DC* (pp. 215-217). Center for Great Lakes Studies, University of Wisconsin-Milwaukee, Milwaukee, WI: Division of Environmental Chemistry, American Chemical Society.

Tobin, P.S. (1994). Risk management through inherently safe chemistry. In *Preprints of Papers Presented at the 208th ACS National Meeting, August 21-25, 1994, Washington, DC* (pp. 268-270). Center for Great

Lakes Studies, University of Wisconsin-Milwaukee, Milwaukee, WI: Division of Environmental Chemistry, American Chemical Society.

Tyler, B.J. (1985). Using the Mond Index to measure inherent hazards. *Plant/Operations Progress 4* (3), 172-75.

van Steen, J. F.J. (1996). *The Inside Project, Phase 1, Task 1f Report: An Assessment of Hurdles and Solutions Concerning the Adoption of Inherently Safer Approaches in Industry*. Apeldoom, Netherlands: TNO.

Velo, E., C. M. Bosch, and F. Recasens (1996). Thermal safety of batch reactors and storage tanks: Development and validation of runaway boundaries. *Ind. Eng. Chem. Res.*, *35*, 1288-1299.

Verwijs, J. W., van den Berg, H., and Westerterp, K.R. (1996). Startup strategy design and safeguarding of industrial adiabatic tubular reactor systems. *AIChE Journal*, *42* (2), 503-515.

Villermaux, J. (1996). New horizons in chemical engineering. In *The 5th World Congress of Chemical Engineering: Technologies Critical to a Changing World, July 14-18, 1996, San Diego, CA, Summary Proceedings Volume* (pp. 16-23). New York: American Institute of Chemical Engineers.

Wallington, T.J., et al. (1994). The environmental impact of CFC replacements – HFCs and HCFCs. *Environmental Science and Technology*, *28* (7), 320-325.

Walsh, F., and Mills, G. (2 August, 1993). Electrochemical techniques for a cleaner environment. *Chemistry and Industry*, 576-579.

Weirick, M.L., Farquhar, S.M., and B.P. Chismar (1994). Spill containment and destruction of a reactive, volatile chemical. *Process Safety Progress, 13* (2), 69-71.

Wells. G.L. and Rose, L.M. (1986). *The Art of Chemical Process Design*, *266*. Amsterdam: Elsevier.

Welter, T.R. (February 4, 1991). The quest for safe substitutes. *Industry Week*, 38-43.

Whiting, M.J.L. (March,1992). The benefits of process intensification for caro's acid production. *Trans. IChemE*, *70*, 195-196.

Wilday, A.J. (1991). The safe design of chemical plants with no need for pressure relief systems. In *IChemE Symposium Series No. 124*, 243-53.

Wilkinson, M., and Geddes, K. (December 1993). An award winning process. *Chemistry in Britain*, 1050-1052.

Wilson, S. (4 April,1994). Peroxygen technology in the chemical industry. *Chemistry and Industry*, 255-258.

Windhorst, J.C.A. (1995). Application of inherently safe design concepts, fitness for use and risk driven design process safety standards to an lpg project. In J.J. Mewis, H.J. Pasman, and E.E. DeRademacker (Eds.). *Loss*

Prevention and Safety Promotion in the Process Industries (543-554). Amsterdam: Elsevier Science B. V.

Wiss, J., Fleury, C. and Fuchs, V. (1995). Modelling and optimization of semi-batch and continuous nitration of chlorobenzene from safety and technical viewpoints. *Journal of Loss Prevention in the Process Industries, 8* (4), 205-213.

Wood, A. (October 14, 1992). DuPont develops waste-free route to terathane. *Chemical Week*, 9.

Wood, A. (August 17, 1994). Mobil cultivates new zeolites to grow process licensing. *Chemical Week*, 34-35.

Wright, M.E., Toplikar, E.G., and Svejda, S.A. (1991). Details concerning the chloromethylation of soluble high molecular weight polystyrene using dimethoxymethane, thionyl chloride and a Lewis acid: A full analysis. *Macromolecules*, *24,* 5879-5880.

Index

B

C

D

T